NEUROMETHODS

Series Editor
Wolfgang Walz
University of Saskatchewan
Saskatoon, SK, Canada

For further volumes:
http://www.springer.com/series/7657

Extracellular Recording Approaches

Edited by

Roy V. Sillitoe

Department of Pathology & Immunology and Neuroscience, Baylor College of Medicine, Houston, TX, USA

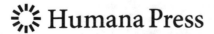 Humana Press

Editor
Roy V. Sillitoe
Department of Pathology & Immunology and Neuroscience
Baylor College of Medicine
Houston, TX, USA

ISSN 0893-2336   ISSN 1940-6045   (electronic)
Neuromethods
ISBN 978-1-4939-8517-3   ISBN 978-1-4939-7549-5   (eBook)
https://doi.org/10.1007/978-1-4939-7549-5

Printed on acid-free paper

This Humana Press imprint is published by Springer Nature
The registered company is Springer Science+Business Media, LLC
The registered company address is: 233 Spring Street, New York, NY 10013, U.S.A.

Preface to the Series

Experimental life sciences have two basic foundations: concepts and tools. The *Neuromethods* series focuses on the tools and techniques unique to the investigation of the nervous system and excitable cells. It will not, however, shortchange the concept side of things as care has been taken to integrate these tools within the context of the concepts and questions under investigation. In this way, the series is unique in that it not only collects protocols but also includes theoretical background information and critiques which led to the methods and their development. Thus it gives the reader a better understanding of the origin of the techniques and their potential future development. The *Neuromethods* publishing program strikes a balance between recent and exciting developments like those concerning new animal models of disease, imaging, in vivo methods, and more established techniques, including, for example, immunocytochemistry and electrophysiological technologies. New trainees in neurosciences still need a sound footing in these older methods in order to apply a critical approach to their results.

Under the guidance of its founders, Alan Boulton and Glen Baker, the *Neuromethods* series has been a success since its first volume published through Humana Press in 1985. The series continues to flourish through many changes over the years. It is now published under the umbrella of Springer Protocols. While methods involving brain research have changed a lot since the series started, the publishing environment and technology have changed even more radically. Neuromethods has the distinct layout and style of the Springer Protocols program, designed specifically for readability and ease of reference in a laboratory setting.

The careful application of methods is potentially the most important step in the process of scientific inquiry. In the past, new methodologies led the way in developing new disciplines in the biological and medical sciences. For example, physiology emerged out of anatomy in the nineteenth century by harnessing new methods based on the newly discovered phenomenon of electricity. Nowadays, the relationships between disciplines and methods are more complex. Methods are now widely shared between disciplines and research areas. New developments in electronic publishing make it possible for scientists that encounter new methods to quickly find sources of information electronically. The design of individual volumes and chapters in this series takes this new access technology into account. Springer Protocols makes it possible to download single protocols separately. In addition, Springer makes its print-on-demand technology available globally. A print copy can therefore be acquired quickly and for a competitive price anywhere in the world.

Saskatoon, SK   *Wolfgang Walz*

Preface

There are now many methods available for studying brain function. Among these are various approaches for recording neurons in vivo. The signals may be collected from populations of neurons using electrode arrays, tetrodes, or single electrodes to collect either local field potentials (LFPs) or action potentials from single cells. These various approaches have been fine-tuned in different species for studying specific behaviors. For example, motor behaviors are heavily studied in mice and rats, but recent work on looming behavior has benefited greatly from studies in locusts, in which brain activity can be examined with precision during execution of the behavior. Moreover, the use of conditional genetics in vertebrates and invertebrates and optogenetics approaches in rodents has increased the power of manipulating neurons and circuits. These different strategies enable researchers to define the causal relationships between neuronal properties in targeted brain areas to distinct behavioral consequences in health and disease.

The chapters in this volume of Neuromethods have brought together experts in the field to discuss extracellular recording techniques with two specific goals in mind. The first is to provide a reasonably detailed outline of the methods, including critical steps for success, advantages of the method, and shortcomings to consider with pointers for how to resolve them. The second is to provide an example application where their specific approach has proven particularly useful, and this may include a discussion of how basic biology has been pushed forward or how translational research and clinical diagnoses or treatments have benefitted from the approaches.

The contributing author list encompasses a unique balance of investigators from different areas of neuroscience who study different regions of the brain, using different model systems. The models include mouse, rat, zebrafish, fly, worm, and even locust. We have also included one chapter describing the use of extracellular recordings in studies of human cognition. The techniques range from single-unit recordings to tetrodes, multi-electrode arrays, and imaging systems. Importantly, we have also included a chapter on recording in perinatal and early developing rodents, highlighting the key advances that studies of critical periods, which shape brain function, have made. Overall, several major sectors of the systems neuroscience field are represented by including brain regions involved in sensory, motor, and cognitive functions.

In summary, this volume will be an invaluable resource because at the heart of the collection is classic extracellular recording, but the authors describe improvements and additions, as well as more modern and novel tools such as optogenetics and multi-channel recording. I have selectively chosen authors who are at the cutting edge of their respective fields. Our wish was to provide readers with the most up-to-date and original applications of the outlined techniques. I am delighted to say that I believe we achieved those goals. With that, I would like to extend my deepest gratitude to each and every author that contributed to what I think is a fantastic and unique collection of chapters. I truly appreciate all the hard work that you put into your chapters, and I know that our readers will share my enthusiasm for your masterfully crafted material.

Houston, TX, USA   *Roy V. Sillitoe*

Contents

Contributors

BENJAMIN R. ARENKIEL • *Department of Molecular & Human Genetics and Neuroscience, Program in Developmental Biology, Baylor College of Medicine, Jan and Dan Duncan Neurological Research Institute of Texas Children's Hospital, Houston, TX, USA*

MARK S. BLUMBERG • *Department of Psychological & Brain Sciences, University of Iowa, Iowa City, IA, USA; DeLTA Center, University of Iowa, Iowa City, IA, USA; Interdisciplinary Graduate Program in Neuroscience, University of Iowa, Iowa City, IA, USA; Department of Biology, University of Iowa, Iowa City, IA, USA*

AMANDA M. BROWN • *Department of Pathology & Immunology and Neuroscience, Baylor College of Medicine, Jan and Dan Duncan Neurological Research Institute of Texas Children's Hospital, Houston, TX, USA*

RICHARD COURTEMANCHE • *Department of Exercise Science & FRQS Groupe de Recherche en Neurobiologie Comportementale/CSBN, Concordia University, Montreal, QC, Canada*

OSCAR H.J. EELKMAN ROODA • *Department of Neuroscience, Erasmus Medical Center, Rotterdam, The Netherlands*

SHANGBANG GAO • *Key Laboratory of Molecular Biophysics of the Ministry of Education, College of Life Science and Technology, Huazhong University of Science and Technology, Wuhan, China*

FABRIZIO GABBIANI • *Department of Neuroscience, Baylor College of Medicine, Houston, TX, USA*

SHANE A. HEINEY • *Department of Neuroscience, Baylor College of Medicine, Houston, TX, USA*

FREEK E. HOEBEEK • *Department of Neuroscience, Erasmus Medical Center, Rotterdam, The Netherlands*

JUI-YI HSIEH • *Circuit Therapeutics, Inc., Menlo Park, CA, USA; Department of Physiology and the Interdepartmental Program in Molecular, Cellular, and Integrative Physiology, David Geffen School of Medicine at UCLA, Los Angeles, CA, USA*

ZHITAO HU • *Queensland Brain Institute, The University of Queensland, Brisbane, QLD, Australia*

LONGWEN HUANG • *Department of Neuroscience, Baylor College of Medicine, Jan and Dan Duncan Neurological Research Institute of Texas Children's Hospital, Houston, TX, USA*

TAYLOR JEFFERSON • *Department of Pathology & Immunology and Neuroscience, Baylor College of Medicine, Jan and Dan Duncan Neurological Research Institute of Texas Children's Hospital, Houston, TX, USA*

DAOYUN JI • *Department of Molecular & Cellular Biology and Neuroscience, Baylor College of Medicine, Houston, TX, USA*

OLIVIA A. KIM • *Department of Neuroscience, Baylor College of Medicine, Houston, TX, USA*

BALAJI KRISHNAN • *Department of Neurology, Mitchell Center for Neurodegenerative Diseases, University of Texas Medical Branch, Galveston, TX, USA*

ERIC J. LANG • *Department of Neuroscience & Physiology, New York University School of Medicine, New York, NY, USA*

MAXIME LÉVESQUE • *Department of Neurology & Neurosurgery, Montreal Neurological Institute, McGill University, Montreal, QC, Canada*

TAO LIN • *Department of Pathology & Immunology and Neuroscience, Baylor College of Medicine, Jan and Dan Duncan Neurological Research Institute of Texas Children's Hospital, Houston, TX, USA*

GARY LIU • *Program in Developmental Biology, Baylor College of Medicine, Jan and Dan Duncan Neurological Research Institute of Texas Children's Hospital, Houston, TX, USA*

ADAM N. MAMELAK • *Department of Neurosurgery, Cedars-Sinai Medical Center, Los Angeles, CA, USA*

JAVIER F. MEDINA • *Department of Neuroscience, Baylor College of Medicine, Houston, TX, USA*

JURI MINXHA • *Computation and Neural Systems Program, California Institute of Technology, Pasadena, CA, USA; Department of Neurosurgery, Cedars-Sinai Medical Center, Los Angeles, CA, USA*

XIANG MOU • *Department of Molecular and Cellular Biology and Neuroscience, Baylor College of Medicine, Houston, TX, USA*

SHOGO OHMAE • *Department of Neuroscience, Baylor College of Medicine, Houston, TX, USA*

SUGI PANNEERSELVAM • *Department of Molecular & Human Genetics, Baylor College of Medicine, Jan and Dan Duncan Neurological Research Institute of Texas Children's Hospital, Houston, TX, USA*

DIANE M. PAPAZIAN • *Department of Physiology and the Interdepartmental Program in Molecular, Cellular, and Integrative Physiology, David Geffen School of Medicine at UCLA, Los Angeles, CA, USA*

BRANDON PEKAREK • *Department of Molecular & Human Genetics, Baylor College of Medicine, Jan and Dan Duncan Neurological Research Institute of Texas Children's Hospital, Houston, TX, USA*

UELI RUTISHAUSER • *Computation and Neural Systems Program, California Institute of Technology, Pasadena, CA, USA; Department of Neurosurgery, Cedars-Sinai Medical Center, Los Angeles, CA, USA; Department of Neurology, Cedars-Sinai Medical Center, Los Angeles, CA, USA*

MARTIJN SCHONEWILLE • *Department of Neuroscience, Erasmus Medical Center, Rotterdam, The Netherlands*

ROY V. SILLITOE • *Department of Pathology & Immunology and Neuroscience, Program in Developmental Biology, Baylor College of Medicine, Jan and Dan Duncan Neurological Research Institute of Texas Children's Hospital, Houston, TX, USA*

GRETA SOKOLOFF • *Department of Psychological & Brain Sciences, University of Iowa, Iowa City, IA, USA; DeLTA Center, University of Iowa, Iowa City, IA, USA*

JESSICA SWANSON • *Department of Molecular & Human Genetics, Baylor College of Medicine, Jan and Dan Duncan Neurological Research Institute of Texas Children's Hospital, Houston, TX, USA*

BURAK TEPE • *Program in Developmental Biology, Baylor College of Medicine, Jan and Dan Duncan Neurological Research Institute of Texas Children's Hospital, Houston, TX, USA*

KEVIN UNG • *Program in Developmental Biology, Baylor College of Medicine, Jan and Dan Duncan Neurological Research Institute of Texas Children's Hospital, Houston, TX, USA*

YOGESH P. WAIRKAR • *Department of Neurology, Mitchell Center for Neurodegenerative Diseases, University of Texas Medical Branch, Galveston, TX, USA*

JOSHUA J. WHITE • *Department of Pathology & Immunology and Neuroscience, Baylor College of Medicine, Jan and Dan Duncan Neurological Research Institute of Texas Children's Hospital, Houston, TX, USA*

BIN WU • *Department of Neuroscience, Erasmus Medical Center, Rotterdam, The Netherlands*

JOY ZHOU • *Department of Pathology & Immunology and Neuroscience, Baylor College of Medicine, Jan and Dan Duncan Neurological Research Institute of Texas Children's Hospital, Houston, TX, USA*

YING ZHU • *Department of Neuroscience, Baylor College of Medicine, Houston, TX, USA; Structural and Computational Biology and Molecular Biophysics Program, Baylor College of Medicine, Houston, TX, USA*

Chapter 1

In Vivo Loose-Patch-Juxtacellular Labeling of Cerebellar Neurons in Mice

Amanda M. Brown, Joshua J. White, Joy Zhou, Taylor Jefferson, Tao Lin, and Roy V. Sillitoe

Abstract

Extracellular recording techniques provide a critical means for measuring neuronal function in vivo. For many experiments, metal electrodes yield spike data that is clean enough for resolving spike waveforms to identify neurons. However, even though some neurons can be distinguished based on their spiking features, it is still not possible to definitively identify most neurons only by their activity. To circumvent this problem, Pinault (J Neurosci Methods 65:113–136, 1996) developed the juxtacellular recording-labeling method to anatomically identify individually recorded neurons. His method utilized glass electrodes to isolate and record single units in vivo. The use of pulled capillary pipettes for recording was key for the success of the method since it allowed him to load the electrode with a tracer that could be delivered across the cell membrane and into the cell. Tracing is achieved by placing the electrode in very close proximity to a cell and then, by delivering pulses of current, pores in the membrane transiently open to allow the tracer to enter. Later studies expanded on the precision of the method by making a loose patch onto the cell before filling it. However, while the juxtacellular method has opened new avenues for relating structure to function at single-cell resolution, the approach remains a challenge to execute because there are several critical steps that are difficult to perform. Here, we provide a step-by-step description for how to perform loose-patch-juxtacellular labeling in vivo in mice. Using the cerebellum as a model system, we outline how to fill Purkinje cells with Neurobiotin. The procedure can be adapted to labeling neurons in any part of the brain, and we discuss its value for unambiguously identifying cells in mutant mice.

Key words Electrophysiology, Spikes, Recording, Iontophoresis, Neurobiotin, Neural tracing, Cerebellum, Purkinje cell, Morphology

---

1 Introduction

There are two overarching goals of neuroscience: to understand brain structure and to resolve brain function. Many methods have been developed to achieve this including neural tract tracing; immunohistochemistry; in situ hybridization; MRI techniques such as fMRI (functional magnetic resonance imaging) and DTI (diffusion tensor imaging); calcium- and voltage-gated channel-mediated imaging; optogenetics; and, of course, a large number of

Roy V. Sillitoe (ed.), *Extracellular Recording Approaches*, Neuromethods, vol. 134,
https://doi.org/10.1007/978-1-4939-7549-5_1, © Springer Science+Business Media, LLC 2018

fundamental electrophysiological approaches. One of the most extensively applied electrophysiology methods is in vivo recording of neurons. Certainly, multielectrode array techniques and tetrode approaches have uncovered critical information on how neuronal populations behave, but to address how specific neurons in distinct regions of the brain function, single cells have to be analyzed. For this, whole-cell methods can be employed for understanding the membrane biophysical properties and the intracellular environment of neurons, or more common extracellular recording methods can be used. But despite the multitude of extracellular recording techniques that have been designed to distinguish the firing activity of individual neurons, the same question often arises. What is the exact morphology of the specific neuron that generates the detected neuronal activity? This is an essential question to answer because the precise cellular architecture of a given neuron, its complete circuit connectivity, and how this cell and its microcircuit behave in vivo will help to elucidate the encoding and execution of sensory, motor, and cognitive behaviors. This, in turn, will provide a major step toward uncovering how these behaviors are impacted in different neurological and neuropsychiatric diseases.

In the mid-1990s, Didier Pinault designed an extracellular recording approach that would help address some of these concerns. He developed an innovative technique that is generally referred to as juxtacellular recording-labeling [1]. The basic principle of this method is that a glass capillary is filled with a neuronal tracer molecule, and then—using the same capillary as a recording electrode—the activity of a single unit is isolated in vivo. Then, using an iontophoresis-based approach again facilitated by the same electrode, pulses of low current are delivered to the cell that is being recorded that simultaneously open pores in the cell's membrane and help drive the tracer into the cell in order to fill it. The brain tissue is then cut and histologically stained to detect the localized deposit of tracer in the exact neuron that was juxtacellularly recorded and labeled. Pinault and his colleagues went on to use his technique to study sensory processing neurons in the somatosensory thalamocortical system of healthy and epileptic rodents [2–6].

Several improvements have been made to the approach, for example, Joshi and Hawken [7] also used glass electrodes, but in their approach the electrode is advanced to the cell membrane to make a "loose-patch" for performing a cell-attached recording. Here, we adapted this loose-patch configuration to label and record cerebellar Purkinje cells in mice. The cerebellum is critical for a number of motor functions including coordination, learning, posture, and balance. Its normal functions are consistent with the behavioral alterations that arise in disorders that affect the cerebellum. Defective cerebellar function is characterized by ataxia,

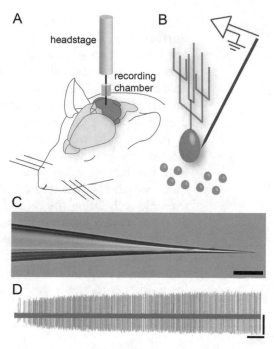

Fig. 1 In vivo electrophysiology of Purkinje cells in the mouse cerebellum. (**a**) Schematic illustrating the general recording setup for juxtacellular labeling of neurons in vivo in an anesthetized adult mouse. (**b**) Schematic showing how a glass electrode is used to record single-unit activity from a Purkinje cell. (**c**) Sample glass electrode that was used for juxtacellular labeling. Note the taper and length of the tip. Scale bar = 100 μm. (**d**) Single-unit recording showing the increase in spike amplitude as the electrode approaches the cell. Scale bar *x*-axis = 2 s, *y*-axis = 10 mV. The schematics in panels **a** and **b** were adapted with permission from [37]

tremor, and dystonia, although there are several non-motor behaviors that are also thought to arise when the cerebellum does not function properly. Indeed, juxtacellular recording has been used to unravel the functions of cerebellar neurons [8–14]. But based on the relatively few studies that have been performed, it is clear that the challenge of labeling cerebellar neurons in vivo has prevented further analysis. We provide a step-by-step protocol for recording and labeling Purkinje cells in vivo and describe in detail how to isolate single neurons and then achieve high-resolution neuronal morphology after labeling the cell with Neurobiotin (Fig. 1). We used Purkinje cells as a model because their physiological properties have been extensively studied in different animal models both in vitro and in vivo, and therefore they are functionally well defined [14–32]. Additionally, their anatomy [33] is one of the most recognizable features in all of neuroscience, stemming from the pioneering works of Johannes Evangelista Purkinje (described in Herndon, 1963 [34]) and Ramon y Cajal [35]. Furthermore, Purkinje cell deficits are a major problem in a growing list of brain diseases [36]. But, when Purkinje cell firing is altered, it becomes difficult to identify them based on their typical properties. To address this challenge, we also provide an example on how the loose-patch-juxtacellular method can be used to gain a deeper appreciation for neuronal structure-function relationships in mutant mice with severely altered cellular activity. We briefly discuss how Purkinje cells can still be identified in vivo after genetically eliminating a key identifying spiking feature, the complex spike, which is typically used to confirm that a recorded cell is a Purkinje cell. Using the cerebellum as a model, we hope that this guide to in vivo juxtacellular labeling will provide a starting point to advance studies of neuronal structure-function properties in other regions of the brain, in the many genetic, pharmacological, and injury-based mouse models of dysfunction, as well as other incoming neuroscience animal models such as the marmoset.

2 Materials

Surgical materials and tools for preparing the mouse for juxtacellular recording

The juxtacellular recording-labeling procedure described below used C57BL6/J inbred and Swiss Webster outbred mice purchased from the Jackson Laboratory or Taconic Biosciences, which were maintained in our animal colony. We bred the mice using timed pregnancies, and we designated noon on the day a vaginal plug was detected as embryonic day (E) 0.5. Mice of both sexes were studied at postnatal day 30 or beyond. All animal studies were carried out under an approved IACUC animal protocol according to the institutional guidelines at Baylor College of Medicine.

The following equipment is required for the described surgical procedure:
Anesthesia system (Parkland Scientific, Coral Springs, FL, USA)

Tank with compressed oxygen

Gas flow regulator (VWR Scientific Cat. # 55850-235)

Isoflurane vaporizer with flowmeter (Cat. # Matrx VIP 3000, Rodent Anesthesia Machine, Midmark, Versailles OH)

Induction box (Midmark, 1.5 L)

Stereotaxic instruments

Model 900 stereotaxic assembly with standard accessories for mice (David Kopf Instruments, Tujunga, CA, USA)

Motorized micromanipulator (MP-225; Sutter Instrument Co., Novato, CA, USA).

Glass bead sterilizer (FST; Steri 205).

Heating pad (Kent Scientific, Torrington, CT, USA; #DCT-15).

Depilatory cream (Nair lotion with body oil, purchased from Walgreens).

Lubricant eye drops (Celluvisc; NDC 0023-4554-30).

70% ethanol pads (No. 5033, Covidien, Mansfield, MA).

Povidone-iodine pads (BETADINE Solution Swab Aids, Purdue Pharma, Stamford, CT, USA; #67618-152-01).

Cotton-tipped applicators (Fisher Scientific; cat no. 23-400-106).

Nonwoven gauze (Fisher Scientific; cat no. 22028559).

Dissection scissors (FST#14082-09; Fine Science Tools, Foster City, CA, USA).

Scalpel

Scalpel blade holder (FST #10003-12; Fine Science Tools, Foster City, CA, USA)

Scalpel blade (Harvard Apparatus, Holliston, MA, USA; #728360)

Forceps

#55 forceps (FST #11255-20; Fine Science Tools, Foster City, CA, USA)

Dumont AA forceps (FST #11210-10; Fine Science Tools, Foster City, CA, USA)

Micro-drill (Ideal Micro-Drill Surgical Drills #726065)

Drill burr (FST #19007-05; Fine Science Tools, Foster City, CA, USA)

Pencil (soft lead, sharpened).
Note: Personal protective equipment should be used for this protocol. Since this is a sterile surgery, we carry out the tracing using sterile technique, which involves using autoclaved tools and wearing sterile personal protective equipment including a gown, gloves, and mask.

2.1 Preparing the Mouse for Surgery

The mouse is weighed and then anesthetized using 2% isoflurane in oxygen. It is then given ketamine/dexmedetomidine (75 mg/kg and 0.5 mg/kg, respectively) via intraperitoneal injection and laid prone on a heating pad on top of a stereotaxic assembly. Isoflurane levels are then reduced to ~1%. The mouse's skull is tightly secured in a stable, prone position by placing the front teeth through the stereotaxic assembly's tooth bar and fitting the ear bars into place.

Place the ear bars into the ear holes with extreme care to avoid puncturing the eardrums. Verify that each ear bar is placed in an equal distance into the ear using the ruler measurement on the ear bars. Secure placement can be checked by lightly pressing on the skull. There should be no movement of the skull laterally or vertically—the head should only pivot forward and backward with a clean roll. *It is imperative for a successful surgery and juxtacellular recording that the head is completely secure and stable.*

2.2 Making the Craniotomy

Fur is removed from the mouse's neck to the edge of the assembly's anesthesia mask using depilatory cream (e.g., Nair) that is then cleaned away using cotton swabs and ethanol pads. The mouse's skin is sterilized with povidone-iodine after the fur has been completely cleared away. An incision is made to reveal bregma and lambda and expose the desired craniotomy site, after which the connective tissues are cleaned from the skull and, for recordings of the cerebellar nuclei and other posterior areas, muscles are carefully teased away. The micromanipulator is then used to find the coordinates of bregma, from which the coordinates of the desired craniotomy site are calculated. The planned recording site is marked with a pencil. The craniotomy over the recording site needs to be wide, as glass electrodes used in juxtacellular recordings are thicker than electrodes for most other methods. Therefore, a long, thin, horizontal oval-shaped craniotomy is opened for recordings of the cerebellum and cerebellar nuclei (please refer to our previously described technique for successful craniotomies [37]). The micro-drill is used to perform the craniotomy by carving in smooth circles around the recording site. Applying light pressure with a cotton or foam swab can quickly abate any bleeding from blood vessels in the skull during the craniotomy. The piece of the skull within the carved circle can be removed with #55 forceps when it appears to easily wiggle when lightly touched. Slowly flip the loosened skull piece away from the craniotomy site (like opening the lid of a box) instead of pulling straight up and away to prevent shearing of the remaining attached tissue. Ripping this tissue can lead to damaged vasculature and "pulsing" in the exposed area of the brain, which makes stable glass recordings a challenge. The craniotomy site can also be briefly soaked with a drop of saline to soften the skull before removal, but this step is not always necessary. Quickly place a drop of saline onto the exposed neural tissue after the skull is removed to prevent it from drying out. A successful craniotomy will have very little or no bleeding or pulsing of the tissue after the skull is removed, and all the visible vasculature should appear intact. After the craniotomy is completed, isoflurane levels may be reduced further to ~0.15–0.25% for the duration of the recording-labeling procedure [29].

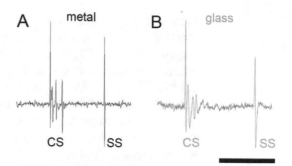

Fig. 2 Purkinje cells fire two types of action potentials that are used to identify the cell type and determine the quality of the signal. (**a**) Purkinje cell recorded with a metal electrode showing a clear complex spike (CS) and a simple spike (SS). (**b**) Purkinje cell recorded using a glass electrode also demonstrating that both action potentials can be isolated before entraining and filling the cells. Scale bar = 20 ms

2.3 In Vivo Firing Properties of Cerebellar Purkinje Cells

One of the advantages of working on Purkinje cells is that they can be identified by their unique electrophysiological properties. Purkinje cells fire two types of action potentials (Fig. 2). The first, called the simple spike, is generated intrinsically within the Purkinje cells [25]. Even though the basal properties of simple spikes are determined intrinsically, sensory-induced afferent activity can influence their firing characteristics through their connections to the granule cells [38–41]. Simple spikes are characterized by a relatively regular pattern and high frequency of firing (~40–50 Hz) [12, 42], although they can fire over a wide range of frequencies as seen in both anesthetized [29, 43–46] and awake preparations [30, 37, 42, 46–51]. The second type of spike is called the "complex" spike. Complex spikes occur at a low frequency (~0.5–1 Hz) and are elicited by climbing fiber input. Complex spikes are distinguished from the simple spikes by their unique waveform consisting of a single large spike followed by three to five smaller spikelets [27]. In adult animals, complex spike firing is thought to modulate simple spike activity [11]. This is important for motor behavior because Purkinje cell firing influences cerebellar nuclear output [52]. Thus, Purkinje cell activity is central to cerebellar function and the control of movement [29, 47, 50, 53–55]. But despite the unique features that greatly help in identifying Purkinje cells, recent work has demonstrated some additional complexity in Purkinje cell behavior. The firing properties of Purkinje cells can differ depending on anatomical location, and the specific rates and patterns of firing fall within a well-known architectural map that is defined by the pattern of zebrinII expression [51, 56]. ZebrinII is a polypeptide antigen located on the aldolase C protein that was used to uncover a remarkable array of stripes called zones [57, 58]. Purkinje cells that express zebrinII (therefore, called positive zones) are positioned into a topographic map that has an alternating pattern with Purkinje cells that do not express zebrinII (therefore, called negative zones) [59, 60]. Work from two different studies, using mouse and rat, showed that not all Purkinje cells fire at the typical 50 Hz [51, 56]. For instance, in mouse they showed that while zebrinII-positive Purkinje cells indeed do fire at around 60 Hz and with a relatively regular pattern, zebrinII-negative Purkinje cells on the other hand fire with a higher rate at around 90 Hz and with a more irregular pattern [51]. Knowing that these differences exist will help build confidence when isolating Purkinje cells in preparation for delivering the current to label them with Neurobiotin. That is, higher rates of firing are not necessarily indicative of a "pinched" cell or an unhealthy dying cell, so as long as the entrainment is maintained for an adequate time period, then perhaps the Purkinje cells with higher rates and relatively irregular patterns might belong to the zebrinII-negative subset. That said care must still be taken when isolating cells in vivo since the electrode can indeed influence the local firing properties.

2.4 Recoding Cerebellar Purkinje Cells In Vivo

For our juxtacellular studies [46], single-unit recordings are attained with 5–13 MΩ glass electrodes (please see below for details) that are controlled by a motorized micromanipulator (MP-225; Sutter Instrument Co., Novato, CA). Signals are band-pass filtered at 0.3–13 kHz, amplified with an ELC-03XS amplifier (NPI Electronic Instruments, Tamm, Germany), and then digitized (CED Power1401) and recorded using Spike2 software (CED, Cambridge, UK). For quantitative analyses of spike rate, patterns (CV and CV2), and rhythms, we collect continuous traces of >300 s and examine spikes with Spike2 (Spike2 software), MS Excel, and MATLAB (MathWorks, Natick, MA). Please refer to our recent work for additional details on electrophysiology and the methods for analysis of spike properties [29, 30, 37, 46].

3 Methods

Method for juxtacellular recording and filling neurons in the mouse cerebellum

The following equipment is used in the described method:

Heating pad (#DCT-15; Kent Scientific, Torrington, CT, USA).

Motorized micromanipulator (MP-225; Sutter Instrument Co., Novato, CA, USA).

Amplifier (ELC-03XS; NPI Electronic Instruments, Tamm, Germany)

Amplifier headstage (ELC-03XS; NPI Electronic Instruments, Tamm, Germany)

Data acquisition interface (CED Power1401; Cambridge Electronic Design Limited, Cambridge, England).

Audio monitor (Grass Technologies AM10; Astro-Med Inc., West Warwick, RI, USA).

Glass pipettes (#30-0057; 100 mm length, inner diameter 0.86 mm, outer diameter 1.5 mm; Harvard Apparatus, Holliston, MA, USA).

Pipette puller (P-1000; Sutter Instrument Co., Novato, CA, USA).

Hamilton Metal Hub Needle (28 gauge, blunt, point style 3; Hamilton Robotics, Reno NV, USA).

Neurobiotin tracer (Vector Laboratories Inc., Burlingame, CA) diluted to 1% in 0.9% sterile saline.

The following recipe is used to pull glass pipettes to the desired specifications.

Line	Heat	Pull	Velocity	Delay	Pressure
1 (×1)	Ramp	0	22	1	600
2 (×1)	Ramp	0	22	1	
3 (×1)	Ramp	0	22	1	
4 (×1)	Ramp	26	22	1	

Note that "Ramp" refers to the safe heat level as determined by a ramp test.

Making glass electrodes:

Glass pipettes can be pulled to 5–13 MΩ resistance with sufficient strength to puncture the dura mater using the recipe described above (Fig. 1c). Rest the pipettes in a container that allows them to be stored either horizontally or with a slight downward angle while preventing the sharp tips from being touched. *Pipettes should be used quickly after being pulled and covered when not in use to prevent dust accumulation. Electrodes should be made "fresh" before each recording and not stored long term.* Place a small drop of 0.9% saline on the back opening of the pipette to allow capillary action to fill the tip of the electrode with saline (it may take a couple of minutes for capillary action to draw the drop into the tip of the electrode). After enough saline has been drawn into the electrode to fill the conical tip, slowly backfill the electrode with 1% Neurobiotin tracer in saline using the infusion needle until the volume is sufficient to submerge the AgCl wire of the electrode interface. There should be no air bubbles in the electrode. Insert the electrode into the headstage and a reference wire and a ground wire into the small pool of saline over the craniotomy. Test the resistance of the electrode by lowering it into the pool of saline and passing a 1 nA current through the electrode. Discard the pipette and repeat the previous steps if the electrode resistance is outside of the 5–13 MΩ range. If the resistance is acceptable, apply positive pressure to the electrode tip by pressing 3 mL of room air behind the electrode.

Isolating single units in vivo:
Slowly lower the electrode past the dura mater and into the brain tissue. Pass a small current into the electrode (we use the manual pulse buzz on the NPI amplifier) to remove any cellular debris that might have accumulated on the electrode after it entered the brain. Continue to slowly advance the electrode through the tissue (for instance, speeds 5–7 on the MP225 micromanipulator) until a cell is encountered. After a cell is encountered, the experimenter should carefully bring the electrode closer to the cell in order to maximize the spike amplitude without damaging the cell (Fig. 1d). Ideally the action potential's positive and negative deflections should both be clearly apparent (Fig. 2b). Release the positive pressure from the electrode when the spike amplitude is optimized in order to form a loose patch onto the cell. The experimenter should observe a marked increase in amplitude after the positive pressure is released. To gain a tighter seal, a small amount of negative pressure can be applied to the electrode if necessary.

Entraining the cell with positive current and then filling it with Neurobiotin:
Once the cell is patched onto, its isolated spiking activity can be recorded for extended periods of time. The authors recommend recording the cell for at least 5 min for analyses and to ensure that the cell was not damaged during the patch. The amplifier should be set for current clamp experiments with a band-pass filter from 0.3 to 13 kHz for these recordings. The experimenter can begin to fill the cell with Neurobiotin tracer after the desired recording duration has been completed. To do this, maintain the patch onto the cell while disrupting the membrane with pulses of positive electrical current: 500 ms square pulse followed by no stimulation for 500 ms and repeat. Start at 0 nA, and slowly increase the current until the cell becomes sustainably entrained to the stimulus (i.e., firing with greater amplitude or more frequently during stimulation and less or not at all between stimulation pulses) (see sample trace in Fig. 3b). The final current should be between 1 and 5 nA. The experimenter may need to adjust the applied current while filling the cell in order to maintain the cell's firing properties. The cell should neither be allowed to fire uncontrollably nor experience depolarization block. The experimenter should use their audio monitor in conjunction with their data acquisition software to detect slight changes in firing properties. If a change is detected, carefully reduce the current until the cell returns to its previous entrained behavior. The experimenter should continue to stimulate the cell for at least 1 h (Fig. 3b), though more time may be needed for lower stimulation currents.

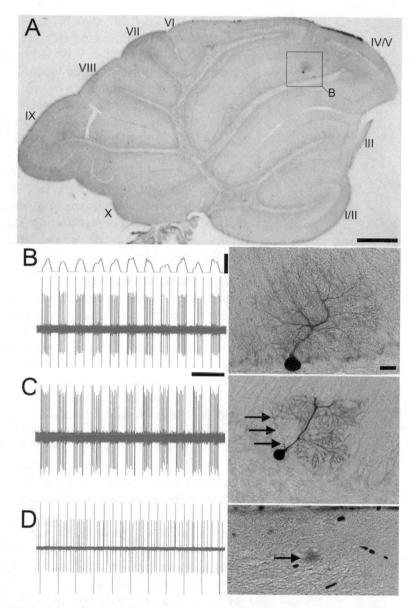

Fig. 3 Juxtacellular labeling of Purkinje cells. (**a**) Low-power view of the cerebellum showing a single-labeled Purkinje cell in lobules IV/V. The black box indicates that the cell is shown at higher power in panel **b**. Scale bar = 500 μm. (**b**) After isolating the Purkinje cell signal, it was entrained with current and filled with Neurobiotin. The cell was almost completely filled with tracer. The pink trace shows the frequency achieved during entrainment. Scale bar for pink trace = 25 Hz, scale bar for green trace = 2 s, and the scale for the staining = 50 μm. (**c**) A partially filled Purkinje cell which was likely the result of a short entrainment. The *black arrows* point to the electrode track as the electrode approaches the Purkinje cell. (**d**) A lightly filled Purkinje cell soma after weak entrainment of the cell (*black arrow*). Note that the large deflections in the green raw traces of electrical activity in **B-D** represent the borders of stimulation pulses being delivered to the tissue while the shorter deflections represent the action potentials

Post-filling recovery of the cell:

Slowly reduce the applied current after filling the cell, and continue to record the cell after 0 nA is reached to ensure the cell's healthy return to non-entrained firing. Then, reapply light positive pressure to the electrode tip (~1 mL room air) to "push" the cell away from the electrode tip. The cell's firing rate and pattern should remain stable, and the experimenter should observe a reduction in spike amplitude as the cell moves away from the electrode. Slowly begin withdrawing the electrode from the tissue. *Do not attempt to continue deeper into the tissue.* If the experimenter wishes to record and/or fill additional cells in the same brain, after the electrode is fully withdrawn from the tissue, remove it from the headstage, choose a new recording site far from the tissue that was displaced from the previous penetration, and begin again with a new electrode.

Perfusion and tissue processing:

Anesthetize the mouse with Avertin or your approved method. Wait until the toe pinch reflex is completely abolished before beginning the perfusion procedure. Flush the blood out of the mouse by driving it with the transcardiac method using PBS pH 7.4 (or saline), and then perfuse the tissue using 4% PFA. Dissect the brain and then postfix it overnight in 4% PFA. The following day put the brain into a 15% sucrose solution diluted in PBS until the brain sinks to the bottom of the container. Next, move the brain into a 30% sucrose solution diluted in PBS, again until the brain sinks. It is now cryoprotected and ready for freezing. Dry the brain gently on a Kimwipe, then mount the brain in OCT and immediately freeze at −80 °C for >1.5 h prior to sectioning. Cut 40 μm sections on a cryostat at −20 °C. For free-floating immunohistochemistry, collect tissue sections in a 24-well plate filled with PBS. The filled cells can then be detected using the VECTASTAIN Elite ABC method (Vector Laboratories Inc.) and visualized with DAB (DAB forms a brown reaction product as seen in Fig. 3). After the desired color intensity is reached, stop the DAB reaction by rinsing the tissue with several rinses of PBS. Float the tissue sections onto glass slides (Fisherbrand ColorFrost Plus microscope slides (Cat. # 12-550-16A, Fisher Scientific)) and then let them dry overnight. Dehydrate the tissue section in a series of ethanol (15%, 50%, 50%, 75%, 100%) and then clear with xylene. Coverslip the tissue using a permanent mounting medium (Entellan mounting media (Electron Microscopy Sciences, Hatfield, PA, USA)) and then allow the slide to dry overnight. The Neurobiotin can also be detected using a fluorescent signal, in which case the tissue is mounted using an aqueous fluorescence-saving medium and the tissue stored in the dark at 4°C.

Imaging the stained tissue:

We capture our bright field photomicrographs of DAB-reacted tissue sections using a Zeiss AxioCam MRc5 camera mounted on a Zeiss Axio Imager M2 microscope. Images of juxtacellularly labeled Neurobiotin-filled Purkinje cells from the stained tissue sections are acquired and analyzed with Zeiss AxioVision software (release 4.8) or Zeiss ZEN software (2012 edition). The raw data are then imported into Adobe Photoshop CS5 and corrected for brightness and contrast levels if necessary.

4 Notes

Stability is one of the most important issues for successfully isolating and recording neurons in vivo [37]. Adequate stability of the animal's head is especially critical for recording with glass electrodes. Therefore, it is necessary to ensure that the mouse is secure throughout the entire process of recording, starting with the proper placement of the ear bars. It is also important to tailor the anesthetic regime, but remember to keep within the recommendations and proper approvals of your local veterinarian and animal welfare office (e.g., IACUC). There are specific measures that one can take to improve the juxtacellular labeling itself. As mentioned before, it is best to entrain and fill the cell for as long as possible, without killing the cell. Although ~15–30 min can be enough time to fill and identify neurons, in our experience labeling for 1 h provides the most robust and complete fill (Fig. 3) [46]. However, after the cell filling is complete, it could be beneficial to wait for several hours after injection before perfusing the animal [61, 62]. Keep in mind that not all neurons are as large as Purkinje cells so shorter periods of filling and post-filling may be fine, especially for smaller interneurons. We advise that the experimenter also fine-tune the glass pulling process. For example, although we have provided parameters that work well for our puller, the ambient humidity (depending on where in the world the lab is located) could impact the quality of the pull. Even though every cell that is recorded is documented with specific coordinates based on where the craniotomy was made and also the specific coordinates that are read on the micromanipulator, after cutting the tissue, it is easy to misjudge where the cell was recorded. Therefore, routinely cutting and saving every tissue section, even if you used coordinates and are confident of the recorded/filled cell's location, is highly recommended. In many cases even a few 10s of microns over- or undershoot increase the risk of missing the cell. If needed, stain and analyze all the tissue sections to ensure that the single section where your cell is located is not missed (or worse, discarded). Once stained, we find it helpful to look for the electrode track, which can be outlined with DAB-stained blood cells. The track is not always obvious, but when it is it can help physically guide you to the stained neuron. We follow the track into the tissue, and often the tip of the track will be near the location of the filled cell (Fig. 3c) [46]. Even with the aid of this protocol, there is much room for improvement because it is not always predictable as to whether a recorded and entrained cell was actually filled to the degree that one might expect or whether the tissue was perfused adequately. For instance, we have shown three examples (Fig. 3): (1) fully filled cell with good staining (good perfusion, although the staining/DAB reaction could have been longer, or perhaps the filling time could have been increased to achieve stronger staining) (Fig. 3a, b), (2) partially filled cell (strong staining, not long enough filling time, or the loose-patch (entrainment) was interrupted during recording) (Fig. 3c), and (3) stained soma (perfusion not so good, which can be predicted by the presence of blood vessels around the cell, current was passed for only 13 min, and while the cell diminished firing frequency during stimulation and therefore to some degree was under the influence of the manipulation, it did not become entrained) (Fig. 3d). Thus, juxtacellular recording requires special attention from the perspective of different disciplines as the behavior of the mouse must be carefully monitored during anesthesia, the electrophysiological properties have to guide the filling, and the anatomical integrity of the tissue determines the quality of the visualized signal.

5   Conclusions

Techniques for injecting dyes and tracers through an electrophysiological recording pipette have been used for more than 50 years [63]. Since then there have been many improvements, although Pinault [1] designed one of the most influential adaptation, the juxtacellular approach. His method sparked interest in developing more consistent methods [7, 64, 65] and also enhanced the research of structure-function relationships in different regions of the brain including the cerebellum [12], hypothalamus [66], striatum [67], subthalamic nucleus [68], various regions of the cortex [69], hippocampus [70], and basolateral amygdala [71]. Juxtacellular labeling has been used in vitro [72] and in vivo in different species including mouse [10], rat [73, 74], and nonhuman primate [7]. Remarkably, some groups have also used the technique for filling neurons that were recorded in behaving animals [75, 76]. Recent applications of the juxtacellular method even include the use of small currents to manipulate single-neuron activity during behavior [77]. Juxtacellular labeling was also recently used to identify and examine the firing properties of Purkinje cells in conditional mutant mice that do not have the complex spike action potential as a signature [46], highlighting the applicability of this technique to probe abnormal cell function and anatomy in mutant and disease states. Together, these different juxtacellular approaches have shed new light on how spike properties of single neurons throughout the central nervous system contribute to circuit function and behavior.

Acknowledgments

This work was supported by funds from Baylor College of Medicine (BCM) and Texas Children's Hospital. R.V.S. received support from the Bachmann-Strauss Dystonia and Parkinson Foundation, Inc., the Caroline Wiess Law Fund for Research in Molecular Medicine, BCM IDDRC U54HD083092, the National Center for Research Resources C06RR029965, and the National Institutes of Neurological Disorders and Stroke (NINDS) R01NS089664. A.M.B. received support from NINDS F31NS101891, and J.J.W. received support from NINDS F31NS092264. The BCM IDDRC Neuropathology Sub-Core performed a portion of the tissue staining (the BCM IDDRC Neurovisualization Core is supported by U54HD083092). The content is solely the responsibility of the authors and does not necessarily represent the official views of the National Center For Research Resources or the National Institutes of Health (NIH).

References

1. Pinault D (1996) A novel single-cell staining procedure performed in vivo under electrophysiological control: morpho-functional features of juxtacellularly labeled thalamic cells and other central neurons with biocytin or Neurobiotin. J Neurosci Methods 65(2):113–136

2. Pinault D, Deschênes M (1998) Anatomical evidence for a mechanism of lateral inhibition in the rat thalamus. Eur J Neurosci 10(11):3462–3469

3. Pinault D, Deschênes M (1998) Projection and innervation patterns of individual thalamic reticular axons in the thalamus of the adult rat: a three-dimensional, graphic, and morphometric analysis. J Comp Neurol 391(2):180–203

4. Pinault D, Vergnes M, Marescaux C (2001) Medium-voltage 5-9-Hz oscillations give rise to spike-and-wave discharges in a genetic model of absence epilepsy: in vivo dual extracellular recording of thalamic relay and reticular neurons. Neuroscience 105(1):181–201

5. Pinault D (2003) Cellular interactions in the rat somatosensory thalamocortical system during normal and epileptic 5-9 Hz oscillations. J Physiol 552(3):881–905

6. Zheng TW, O'Brien TJ, Morris MJ et al (2012) Rhythmic neuronal activity in S2 somatosensory and insular cortices contribute to the initiation of absence-related spike-and-wave discharges. Epilepsia 53(11):1948–1958

7. Joshi S, Hawken MJ (2006) Loose-patch-juxtacellular recording in vivo—a method for functional characterization and labeling of neurons in macaque V1. J Neurosci Methods 156(1–2):37–49

8. Simpson JI, Hulscher HC, Sabel-Goedknegt E et al (2005) Between in and out: linking morphology and physiology of cerebellar cortical interneurons. Prog Brain Res 148:329–340

9. Holtzman T, Rajapaksa T, Mostofi A et al (2006) Different responses of rat cerebellar Purkinje cells and Golgi cells evoked by widespread convergent sensory inputs. J Physiol 574(2):491–507

10. Barmack NH, Yakhnitsa V (2008) Functions of interneurons in mouse cerebellum. J Neurosci 28(5):1140–1152

11. Barmack NH, Yakhnitsa V (2011) Microlesions of the inferior olive reduce vestibular modulation of Purkinje cell complex and simple spikes in mouse cerebellum. J Neurosci 31(27):9824–9835

12. Ruigrok TJH, Hensbroek RA, Simpson JI (2011) Spontaneous activity signatures of morphologically identified interneurons in the vestibulocerebellum. J Neurosci 31(2):712–724

13. Barmack NH, Yakhnitsa V (2013) Modulated discharge of Purkinje and stellate cells persists after unilateral loss of vestibular primary afferent mossy fibers in mice. J Neurophysiol 110(10):2257–2274

14. Hensbroek RA, Belton T, van Beugen BJ et al (2014) Identifying Purkinje cells using only their spontaneous simple spike activity. J Neurosci Methods 232:173–180

15. Eccles JC, Llinás R, Sasaki K (1966) The action of antidromic impulses on the cerebellar Purkinje cells. J Physiol 182(2):316–345

16. Eccles JC, Llinás R, Sasaki K (1966) The excitatory synaptic action of climbing fibres on the Purkinje cells of the cerebellum. J Physiol 182(2):268–296

17. Eccles JC, Llinás R, Sasaki K (1966) Intracellularly recorded responses of the cerebellar Purkinje cells. Exp Brain Res 1(2):161–183

18. Eccles JC, Sasaki K, Strata P (1966) The profiles of physiological events produced by a parallel fibre volley in the cerebellar cortex. Exp Brain Res 2(1):18–34

19. Eccles JC, Llinás R, Sasaki K et al (1966) Interaction experiments on the responses evoked in Purkinje cells by climbing fibres. J Physiol 182(2):297–315

20. Eccles JC, Llinás R, Sasaki K (1966) Parallel fibre stimulation and the responses induced thereby in the Purkinje cells of the cerebellum. Exp Brain Res 1(1):17–39

21. Llinás R, Sugimori M (1980) Electrophysiological properties of in vitro Purkinje cell somata in mammalian cerebellar slices. J Physiol 305:171–195

22. Llinás R, Sugimori M (1980) Electrophysiological properties of in vitro Purkinje cell dendrites in mammalian cerebellar slices. J Physiol 305:197–213

23. Desclin JC, Colin F, Manil J (1981) Morphological correlates of cerebellar Purkinje cell activity. Prog Clin Biol Res 59A:269–277

24. Lang EJ, Sugihara I, Welsh JP et al (1999) Patterns of spontaneous purkinje cell complex spike activity in the awake rat. J Neurosci 19(7):2728–2739

25. Cerminara NL, Rawson JA (2004) Evidence that climbing fibers control an intrinsic spike generator in cerebellar Purkinje cells. J Neurosci 24(19):4510–4517

26. Davie JT, Clark BA, Hausser M (2008) The origin of the complex spike in cerebellar Purkinje cells. J Neurosci 28(30):7599–7609

27. Schmolesky MT, Weber JT, De Zeeuw CI et al (2002) The making of a complex spike: ionic composition and plasticity. Ann N Y Acad Sci 978:359–390

28. Yang Y, Lisberger SG (2014) Purkinje-cell plasticity and cerebellar motor learning are graded by complex-spike duration. Nature 510(7506):529–532

29. White JJ, Arancillo M, Stay TL et al (2014) Cerebellar zonal patterning relies on Purkinje cell neurotransmission. J Neurosci 34(24):8231–8245

30. Arancillo M, White JJ, Lin T et al (2015) In vivo analysis of Purkinje cell firing properties during postnatal mouse development. J Neurophysiol 113(2):578–591

31. Hewitt AL, Popa LS, Ebner TJ (2015) Changes in Purkinje cell simple spike encoding of reach kinematics during adaption to a mechanical perturbation. J Neurosci 35(3):1106–1124

32. Ohmae S, Medina JF (2015) Climbing fibers encode a temporal-difference prediction error during cerebellar learning in mice. Nat Neurosci 18(12):1798–1803

33. Cerminara NL, Lang EJ, Sillitoe RV et al (2015) Redefining the cerebellar cortex as an assembly of non-uniform Purkinje cell microcircuits. Nat Rev Neurosci 16(2):79–93

34. Herndon RM (1963) The fine structure of the Purkinje cell. J Cell Biol 18(1):167–180

35. Ramón y Cajal S. (1909) Histologie du système nerveux de l'homme. Paris

36. Reeber SL, Otis TS, Sillitoe RV (2013) New roles for the cerebellum in health and disease. Front Syst Neurosci 7:83

37. White JJ, Lin T, Brown AM et al (2016) An optimized surgical approach for obtaining stable extracellular single-unit recordings from the cerebellum of head-fixed behaving mice. J Neurosci Methods 262:21–31

38. Armstrong DM, Rawson JA (1979) Responses of neurones in nucleus interpositus of the cerebellum to cutaneous nerve volleys in the awake cat. J Physiol 289:403–423

39. Armstrong DM, Rawson JA (1979) Activity patterns of cerebellar cortical neurones and climbing fibre afferents in the awake cat. J Physiol 289:425–448

40. Thach WT (1967) Somatosensory receptive fields of single units in cat cerebellar cortex. J Neurophysiol 30(4):675–696

41. Wise AK, Cerminara NL, Marple-Horvat DE et al (2010) Mechanisms of synchronous activity in cerebellar Purkinje cells. J Physiol 588(Pt 13):2373–2390

42. Shin S-L, Hoebeek FE, Schonewille M et al (2007) Regular patterns in cerebellar Purkinje cell simple spike trains. PLoS One 2(5):e485

43. Bosman LWJ, Koekkoek SKE, Shapiro J et al (2010) Encoding of whisker input by cerebellar Purkinje cells. J Physiol 588(Pt 19):3757–3783

44. de Solages C, Szapiro G, Brunel N et al (2008) High-frequency organization and synchrony of activity in the purkinje cell layer of the cerebellum. Neuron 58(5):775–788

45. White JJ, Arancillo M, King A et al (2016) Pathogenesis of severe ataxia and tremor without the typical signs of neurodegeneration. Neurobiol Dis 86:86–98

46. White JJ, Sillitoe RV (2017) Genetic silencing of olivocerebellar synapses causes dystonia-like behavior in mice. Nat Commun 8:14912. https://doi.org/10.1038/ncomms14912

47. Chaumont J, Guyon N, Valera AM et al (2013) Clusters of cerebellar Purkinje cells control their afferent climbing fiber discharge. Proc Natl Acad Sci U S A 110(40):16223–16228

48. Cheron G, Sausbier M, Sausbier U et al (2009) BK channels control cerebellar Purkinje and Golgi cell rhythmicity in vivo. PLoS One 4(11):e7991

49. Goossens HH, Hoebeek FE, Van Alphen AM et al (2004) Simple spike and complex spike activity of floccular Purkinje cells during the optokinetic reflex in mice lacking cerebellar long-term depression. Eur J Neurosci 19(3):687–697

50. Witter L, Canto CB, Hoogland TM et al (2013) Strength and timing of motor responses mediated by rebound firing in the cerebellar nuclei after Purkinje cell activation. Front Neural Circuits 7:133

51. Zhou H, Lin Z, Voges K et al (2014) Cerebellar modules operate at different frequencies. elife 3:e02536

52. Person AL, Raman IM (2012) Synchrony and neural coding in cerebellar circuits. Front Neural Circuits 6:97

53. Badura A, Schonewille M, Voges K et al (2013) Climbing fiber input shapes reciprocity of Purkinje cell firing. Neuron 78(4):700–713

54. Medina JF, Lisberger SG (2008) Links from complex spikes to local plasticity and motor learning in the cerebellum of awake-behaving monkeys. Nat Neurosci 11(10):1185–1192

55. Popa L.S., Hewitt A.L., Ebner T.J. (2013) Purkinje cell simple spike discharge encodes error signals consistent with a forward internal model. Cerebellum 12(3), 331–3

56. Xiao J, Cerminara NL, Kotsurovskyy Y et al (2014) Systematic regional variations in Purkinje cell spiking patterns. PLoS One 9(8):e105633

57. Brochu G, Maler L, Hawkes R (1990) Zebrin II: a polypeptide antigen expressed selectively by Purkinje cells reveals compartments in rat and fish cerebellum. J Comp Neurol 291(4):538–552

58. Ahn AH, Dziennis S, Hawkes R et al (1994) The cloning of zebrin II reveals its identity with aldolase C. Development 120(8):2081–2090

59. Sillitoe RV, Hawkes R (2002) Whole-mount immunohistochemistry: a high-throughput screen for patterning defects in the mouse cerebellum. J Histochem Cytochem 50(2):235–244

60. Apps R, Hawkes R (2009) Cerebellar cortical organization: a one-map hypothesis. Nat Rev Neurosci 10(9):670–681

61. Bevan MD, Booth PA, Eaton SA et al (1998) Selective innervation of neostriatal interneurons by a subclass of neuron in the globus pallidus of the rat. J Neurosci 18(22):9438–9452

62. Sadek AR, Magill PJ, Bolam JP (2007) A single-cell analysis of intrinsic connectivity in the rat globus pallidus. J Neurosci 27(24):6352–6362

63. Stretton AO, Kravitz EA (1968) Neuronal geometry: determination with a technique of intracellular dye injection. Science 162(3849):132–134

64. Wilson C.J., Sachdev R.N.S. (2004) Intracellular and juxtacellular staining with biocytin. Curr Protoc Neurosci Chapter 1, Unit 1.12

65. Duque A, Zaborszky L (2006) Juxtacellular labeling of individual neurons in vivo: from electrophysiology to synaptology. In: Neuroanatomical tract-tracing 3. Springer US, New York, pp 197–236

66. Hassani OK, Henny P, Lee MG et al (2010) GABAergic neurons intermingled with orexin and MCH neurons in the lateral hypothalamus discharge maximally during sleep. Eur J Neurosci 32(3):448–457

67. Inokawa H, Yamada H, Matsumoto N et al (2010) Juxtacellular labeling of tonically active neurons and phasically active neurons in the rat striatum. Neuroscience 168(2):395–404

68. Hartung H, Tan SKH, Steinbusch HMW et al (2011) High-frequency stimulation of the subthalamic nucleus inhibits the firing of juxtacellular labelled 5-HT-containing neurons. Neuroscience 186:135–145

69. Herfst L, Burgalossi A, Haskic K et al (2012) Friction-based stabilization of juxtacellular recordings in freely moving rats. J Neurophysiol 108(2):697–707

70. Varga C, Golshani P, Soltesz I (2012) Frequency-invariant temporal ordering of interneuronal discharges during hippocampal oscillations in awake mice. Proc Natl Acad Sci U S A 109(40):E2726–E2734

71. Sun Y-N, Li L-B, Zhang Q-J et al (2013) The response of juxtacellular labeled GABA interneurons in the basolateral amygdaloid nucleus anterior part to 5-HT$_2$A/$_2$C receptor activation is decreased in rats with 6-hydroxydopamine lesions. Neuropharmacology 73:404–414

72. Daniel J, Polder HR, Lessmann V et al (2013) Single-cell juxtacellular transfection

and recording technique. Pflugers Arch 465(11):1637–1649

73. Tang Q, Ebbesen CL, Sanguinetti-Scheck JI et al (2015) Anatomical organization and spatiotemporal firing patterns of layer 3 neurons in the rat medial entorhinal cortex. J Neurosci 35(36):12346–12354

74. Dempsey B, Turner AJ, Le S et al (2015) Recording, labeling, and transfection of single neurons in deep brain structures. Phys Rep 3(1):e12246–e12246

75. Tang Q, Brecht M, Burgalossi A (2014) Juxtacellular recording and morphological identification of single neurons in freely moving rats. Nat Protoc 9(10):2369–2381

76. Cazakoff BN, Lau BYB, Crump KL et al (2014) Broadly tuned and respiration-independent inhibition in the olfactory bulb of awake mice. Nat Neurosci 17(4):569–576

77. Doron G, von Heimendahl M, Schlattmann P et al (2014) Spiking irregularity and frequency modulate the behavioral report of single-neuron stimulation. Neuron 81(3):653–663

Chapter 2

Targeted Electrophysiological Recordings In Vivo in the Mouse Cerebellum

Bin Wu and Martijn Schonewille

Abstract

Single-unit recordings in vivo are the unitary elements in the processing of the brain and as such essential in systems physiology to understand brain functioning. In the cerebellum, a structure with high levels of intrinsic activity, studying these elements in vivo in an awake animal is imperative to obtain information regarding the processing features of these units in action. In this chapter we address the rationale and the approach of recording electrophysiological activity in the cerebellum, particularly that of Purkinje cells, in vivo in the awake, active animal. In line with the developing appreciation for the diversity within populations of the cells of the same type, there is a growing interest in the differentiation within the population of Purkinje cells. Here we describe a successful approach to analyzing the activity of two populations of Purkinje cells, which differ in connectivity and the expression of several genes. By driving the expression of a fluorescent marker with the promotor of one of the differentiating genes, the presence of a fluorescence signal could be used to recognize and approach Purkinje cells, while the intensity of the signal can be used as a marker to identify the two subpopulations. Finally, the drawbacks and the advantages of this technique are discussed and placed into a future perspective.

Key words Purkinje cells, Extracellular recordings, Two-photon microscopy, Zebrin/aldolase C, Mouse

1 Introduction

1.1 Single-Unit Recordings In Vivo

Understanding what goes on in the most studied black box we know, the brain, is the essence of neuroscience. Brain physiology traditionally has been studied at several levels, varying from broad view approaches using EEGs or fMRI to that of individual cell structures such as axons, dendrites, or even synapses using patch-clamp recordings. Analysis at this level of detail has provided a wealth of knowledge regarding the cellular activity underlying concepts such as sensory processing, motor coding, and learning. Top-down, noninvasive approaches have taught us about functionalities, about somatotopy, or about temporal aspects of activity and about the connectivity between areas. To bridge the gap between molecular processes occurring at the (sub)cellular level and the activity observed using noninvasive techniques,

Roy V. Sillitoe (ed.), *Extracellular Recording Approaches*, Neuromethods, vol. 134,
https://doi.org/10.1007/978-1-4939-7549-5_2, © Springer Science+Business Media, LLC 2018

several electrophysiological approaches are used. Local field potential recordings indicate activity in a particular volume of recording area, while imaging—at the expense of temporal resolution—adds or increases spatial resolution. These forms of recording population code will help us understand the information passing through and processed by the ensemble but are less informative regarding the individual units. For detailed information on the specific activity of individual units, the neurons, single-unit recordings are required. Neuronal activity presents itself predominantly in the form of rapid fluctuation of the electrical membrane potential, also known as action potentials. Whereas the membrane potential can only be recorded inside the cell with patch-clamp recordings or voltage-sensitive dyes, the currents underlying the action potential are significantly large enough to reliably be recorded in the extracellular space. The potential of this technique has been known and used for a long time, including, for instance, in the landmark scientific discovery of the direction selectivity of neurons in the cat striatum [1]. But even today the technique is still used in many key studies [2, 3], in part driven by the revival of systems physiology. New technical developments such as optogenetic manipulations of cellular activity require, to assure appropriate manipulation, a cell-specific readout of the change in the firing rate induced by the light.

1.2 Targeted Recordings In Vivo: Why and How?

Although the local environment of neurons in slice preparations is thought to remain intact, features such as sensory input, extracellular chemical configuration, and axons of projecting neurons are typically disrupted. Switching to recording in vivo circumvents these issues but at the expense of the possibility to use visual information for guidance. In many instances recordings can be somewhat directed based on neuroanatomy or known cellular activity patterns or are simply not needed. If information on cell type is essential but not needed to direct sampling, cells can also be labeled and classified postmortem. In all other situations, whether it is to minimize variation or to zoom in on the properties of specific cell types, the ability to target recordings can be essential. This technique was pioneered by Margrie and colleagues, who performed in vivo electrophysiological recordings based on images generated by two-photon excitation laser-scanning microscopy [4], a technique they named two-photon targeted patching or TPTP [5]. Transgenic mice expressing enhanced green fluorescent protein (eGFP) under control of the parvalbumin promotor to label cortical interneurons were anesthetized and placed under a two-photon microscope. With the cells labeled in green and with the pipette filled with red Alexa-594, the approach could be visualized and directed, and the success rate of patching was >50% per animal and over >10% per penetration [4]. Following the

TPTP approach described above, an alternative, inverse method was developed. Kitamura and colleagues demonstrated that by labeling the extracellular space with a fluorescent dye, the neurons can be "visualized" as a negative image [6]. By perfusing the extracellular space with the same Alexa-594, which is not taken up by neurons, the authors were able to identify neurons as shadows and either electroporate or recorded from them. The success rate of obtaining a giga-seal and whole-cell configuration was >60% for pyramidal cells and interneurons per attempt. Despite promising success rates, both approaches are, however, used quite rarely, as whole-cell recordings in vivo remain labor intensive with relatively low success rates and hence their use in today's neuroscience is limited.

1.3 Cerebellar Wiring and Physiology

The cerebellum holds a special place in neuroscience, not in the least due to its electrophysiological properties. The "little brain" presents a combination of central nuclei surrounded by a highly organized, crystalline cortical structure (Fig. 1), with a sole principal output neuron, the Purkinje cell, that uses GABA as a neurotransmitter and has an immense, two-dimensional dendritic tree [7]. Information enters the cerebellum predominantly via two sources: mossy fibers and climbing fibers. Mossy fibers carry information about sensory stimulation and ongoing motor activities in the form of efference copies, or corollary discharges [8], and contact excitatory granule cells and inhibitory Golgi cells [9, 10]. In contrast, single climbing fibers provide excitatory input to each Purkinje cell resulting in a complex spike [11, 12] and are commonly thought to carry error signals [13, 14], but other potential functions have also been described [15, 16]. Granule cells send their axon into the molecular layer, where they bifurcate and provide a direct excitatory drive to the Purkinje cell's extensive dendritic tree via parallel fibers and an indirect inhibitory input via the molecular layer interneurons, stellate cells, and basket cells, which predominantly contact the Purkinje cell's dendrite and the soma, respectively [17]. Granule cells typically fire action potentials in bursts [18]. Purkinje cells are intrinsically active and integrate excitatory and inhibitory input into a GABAergic output onto cerebellar nuclear neurons. Most inhibitory interneurons have an intermediate firing rate, with less pronounced differences among them [19]. Together the mossy fiber and climbing fiber system create a matrix-like configuration. Parallel fibers run horizontally for millimeters, contacting spines on the dendrites of numerous Purkinje cells in the coronal plane. In contrast, the Purkinje cell dendrite is oriented sagittally, and the inferior olive with its climbing fibers connects to multiple Purkinje cells in a sagittal plane, with each Purkinje cell receiving hundreds of inputs from only a single climbing fiber in the adult stage (Fig. 1c). This typical configuration

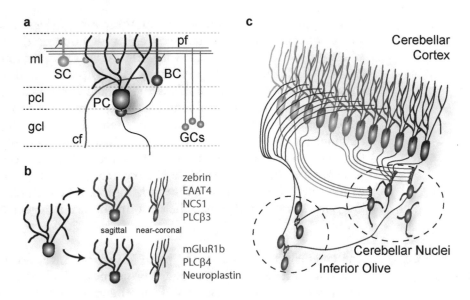

Fig. 1 The olivocerebellar circuit is constructed of modules with distinct cytochemical signatures. (**a**) Schematic drawing of the main cortical cell types, including granule (GC), stellate (SC), basket (BC), and Purkinje cells (PC) with parallel (pf) and climbing fibers (cf) in their respective layers: gcl, pcl, and molecular layer. (**b**) Purkinje cells can be subdivided based on the expression pattern of particular proteins, of which the most well known is zebrin or aldolase c. Purkinje cell dendritic trees are virtually two dimensional; in the vermis they can be fully visualized in a sagittal plane. (**c**) Schematic representation of the three-dimensional configuration and the inter-modular connectivity between the inferior olive, cerebellar cortex, and cerebellar nuclei. Inferior olivary neurons typically project to Purkinje cells of the same zebrin identity in similar modules, which in turn commonly project to the same regions in the cerebellar nuclei. As a result, the cerebellar nuclei can be divided into zebrin-positive and zebrin-negative parts too, based on zebrin-stained Purkinje cell axons. In contrast to this organization in the sagittal plane, parallel fibers (not depicted) run horizontally, i.e., perpendicular to the orientation of the virtually flat Purkinje cell dendritic tree

sparked the interest of computational neuroscientists to generate a theory for cerebellar learning [20, 21], which is still the topic of discussion today [22–27].

1.4 Electrophysiological Recordings in the Cerebellum In Vivo

The cerebellum is characterized by several features that facilitate convenient, reliable in vivo electrophysiological recordings. Firstly there is the location; as the cerebellum is situated in the back of the skull, there is ample space to place an immobilizing construct on the more frontal parts, assuring a preparation that is solid enough for chronic experiments. Secondly, the layered, cytoarchitectural organization, due to its predictable pattern, allows for online reconstructions of the trajectory and thereby makes it possible to determine in which lobules recordings are made, even without postmortem analysis (Fig. 3a). Moreover, due to the two types of inputs, parallel fibers and climbing fibers, the cells that integrate all inputs and form the only output of the cerebellar cortex, the Purkinje cells, can be readily identified by their two types of action potentials. Activation of the hundreds of synapses from a single climbing fiber onto a Purkinje cell results in a massive, prolonged depolarization, also known as the "complex" spike [12]. This signature response can also be observed in extracellular recordings, here too as a complex shape (Fig. 3a), and as such forms a landmark feature to identify cerebellar Purkinje cells. Hence, it has been possible for decades, and thus since long before two-photon imaging became available, to perform cell-type-specific targeted recordings of Purkinje cells. The "visualization" here, however, comes in the form of recorded electrical signal, which is commonly converted into an auditory signal that allows the experimenter to "hear" Purkinje cells, rather than see them. Finally, combining the advantages elements described above, the ability to record cell-specific activity at relatively deep locations in the cerebellum with long-tapered pipettes brings the advantage of increased stability due to the intimate and extended contact between the pipette and brain. The relevance of the technique in neuroscience is substantiated by the list of animal types (and labs) that have been used to perform cerebellar extracellular recordings in vivo. Originally technical limitation prohibited the use of smaller animals, and experiments were done in larger mammals, e.g., in monkeys [28–31], cats [32–35], and rabbits [36–39], and less conventional animal models including fish [40, 41], birds [42], and amphibians [43]. More recent work has been done on rats [19, 44, 45] and ferrets [46, 47], but the mouse has rapidly moved into the spot of the most popular animal for extracellular recordings in vivo, also in the cerebellum [3, 13, 48–51].

1.5 Cerebellar Genetics and Connectivity

With its crystalline architecture, the cerebellum has attracted the attention from neuroanatomists since the early days of neuroscience [52, 53]. Particularly the cerebellar cortex attracted attention, with its apparent highly homogenously present layered organization (Fig. 1c). Over the years, or even decades, this image of a homogeneous structure has been contested at different levels. The first evidence against the uniform nature is the description of a patterned expression of 5′-nucleotidase, more than half a century ago [54]. Since then, numerous proteins have been found to be expressed in a particular patterns, with the most well-known example being, identified in rats, the zebrins: zebrins I and II [55, 56]. Zebrins owe their name to the peculiar pattern of expression, consisting of sagittal stripes of stained, "positive" and non-stained, "negative" Purkinje cells (Fig. 1b). This pattern of stripes of varying width is not unique to rats but has been observed in various species of different classes ranging from pigeons [57], via, e.g., hedgehogs [58], to humans. It has been observed for numerous proteins since, including, mGlur1b [59], EAAT4 [60], PLCβ3 and β4 [61], PKCδ [62], NCS-1 [63], IP3R1 [64], and neuroplastin [65] (Fig. 1b). Interestingly, the differentiation does not only exist in the genetic profile but also in morphology and connectivity. For instance, Altman and Bayer in 1977 described a cell type that is abundantly present but almost only in the granule cell layer of vestibule-cerebellum [66], later to be recognized and named by its peculiar shape as the unipolar brush cell (UBC) [67]. This discovery signifies the more general notion that despite its homogeneous appearance, the cerebellar cortex has regional variations in composition.

Except for a few variations, cumulative research indicates that the connectivity of the cerebellum adheres to particular patterns as well. PCs with the same zebrin II identity form a cerebellar module, in that they receive CF inputs from the same subnucleus of the inferior olive and send their output to the parts of the cerebellar nuclei (Fig. 1c) [68–73]. In fact, even mossy fibers can adhere to the zebrin signature [68], although there is integration of different modalities at the input stage already [74]. Hence, one could argue that zebrin is a marker identifying cerebellar cortical regions, or modules, that form a functional unit that integrates particular mossy and climbing information [68, 70, 75].

1.6 Differential Cerebellar Physiology

The presence of differential gene expression profiles between distinct Purkinje cell populations could implicate that there are related differences in physiology as well. The first evidence supporting this concept came from Wadiche and Jahr, who demonstrated that the presence of EAAT4, which has an expression pattern very similar to that of zebrin, inhibits the activation of metabotropic glutamate receptors. As a result, EAAT4-negative (and thus zebrin-negative) Purkinje cells in lobule III exhibit long-term depression of the PF-PC synapse (PF-PC LTD), while zebrin−/EAAT4-positive Purkinje cells in lobule X do not [76]. Similar differences between anterior and posterior lobules have been described for a depolarization-induced slow current, a current related to postsynaptic dopamine, released in an autocrine manner [77, 78]. In vitro experiments also revealed differences in general excitability between Purkinje cells in anterior and posterior lobules and related them to differences in specific K^+ and Na^+ currents [79]. Even the climbing fiber input itself has been suggested to be physiologically different between zebrin-positive and zebrin-negative Purkinje cells, although is not clear if these differences work in conjunction with or against those observed in Purkinje cells [80].

Taken together these results sparked our interest in the activity of individual zebrin-positive and zebrin-negative Purkinje cells in the awake, active mouse. This scientific question required exactly those recordings conditions that allow insight not only into the identity of the recorded cell in terms of cell type but also that in terms of genetic "identity," a feature that can only be visualized by immunohistochemistry or genetically encoded fluorescent markers. To determine the impact of these features or other differences in gene expression profiles, hundreds of Purkinje cells were recorded throughout the cerebellar cortex, and the activity patterns were correlated to their location and zebrin identity, either by postmortem analysis or by online identification based on two-photon imaging (Fig. 3b, c) [81, 82], results that were confirmed by others [44].

2   Methods

By preparing mice for in vivo recordings in the awake animal with the option for chronic recordings over several days, one can avoid side effects of anesthetics or analgesics (see, e.g., Schonewille et al. [48]). In this section we provide a detailed protocol for the surgical preparation of a craniotomy to access the cerebellum, followed by an overview of the experimental setup and the protocol for in vivo electrophysiological recording in an awake mouse and the data analysis. These methods have been successfully developed and applied in our laboratory over the last decade to record neuronal activities in an anesthetized [19, 48], alert [81, 83, 84], or behaving mouse [85, 86].

2.1   Surgical Procedures

Mice are commonly prepared for experiments with a preparational surgery, as also described by White et al. [87]. During the surgery a "pedestal" for immobilization of the head and a chamber for a single or multiple electrophysiological recording session(s) will be placed as follows (see also Fig. 2a). Mice are anesthetized by a mixture of isoflurane and oxygen (initial concentration, 4% V/V in O_2; maintenance concentration, 1.5–2% V/V in O_2) in a gas chamber. Once the mouse no longer has a foot pinching reflex, the mouth is opened, and the tongue is pulled out with tweezers. Next, the bar of a stereotaxic apparatus (Bilaney, Germany) is gently inserted into the mouth. When in the proper position, the screw is fastened to fix the head horizontally. To assure that body temperature is maintained at 35–37 °C, the anal thermosensor is carefully placed and connected to a heating pad (FHC, Bowdoinham, ME). The eyes are protected by covering them with an eye ointment (Duratears, Alcon, Belgium). The dorsal cranial fur is shaven with soap and a scalpel, a midline incision is made to expose the skull, and the periost is cleared and cleaned with cotton swabs (Fig. 2a). A drop of OptiBond All-In-One (Kerr, Salerno, Italy) is applied to the dorsal cranial surface of the skull from bregma to lambda and cured with light for 60 s. The adhesive layer is covered with a thin layer of Charisma composite (Heraeus Kulzer, Germany), and the pedestal is embedded in the composite and cured it immediately. Additional layers of composite are added to surround the base of

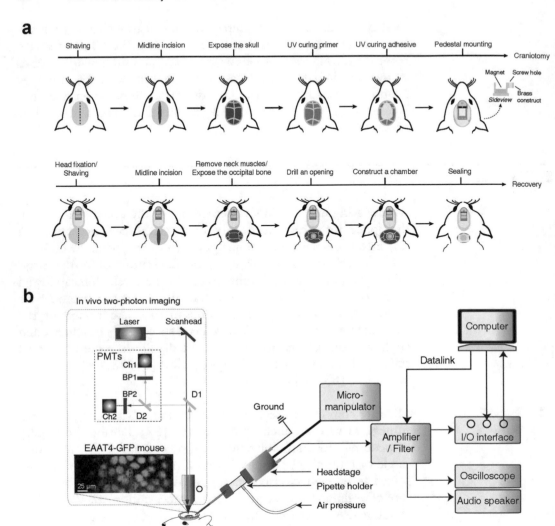

Fig. 2 Outline of preparational surgery and recording setup. (**a**) Diagram of procedures for installation of pedestal and craniotomy surgeries in anesthetized mice. (**b**) Schematic drawing of the configuration for combined electrophysiological (*right*) and two-photon imaging to allow targeted recordings in vivo (*left*). In short, the mouse is head-fixed in the restrainer. Electrical signals are recorded by a microelectrode and amplified by the headstage, which is attached to a three-axis micromanipulator, and forwarded to the main amplifier. The signal is then sent to a sound monitor and/or oscilloscope for rapid monitoring of the recorded signals, digitized by the I/O interface and stored on the computer controlling the recording settings and stimulations. *Left*, scheme of two-photon microscope for imaging the eGFP-labeled neurons and the electrode. *PMT* photomultiplier tube, *D1* dichroic 1 (695 nm split), *D2* dichroic 2 (585 nm split), *BP1* band-pass filter 1 (500–550 nm), *BP2* band-pass filter 2 (584–676 nm), *O* water immersion objective. *Inset with image*: a glass pipette containing Alexa Fluor 594 (10 μM in 2 M NaCl, depicted in *red*) is targeted toward Purkinje cells (*green circles*) using the eGFP signal from the EAAT4-positive cells, and cell population-specific extracellular recordings are obtained

the pedestal and cured with light, to complete the pedestal placement part of the surgery (Fig. 2a).

From here on, the mouse can be fixed by either the pedestal or the mouth bar. The skin over the skull is opened until the foramen magnum and the three layers of medial neck muscles overlying the occipital bone and foramen magnum are removed. A craniotomy is made by drilling an opening (max. Ø < 3 mm) in the interparietal or occipital bone, overlying the cerebellar part of interest. Size and location of the craniotomy can be adapted to the experimental approach. Smaller craniotomies are preferable for reasons of stability, and the dura mater should be preserved to keep the brain intact and healthy, if possible (Fig. 2b). A recording chamber is constructed around the craniotomy with Charisma composite again and sealed with self-curing silicone composite (Twinsil speed 22, Picodent, Germany). The mouse preferably is allowed to recover in its home cage for at least 3 days after the surgery before starting electrophysiological recordings.

2.2 Anesthesia, Analgesia, and Anti-inflammatory Drugs for Experimental Mice

Discomfort to the mice is limited by the use of several anesthetics, analgesics, and anti-inflammatory drugs given pre-, peri-, and postsurgery. A balanced combination could include isoflurane (4% induction, 1.5–2% maintenance), lidocaine (in Xylocaine 10%, drop of 100 mg/mL), bupivacaine (Bupivacaine Actavis, drop of 2.5 mg/mL), buprenorphine (Temgesic, 0.3 mg/mL), and Rimadyl (carprofen, 50 mg/mL). Isoflurane is given as indicated in the experimental approach, lidocaine and bupivacaine are administered locally as analgesic on the skull, and buprenorphine (diluted to 0.02 mg/kg) and Rimadyl (5 mg/kg) are injected intraperitoneally for postoperative analgesia and to prevent inflammation.

2.3 The Equipment and Recording Procedures

There is no golden standard for in vivo electrophysiological recordings. Below the method that has been developed over several years of using this technique is detailed. First, to assure that the animal is comfortable with the experimental conditions including the restraining of the head, mice are habituated by placing them in the experimental setup for 30–60 min for 1 or 2 days before the first experimental session. For habituation and experiments, the mice are immobilized using a custom-made restrainer, by bolting the head holder to a head fixation post. Prior to experiments, the silicone mixture should be removed to expose the craniotomy. Tissue can accumulate on top of the dura, which needs to be removed to allow a clean entrance and easy penetration for the recording and stimulation probes. Using a hypodermic needle, an incision is made in the dura to facilitate the penetration of glass pipettes (only needed for sharper tips), and the exposed cortex is rinsed with saline to remove blood. During the recording, the brain surface is covered with saline or agar in order to prevent dehydration. Glass pipettes (OD 1.5 mm, ID 0.86 mm, borosilicate, Sutter Instruments, USA; 1–2 µm tips, 4–8 MΩ) are pulled (P-1000, Sutter Instruments, USA) and filled with 2 M NaCl solution and Alcian Blue (for pressure injections) or 2–3% solution of biotin (Vector Laboratories). The pipette is lowered slowly into the cerebellar cortex by either an analogue, hydraulic microdrive (Trent Wells, TX, USA) or a digital microdrive (Luigs & Neumann, Ratingen, Germany) (in the final approach set to 2–4-µm step size, step every 2–3 s). The recorded signals are preamplified (custom-made preamplifier, 1000× DC), filtered and digitized (either CyberAmp320, Molecular Devices, with Power1401, CED, Cambridge, UK, or Axon MultiClamp 700B with Digidata 1440A, Molecular Devices, Sunnyvale, CA, USA), and stored on a disk (using Spike2, CED, or pClamp10, Molecular Devices) for offline analysis (Fig. 2c). To mark the recording location, brief air pulses (MPPI-3, ASI, USA) or iontophoresis (custom-made device, 4–8 µA, 7 s on–7 s off) can be used to locally deliver the Alcian Blue or biotin, respectively. After the recording session, the animal will be sacrificed if the experiment is completed or the brain is covered with an ointment (Duratears, Alcon, Belgium), and the chamber is sealed using self-curing silicone composite (Twinsil speed 22, Picodent, Germany) to continue the next day.

2.4 Targeted Recordings Using Two-Photon Imaging

Described above is the general surgical preparation for in vivo recording in awake mice. For in vivo two-photon targeted recordings, the approach is similar to these procedures, but the construction for immobilization and craniotomy is adapted to the requirements of the microscope, and the experiment is performed as soon as the animal has recovered from anesthesia. For these experiments a rectangular metal plate is used and placed perpendicular to the midline of the mouse, with grooves for solid fixation and a larger opening for imaging and recordings.

Images are acquired using a TriM Scope II (LaVision BioTec, Bielefeld, Germany) attached to an upright microscope with a 40×/0.8 NA water-immersion objective (Olympus, Tokyo, Japan). Laser illumination is provided by a Chameleon Ultra titanium sapphire laser (Coherent, Santa Clara, CA), and to detect fluorescent emissions, two photomultiplier tubes (Hamamatsu, Iwata City, Japan) are used. The recording pipette is filled with Alexa-594 (10 µM in 2 M NaCl; Life Technologies, Carlsbad, CA) and visualized with an excitation wavelength of 800 nm. Images from eGFP and Alexa-594 are filtered using a Gaussian kernel, contrast-optimized, and subsequently merged in Photoshop (Adobe, San Jose, CA). A typical recording sampled 40 × 200 µM with a frame rate of approximately 25 Hz [81] (Fig. 2c).

2.5   Selecting Cells for Recording

In contrast to *in vitro techniques*, in vivo approaches can be used to study neurons in physiological conditions with normal inputs and intact axon and in their virtually natural ambience of neuronal circuits, thus resulting in more realistic characterization that can be more easily extrapolated than in vitro recordings. However, in an intact animal, it is not easy to determine the type and location of the neuron that is being recorded. Fortunately, there are several ways to identify the cell type and location in the cerebellum.

Spike waveform. Even when recorded extracellularly, different types of cells present distinct spike waveforms. For instance, Purkinje cells are identified by the presence of simple and complex spikes and are confirmed to be a single unit by the presence of a pause in simple spikes after each complex spike [88]. Interestingly, the signal recorded when approaching a Purkinje cell depends on the location relative to the orientation of the cell. When approached from the molecular layer, the first signal that can be identified visually (and heard if the recording signals are converted to audio) is that of the complex spikes. These large, slow spikes (lower tones) appear as negative deflections from the baseline, at ~1 Hz (Fig. 3a, top). Continuing in the direction of the soma will slowly reveal the ~50–100 Hz firing rate (a soft hum) of the simple spikes, also as negative deflections, while the complex spike signal slowly changes into a predominantly positive signal (see Fig. 3a, middle). Close to the soma, the spike waveform indicates the distance between the tip of pipette and neuron. In extracellular recordings, the tip of an electrode is positioned adjacent to but outside of a neuron, as depicted in Fig. 3a (upper middle). With a clean tip one can, in optimal circumstances and—if needed—aided by a little negative pressure, establish a juxtacellular recording (Fig. 3a, lower middle). The ultimate step would be to proceed to whole-cell patch-clamp recording configuration, the most commonly used patch-clamp mode where the membrane patch is disrupted by to establish electrical and molecular access to the intracellular space. It is the gold standard for high-fidelity analysis of the electrical properties and functional connectivity of neurons, but its success rate is much lower than extracellular and juxtacellular recording (Fig. 3a, bottom) [2, 3, 89].

Firing pattern. In the cerebellar cortex, other cell types can be identified, apart from Purkinje cells. Separating these is more complicated but can be done quite reliably based on waveforms and firing characteristics under anesthesia as indicated by Ruigrok et al. [19], although this is debated [90]. In this manner, granule cells can be identified by their high irregularity and low average firing rate; unipolar brush cells have a signature low CV2 (see "data analysis"); molecular layer interneurons typically stand out by their irregularity and intermediate to high firing rate, while Golgi cells

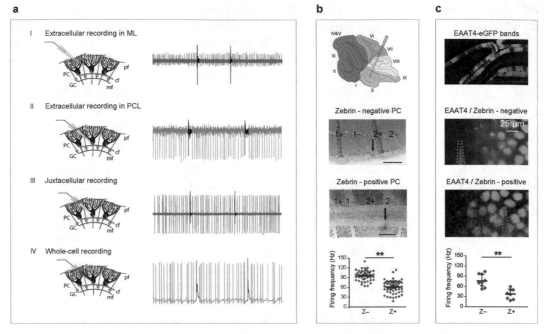

Fig. 3 In vivo recording configurations and example data from labeled and targeted Purkinje cell recordings. (**a**) The position of the recording electrode relative to the Purkinje cell determines the observed signals. Liquid-filled glass pipette electrodes are used to record single-cell activity from PCs. In the molecular layer, complex spikes are most obvious, seen as slow, negative deflections (*I*). Toward the soma complex spike, signals become predominantly positive with slower elements, while simple spikes appear as negative or more bidirectional deflections (*II*). When the tip of the electrode is placed even more intimately with the soma, this is referred to as juxtacellular, a configuration that allows cell-specific labeling (*III*). Ultimately, using suction, followed by sealing and breaking in, the configuration can be converted into a whole-cell patch-clamp recording (*IV*). (**b**) Schematic representation in sagittal plane of the electrode path. Micrographs show coronal sections of recorded Purkinje cells labeled with Alcian Blue after staining against zebrin; *top*, zebrin-negative Purkinje cell in the anterior cerebellum; *bottom*, zebrin positive in the posterior cerebellum. The *bottom graph* depicts the difference in simple spike firing rate between modules observed using this technique. (**c**) Confocal image of a coronal section of an EAAT4-eGFP brain, showing the typical EAAT4/zebrin pattern. Two-photon images (approx. The same location, shifted about 25 μm) with the pipette recording a negative (*top*) and a positive (*bottom*) Purkinje cell. The bottom graph shows how the targeted recordings confirmed the difference observed in b, with a lower number of cells. Results in **b** and **c** are adapted from Zhou et al., 2014. *PC* Purkinje cell, *GC* granule cell, *PF* parallel fiber, *CF* climbing fiber, *MF* mossy fiber

have an intermediate firing rate and are more regular [19]. Due to the poor level of somatotopy in the cerebellar cortex, the potential to identify the recording area based on responses to sensory input is limited [91]. Nonetheless there are several exceptions of well-studied areas including the (para)flocculus, which responds with reciprocal modulation of simple and complex spikes to visual and vestibular input [28, 85], the saccade-related area in lobules VI and VII in which Purkinje cells respond to saccadic eye movements [29], and a designated area in the transitions between the vermis and the hemispheres and (the hemispheral parts of) lobules

IV–V and lobule VI—simple lobule related to eyeblink conditioning [24, 35].

Labeling. To know the exact recording location when recording deeper in the brain, it is imperative to label the cell individually after completing the recording. The two most common forms of labeling with glass pipettes are iontophoresis and pressure injections, whereas with metal electrodes often lesions are placed to mark recording location. Iontophoresis results in minimal tissue damage but requires a charged tracer or dye and has limited control over injection volume or concentration. Several tracers are very suitable for iontophoresis, including biotinylated dextran amines (i.e., BDA-3000), as they can be used for visualization of long-range axonal projections and hence require intact structures and up to a few days of survival after labeling [92, 93]. Alternatively, an uncharged dye can be injected with controlled back pressure on the glass pipette, such as Alcian Blue (0.1–2% solution in saline, Sigma-Aldrich, USA) by air pressure [81, 94]. Ideal for these injections are double barrel or septum borosilicate glass pipettes (e.g., theta septum, 1.5 OD, 1.02 ID, USA), one half of the pipette is filled with 2 M NaCl for recording, and the other half barrel is filled with a blue dye for labeling. Although the immunohistological procedures are time-consuming, iontophoresis-based injections allow for the most controlled small injections that, by controlling the amount of current and duration, ideally label only one cell.

Imaging. With the growing understanding of the complexity of the brain, there is also a growing appreciation for the diversity present, even within populations of a particular cell type. To be able to characterize this diversity in vivo, tools are required to identify this subdivision within cell types, either offline with immunohistochemistry or online with fluorescent markers. Hence, to be able to identify the different cerebellar modules, based on the expression of zebrin, we used EAAT4 promotor-driven eGFP-expressing, or EAAT4-eGFP, mice [95]. EAAT4, or excitatory amino acid transporter 4, is expressed in a pattern similar to that of zebrin [96]. Alternatively, one could also use a zebrin (aldolase c) promotor-driven expression of a fluorescent marker [97]. The strong expression of eGFP in both the dendrite and the soma readily identifies zebrin−/EAAT4-positive Purkinje cells, but the weak expression in zebrin−/EAAT4-negative cells helps localizing also those, and thus facilitates targeted recordings of both types, with a single genetic manipulation [81]. By visualizing also the recording pipette with Alexa-594, both the pipette and the preferred area can be made visible, allowing for selection of and careful, precise approaches to the chosen cells.

2.6   Data Analysis

Several commercial software packages are available to analyze the recorded traces. Offline analysis is performed using SpikeTrain (Neurasmus B.V., Rotterdam, the Netherlands, www.neurasmus.com), running under MATLAB (MathWorks, MA, USA). SpikeTrain is an object-oriented program code that can use either superparamagnetic clustering [98] or principal component analysis. After automated SS and CS detection and discrimination of simple spike and complex spike, the assigned codes can be manually checked and corrected, if needed. Histograms of SS trigger on the occurrence of a CS (bin width 1 ms) can be used to verify that each isolated Purkinje cell shows a clean climbing fiber pause (i.e., no simple spikes for the duration of each complex spike; see also Fig. 3a). The absence of a climbing fiber pause, the minimum duration between a complex spike and the following simple spike, is taken to indicate that the isolation is imperfect and there is a second unit present. For each cell several parameters for simple and complex spikes can be calculated, including firing rate, CV and mean CV2, as well as the climbing fiber pause [84]. CV is the standard deviation of inter-spike intervals (ISI) divided by the mean; the mean CV2 is calculated as the mean of $2 \cdot |(ISI)n + 1 - ISIn|/(ISIn + 1 + ISIn)$ [99]. Both are measures for the regularity of the firing, also referred to as precision [50], with CV reflecting that of the entire recording and mean CV2 that of adjacent intervals, making the latter a measure of regularity on small timescales. In addition to auto-correlograms, which also illustrates the regularity, the presence of complex spikes also allows for cross-correlograms [100], of which the shape correlates with the cerebellar region [81, 82].

2.7 Technical Considerations

Several aspects need to be taken into account when performing recordings in vivo and when using fluorescent signals. First, when experimental procedures and ethical considerations allow to do so, recording in awake animals is preferable. Generalize anesthesia significantly reduces the discomfort of the animal and enhances the stability of the preparation, especially if the slower and deeper breathing can be properly controlled to avoid movement of the brain. However, anesthesia significantly impacts, for instance, Purkinje cells and changes various properties of its firing activity, depending also on the type of anesthetic [48]. The effects of anesthetics are comprehendible, as the cerebellum is a sensorimotor integrator [101] and under anesthesia at least the motor part is disrupted. Isoflurane had a more pronounced effect on Purkinje cell activity then ketamine/xylazine in that it increased the pausing time and other parameters related to inter-spike intervals more severely [48].

Moreover, both in anesthetized and in awake animals, it is essential to consider the role of temperature. It is common knowledge that in in vitro experiments temperature has a substantial influence on activity and plasticity. Similar effects can be predicted in vivo, but it is less clear to what extent they are present [102].

Using a miniature probe in vivo, we found that the temperature drops dramatically at the surface of the craniotomy in the absence of any form of heating. In fact, we observed a temperature difference of up to 5 °C compared to deeper parts of the cerebellum that correlated with a lower Purkinje cell simple spike firing rate in the first 0.5–1 mm of tissue [82]. Two-photon imaging only reaches to first couple of hundred micrometers, implicating that recordings under these conditions can be affected (compare Fig. 3b vs. Fig. 3c). A perfusion system, similar to that used in in vitro recordings, could potentially solve the issue, which is of particular relevance in studies that also include a behavioral component [103]. Conversely, during two-photon imaging, photo-stimulation can hamper the success rate of obtaining patch recordings in vivo, possibly due to photodamage or interactions with the pipette solution [4].

3   Conclusions

In this chapter we describe the use of extracellular recordings in mice during quiet wakefulness combined with methods to target specific subpopulations. We focus on cerebellar Purkinje cells and describe the approach to targeted recordings from specific subtypes of Purkinje cells. In addition to postmortem labeling, we detailed the approach to imaging-based recordings of specific Purkinje cells. In mice expressing a fluorescent protein under a cell-subtype-specific promotor (so-called zebrin-positive or zebrin-negative neurons), we used two-photon imaging to identify the target neurons and subdivide them based on the intensity of fluorescence. By visualizing the pipette tip with a fluorescent dye, we assure an optimal approach to the identified neuron and create the potential to even do dual or multiple, side-by-side recordings. In this time of increasing appreciation for the heterogeneity within particular cell types, the ability to make targeted recordings of specific subtypes will become more and more relevant.

References

1. Hubel DH, Wiesel TN (1959) Receptive fields of single neurons in the cat's striate cortex. J Physiol 148:574–591

2. van Welie I, Roth A, Ho SS, Komai S, Hausser M (2016) Conditional spike transmission mediated by electrical coupling ensures millisecond precision-correlated activity among interneurons in vivo. Neuron 90:810–823

3. Chen S, Augustine GJ, Chadderton P (2016) The cerebellum linearly encodes whisker position during voluntary movement. elife 5:e10509

4. Margrie TW et al (2003) Targeted whole-cell recordings in the mammalian brain in vivo. Neuron 39:911–918

5. Komai S, Denk W, Osten P, Brecht M, Margrie TW (2006) Two-photon targeted patching (TPTP) in vivo. Nat Protoc 1:647–652

6. Kitamura K, Judkewitz B, Kano M, Denk W, Hausser M (2008) Targeted patch-clamp recordings and single-cell electroporation of unlabeled neurons in vivo. Nat Methods 5:61–67

7. Palay SL, Chan-Palay V (1974) Cerebellar cortex: cytology and organization. Springer, Berlin, pp 180–336

8. Kennedy A et al (2014) A temporal basis for predicting the sensory consequences of motor commands in an electric fish. Nat Neurosci 17:416–422

9. Eccles J, Llinas R, Sasaki K (1964) Golgi cell inhibition in the cerebellar cortex. Nature 204:1265–1266

10. Eccles JC, Llinas R, Sasaki K (1966) The mossy fibre-granule cell relay of the cerebellum and its inhibitory control by Golgi cells. Exp Brain Res 1:82–101

11. Eccles JC, Llinas R, Sasaki K (1966) The excitatory synaptic action of climbing fibres on the Purkinje cells of the cerebellum. J Physiol 182:268–296

12. Schmolesky MT, Weber JT, De Zeeuw CI, Hansel C (2002) The making of a complex spike: ionic composition and plasticity. Ann N Y Acad Sci 978:359–390

13. Najafi F, Medina JF (2013) Beyond "all-or-nothing" climbing fibers: graded representation of teaching signals in Purkinje cells. Front Neural Circuits 7:115

14. Ito M (2001) Cerebellar long-term depression: characterization, signal transduction, and functional roles. Physiol Rev 81:1143–1195

15. Winkelman B, Frens M (2006) Motor coding in floccular climbing fibers. J Neurophysiol 95:2342–2351

16. Yarom Y, Cohen D (2002) The olivocerebellar system as a generator of temporal patterns. Ann N Y Acad Sci 978:122–134

17. Andersen P, Eccles J, Voorhoeve PE (1963) Inhibitory synapses on somas of purkinje cells in the cerebellum. Nature 199:655–656

18. van Beugen BJ, Gao Z, Boele HJ, Hoebeek F, De Zeeuw CI (2013) High frequency burst firing of granule cells ensures transmission at the parallel fiber to purkinje cell synapse at the cost of temporal coding. Front Neural Circuits 7:95

19. Ruigrok TJ, Hensbroek RA, Simpson JI (2011) Spontaneous activity signatures of morphologically identified interneurons in the vestibulocerebellum. J Neurosci 31:712–724

20. Albus JS (1971) A theory of cerebellar function. Math Biosci 10:25–61

21. Marr D (1969) A theory of cerebellar cortex. J Physiol 202:437–470

22. Ito M, Sakurai M, Tongroach P (1982) Climbing fibre induced depression of both mossy fibre responsiveness and glutamate sensitivity of cerebellar Purkinje cells. J Physiol 324:113–134

23. Schonewille M et al (2011) Reevaluating the role of LTD in cerebellar motor learning. Neuron 70:43–50

24. ten Brinke MM et al (2015) Evolving models of pavlovian conditioning: cerebellar cortical dynamics in awake behaving mice. Cell Rep 13:1977–1988

25. Boyden ES, Katoh A, Raymond JL (2004) Cerebellum-dependent learning: the role of multiple plasticity mechanisms. Annu Rev Neurosci 27:581–609

26. Gao Z, van Beugen BJ, De Zeeuw CI (2012) Distributed synergistic plasticity and cerebellar learning. Nat Rev Neurosci 13:619–635

27. Ito M, Yamaguchi K, Nagao S, Yamazaki T (2014) Long-term depression as a model of cerebellar plasticity. Prog Brain Res 210:1–30

28. Lisberger SG, Fuchs AF (1974) Response of flocculus Purkinje cells to adequate vestibular stimulation in the alert monkey: fixation vs. compensatory eye movements. Brain Res 69:347–353

29. Thier P, Dicke PW, Haas R, Barash S (2000) Encoding of movement time by populations of cerebellar Purkinje cells. Nature 405:72–76

30. Pasalar S, Roitman AV, Durfee WK, Ebner TJ (2006) Force field effects on cerebellar Purkinje cell discharge with implications for internal models. Nat Neurosci 9:1404–1411

31. Roy JE, Cullen KEA (1998) Neural correlate for vestibulo-ocular reflex suppression during voluntary eye-head gaze shifts. Nat Neurosci 1:404–410

32. Sato Y, Miura A, Fushiki H, Kawasaki T, Watanabe Y (1993) Complex spike responses of cerebellar Purkinje cells to constant velocity optokinetic stimuli in the cat flocculus. Acta Otolaryngol Suppl 504:13–16

33. Jorntell H, Ekerot CF (2002) Reciprocal bidirectional plasticity of parallel fiber receptive fields in cerebellar Purkinje cells and their afferent interneurons. Neuron 34:797–806

34. Yartsev MM, Givon-Mayo R, Maller M, Donchin O (2009) Pausing purkinje cells in the cerebellum of the awake cat. Front Syst Neurosci 3:2

35. Hesslow G (1994) Inhibition of classically conditioned eyeblink responses by stimulation of the cerebellar cortex in the decerebrate cat. J Physiol Lond 476:245–256

36. Ekerot CF, Kano M (1989) Stimulation parameters influencing climbing fibre induced long-term depression of parallel fibre synapses. Neurosci Res 6:264–268

37. Simpson JI, Alley KE (1974) Visual climbing fiber input to rabbit vestibulo-cerebellum: a source of direction-specific information. Brain Res 82:302–308

38. Yagi N, Chikamori Y, Matsuoka I (1977) Response of single Purkinje neurons in the flocculus of albino rabbits to caloric stimulation. Acta Otolaryngol 84:98–104

39. Miyashita Y (1984) Eye velocity responsiveness and its proprioceptive component in the floccular Purkinje cells of the alert pigmented rabbit. Exp Brain Res 55:81–90

40. Yoshida M, Kondo H (2012) Fear conditioning-related changes in cerebellar Purkinje cell activities in goldfish. Behav Brain Funct 8:52

41. Sawtell NB, Williams A, Bell CC (2007) Central control of dendritic spikes shapes the responses of Purkinje-like cells through spike timing-dependent synaptic plasticity. J Neurosci 27:1552–1565

42. Wylie DR, Frost BJ (1991) Purkinje cells in the vestibulocerebellum of the pigeon respond best to either translational or rotational wholefield visual motion. Exp Brain Res 86:229–232

43. Llinas R, Bloedel JR, Hillman DE (1969) Functional characterization of neuronal circuitry of frog cerebellar cortex. J Neurophysiol 32:847–870

44. Xiao J et al (2014) Systematic regional variations in Purkinje cell spiking patterns. PLoS One 9:e105633

45. Shin SL et al (2007) Regular patterns in cerebellar Purkinje cell simple spike trains. PLoS One 2:e485

46. Hesslow G, Ivarsson M (1994) Suppression of cerebellar Purkinje cells during conditioned responses in ferrets. Neuroreport 5:649–652

47. Lou JS, Bloedel JR (1986) The responses of simultaneously recorded Purkinje cells to the perturbations of the step cycle in the walking ferret: a study using a new analytical method—the real-time postsynaptic response (RTPR). Brain Res 365:340–344

48. Schonewille M et al (2006) Purkinje cells in awake behaving animals operate at the upstate membrane potential. Nat Neurosci 9:459–461; author reply 461

49. Arancillo M, White JJ, Lin T, Stay TL, Sillitoe RV (2015) In vivo analysis of Purkinje cell firing properties during postnatal mouse development. J Neurophysiol 113:578–591

50. Walter JT, Alvina K, Womack MD, Chevez C, Khodakhah K (2006) Decreases in the precision of Purkinje cell pacemaking cause cerebellar dysfunction and ataxia. Nat Neurosci 9:389–397

51. Barmack NH, Yakhnitsa V (2008) Functions of interneurons in mouse cerebellum. J Neurosci 28:1140–1152

52. Cajal SR y (1911) Histologie du Système Nerveux de l'Homme et des Vertébrés. Vol. I–II

53. Henle J (1879) Handbuch der Nervenlehre des Menschen. Fachbuchverlag, Dresden

54. Scott TG (1963) A unique pattern of localization within the cerebellum. Nature 200:793

55. Leclerc N, Dore L, Parent A, Hawkes R (1990) The compartmentalization of the monkey and rat cerebellar cortex: zebrin I and cytochrome oxidase. Brain Res 506:70–78

56. Brochu G, Maler L, Hawkes R (1990) Zebrin II: a polypeptide antigen expressed selectively by Purkinje cells reveals compartments in rat and fish cerebellum. J Comp Neurol 291:538–552

57. Graham DJ, Wylie DR (2012) Zebrin-immunopositive and -immunonegative stripe pairs represent functional units in the pigeon vestibulocerebellum. J Neurosci 32:12769–12779

58. Sillitoe RV, Kunzle H, Hawkes R (2003) Zebrin II compartmentation of the cerebellum in a basal insectivore, the Madagascan hedgehog tenrec Echinops telfairi. J Anat 203:283–296

59. Grandes P, Mateos JM, Ruegg D, Kuhn R, Knopfel T (1994) Differential cellular localization of three splice variants of the mGluR1 metabotropic glutamate receptor in rat cerebellum. Neuroreport 5:2249–2252

60. Nagao S, Kwak S, Kanazawa I (1997) EAAT4, a glutamate transporter with properties of a chloride channel, is predominantly localized in Purkinje cell dendrites, and forms parasagittal compartments in rat cerebellum. Neuroscience 78:929–933

61. Sarna JR, Marzban H, Watanabe M, Hawkes R (2006) Complementary stripes of phospholipase Cbeta3 and Cbeta4 expression by Purkinje cell subsets in the mouse cerebellum. J Comp Neurol 496:303–313

62. Barmack NH, Qian Z, Yoshimura J (2000) Regional and cellular distribution of protein kinase C in rat cerebellar purkinje cells [in process citation]. J Comp Neurol 427:235–254

63. Jinno S, Jeromin A, Roder J, Kosaka T (2003) Compartmentation of the mouse cerebellar cortex by neuronal calcium sensor-1. J Comp Neurol 458:412–424

64. Furutama D et al (2010) Expression of the IP3R1 promoter-driven nls-lacZ transgene in Purkinje cell parasagittal arrays of developing mouse cerebellum. J Neurosci Res 88:2810–2825

65. Marzban H et al (2003) Expression of the immunoglobulin superfamily neuroplastin adhesion molecules in adult and developing mouse cerebellum and their localisation to parasagittal stripes. J Comp Neurol 462:286–301

66. Altman J, Bayer SA (1977) Time of origin and distribution of a new cell type in the rat cerebellar cortex. Exp Brain Res 29:265–274

67. Harris J, Moreno S, Shaw G, Mugnaini E (1993) Unusual neurofilament composition in cerebellar unipolar brush neurons. J Neurocytol 22:1039–1059

68. Pijpers A, Apps R, Pardoe J, Voogd J, Ruigrok TJ (2006) Precise spatial relation-

ships between mossy fibers and climbing fibers in rat cerebellar cortical zones. J Neurosci 26:12067–12080

69. Sugihara I, Shinoda Y (2007) Molecular, topographic, and functional organization of the cerebellar nuclei: analysis by three-dimensional mapping of the olivonuclear projection and aldolase C labeling. J Neurosci 27:9696–9710

70. Voogd J, Ruigrok TJ (2004) The organization of the corticonuclear and olivocerebellar climbing fiber projections to the rat cerebellar vermis: the congruence of projection zones and the zebrin pattern. J Neurocytol 33:5–21

71. Apps R, Hawkes R (2009) Cerebellar cortical organization: a one-map hypothesis. Nat Rev Neurosci 10:670–681

72. Sugihara I (2011) Compartmentalization of the deep cerebellar nuclei based on afferent projections and aldolase C expression. Cerebellum 10:449–463

73. Sugihara I et al (2009) Projection of reconstructed single Purkinje cell axons in relation to the cortical and nuclear aldolase C compartments of the rat cerebellum. J Comp Neurol 512:282–304

74. Huang CC et al (2013) Convergence of pontine and proprioceptive streams onto multimodal cerebellar granule cells. elife 2:e00400

75. Ruigrok TJ (2011) Ins and outs of cerebellar modules. Cerebellum 10:464–474

76. Wadiche JI, Jahr CE (2005) Patterned expression of Purkinje cell glutamate transporters controls synaptic plasticity. Nat Neurosci 8:1329–1334

77. Shin JH, Kim YS, Linden DJ (2008) Dendritic glutamate release produces autocrine activation of mGluR1 in cerebellar Purkinje cells. Proc Natl Acad Sci U S A 105:746–750

78. Kim YS, Shin JH, Hall FS, Linden DJ (2009) Dopamine signaling is required for depolarization-induced slow current in cerebellar Purkinje cells. J Neurosci 29:8530–8538

79. Kim CH et al (2012) Lobule-specific membrane excitability of cerebellar Purkinje cells. J Physiol 590:273–288

80. Paukert M, Huang YH, Tanaka K, Rothstein JD, Bergles DE (2010) Zones of enhanced glutamate release from climbing fibers in the mammalian cerebellum. J Neurosci 30:7290–7299

81. Zhou H et al (2014) Cerebellar modules operate at different frequencies. elife 3:e02536

82. Zhou H, Voges K, Lin Z, Ju C, Schonewille M (2015) Differential Purkinje cell simple spike activity and pausing behavior related to cerebellar modules. J Neurophysiol 113:2524–2536

83. Peter S et al (2016) Dysfunctional cerebellar Purkinje cells contribute to autism-like behaviour in Shank2-deficient mice. Nat Commun 7:12627

84. Goossens J et al (2001) Expression of protein kinase C inhibitor blocks cerebellar long-term depression without affecting Purkinje cell excitability in alert mice. J Neurosci 21:5813–5823

85. Schonewille M et al (2006) Zonal organization of the mouse flocculus: physiology, input, and output. J Comp Neurol 497:670–682

86. Badura A et al (2013) Climbing fiber input shapes reciprocity of Purkinje cell firing. Neuron 78(4):700–713

87. White JJ et al (2016) An optimized surgical approach for obtaining stable extracellular single-unit recordings from the cerebellum of head-fixed behaving mice. J Neurosci Methods 262:21–31

88. Simpson JI, Wylie DR, De Zeeuw CI (1996) On climbing fiber signals and their consequence(s). Behav Brain Sci 19:380–394

89. Bengtsson F, Jorntell H (2014) Specific relationship between excitatory inputs and climbing fiber receptive fields in deep cerebellar nuclear neurons. PLoS One 9:e84616

90. Haar S, Givon-Mayo R, Barmack NH, Yakhnitsa V, Donchin O (2015) Spontaneous activity does not predict morphological type in cerebellar interneurons. J Neurosci 35:1432–1442

91. Manni E, Petrosini LA (2004) Century of cerebellar somatotopy: a debated representation. Nat Rev Neurosci 5:241–249

92. Pinault D (1996) A novel single-cell staining procedure performed in vivo under electrophysiological control: morpho-functional features of juxtacellularly labeled thalamic cells and other central neurons with biocytin or Neurobiotin. J Neurosci Methods 65:113–136

93. Boele HJ, Koekkoek SK, De Zeeuw CI, Ruigrok TJ (2013) Axonal sprouting and formation of terminals in the adult cerebellum during associative motor learning. J Neurosci 33:17897–17907

94. Hoebeek FE et al (2005) Increased noise level of purkinje cell activities minimizes impact of their modulation during sensorimotor control. Neuron 45:953–965

95. Gincel D et al (2007) Analysis of cerebellar Purkinje cells using EAAT4 glutamate transporter promoter reporter in mice generated via bacterial artificial chromosome-mediated transgenesis. Exp Neurol 203:205–212

96. Dehnes Y et al (1998) The glutamate transporter EAAT4 in rat cerebellar Purkinje cells: a glutamate-gated chloride channel concen-

trated near the synapse in parts of the dendritic membrane facing astroglia. J Neurosci 18:3606–3619

97. Fujita H et al (2014) Detailed expression pattern of aldolase C (Aldoc) in the cerebellum, retina and other areas of the CNS studied in Aldoc-Venus knock-in mice. PLoS One 9:e86679

98. Quiroga RQ, Nadasdy Z, Ben-Shaul Y (2004) Unsupervised spike detection and sorting with wavelets and superparamagnetic clustering. Neural Comput 16:1661–1687

99. Holt GR, Softky WR, Koch C, Douglas RJ (1996) Comparison of discharge variability in vitro and in vivo in cat visual cortex neurons. J Neurophysiol 75:1806–1814

100. De Zeeuw CI, Wylie DR, Stahl JS, Simpson JI (1995) Phase relations of Purkinje cells in the rabbit flocculus during compensatory eye movements. J Neurophysiol 74:2051–2063

101. De Zeeuw CI et al (2011) Spatiotemporal firing patterns in the cerebellum. Nat Rev Neurosci 12:327–344

102. Kalmbach AS, Waters J (2012) Brain surface temperature under a craniotomy. J Neurophysiol 108:3138–3146

103. Long MA, Fee MS (2008) Using temperature to analyse temporal dynamics in the songbird motor pathway. Nature 456:189–194

Chapter 3

Single-Unit Extracellular Recording from the Cerebellum During Eyeblink Conditioning in Head-Fixed Mice

Shane A. Heiney, Shogo Ohmae, Olivia A. Kim, and Javier F. Medina

Abstract

This chapter presents a method for performing in vivo single-unit extracellular recordings and optogenetics during an associative, cerebellum-dependent learning task in head-fixed mice. The method uses a cylindrical treadmill system that reduces stress in the mice by allowing them to walk freely, yet it provides enough stability to maintain single-unit isolation of neurons for tens of minutes to hours. Using this system, we have investigated sensorimotor coding in the cerebellum while mice perform learned skilled movements.

Key words Cerebellum, Classical conditioning, Electrophysiology, Behavior, Motor learning

1   Introduction

Single-unit extracellular recordings have a rich history of use for the investigation of neural coding during behavior in many animal models [1]. Despite exciting advances in multi-neuron recording technologies such as tetrodes and silicon linear or matrix arrays, single-unit recording with metal or glass electrodes is still the gold standard for interrogating neural spiking, especially in deep brain areas with densely packed cells such as the cerebellum. The method allows the investigator to relate the spiking of individual neurons to sensory events and behavior, and thus allows one to examine the "neural code" with unparalleled fidelity.

Despite its advantages, single-unit extracellular recording in the cerebellum has only recently begun to be used in experiments on awake, behaving mice [2–7]. Unlike other model species, the restraint required for stable, acute single-unit recordings in the cerebellum typically elicits a stress response in mice, which can prevent effective task performance or engage fear-related circuits that mask or interfere with the behavior under study [8–10]. However, we have found that mice tolerate head restraint when permitted to walk freely on a treadmill. The ability to walk at will reduces stress and allows the mice to learn and perform a

Roy V. Sillitoe (ed.), *Extracellular Recording Approaches*, Neuromethods, vol. 134,
https://doi.org/10.1007/978-1-4939-7549-5_3, © Springer Science+Business Media, LLC 2018

cerebellum-dependent associative motor learning task [11, 12]. Head fixation allows the investigator to exert fine control over stimulus presentation, collect high-quality extracellular single-unit recordings, and precisely measure behavior. Thus, the treadmill system facilitates the collection of datasets that can be used to correlate neural activity with stimulus presentation and behavior. When used in combination with techniques such as optogenetics or pharmacology [7, 12], the system also permits investigators to test the causal relationships between neural activity and behavior.

While it can be used in many different tasks, we use the head-fixed treadmill system for an associative motor learning task called eyeblink conditioning (EBC). During EBC, the animal is presented with repeated trials in which a neutral stimulus like a flash of light (conditioned stimulus, CS) precedes a blink-eliciting air puff to the cornea (unconditioned stimulus, US) [13, 14]. After many pairings of light and puff presentation, the animal learns to blink (conditioned response, CR) when it sees the light, such that the eye is closed by the time the puff arrives. EBC in mice requires an intact cerebellum [12, 15], and work in other animal species indicates that the memory for the association is formed and maintained within cerebellar circuits [16, 17]. Further, studies in mutant mice have identified specific cerebellar plasticity mechanisms that are required for EBC [15, 18–20]. Thus, EBC is an excellent model behavior for studying cerebellar mechanisms of learning and memory.

This chapter will present methods we have developed that specifically allow for the recording and optogenetic manipulation of multiple cell types within the cerebellar circuit during EBC, including Purkinje cells (PCs) and deep cerebellar nuclei (DCN) cells. However, many aspects of the described techniques should be useful for anyone wishing to make extracellular single-unit recordings in awake, behaving mice.

2   Materials

The apparatus we use for EBC and in vivo electrophysiology consists of mostly off-the-shelf components and a few custom parts and software. Below we describe the materials needed to assemble an apparatus like ours and give some assembly instructions. We also provide a list of materials required for the surgeries needed to prepare mice for these experiments. The heart of the EBC system is a cylindrical treadmill and infrared-sensitive high-speed camera that are mounted on an optical breadboard. Various components for stimulus delivery and electrophysiological recording are then mounted around the treadmill and camera systems, as depicted in Fig. 1.

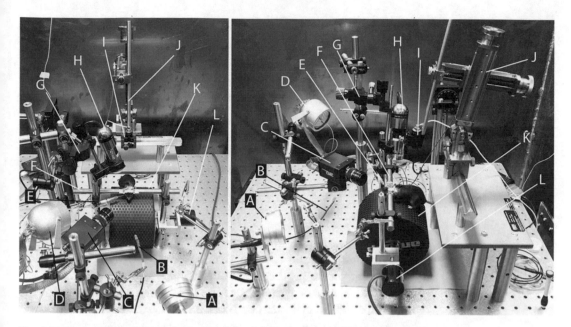

Fig. 1 Apparatus for head-fixed electrophysiology, optogenetics, and behavior experiments. The two photographs show a top view (*left*) and side view (*right*) of the assembled components. Aluminum foil electrical shields have been removed from equipment (e.g., infrared light (**d**)) for better visibility of materials. Labels reference components described in the materials section. Briefly: (**a**) Multi-Field Magnetic Speakers (TDT MF1). (**b**) BNC to LED cable (custom). (**c**) High-speed camera (AVT GE680). (**d**) Infrared light (Towallmark Crazy Cart 48-LED IR Infrared Night Vision Illuminator). (**e**) Blunt needle (23G), adapter (WPI 13160), and tubing for delivering air puff (Clippard URH1-0402-CLT). (**f**) Bars for holding headplate (custom). (**g**) Three-axis micromanipulator (Narishige UMC3). (**h**) Handheld microscope camera (Dino-Lite AD-4013 MTL). (**i**) One-axis micromanipulator (Narishige MMO-220A modified to remove ball joint). (**j**) Stereotaxic micromanipulator (Narishige SMM-100 with custom shaft (EDMS16-054b) and modified MA-2 micromanipulator stand (SDMS16-033a)). (**k**) Treadmill (custom). (**l**) Rotary encoder (Karlsson Robotics E6C2-CWZ3E 1024 count/revolution quadrature encoder)

2.1 General Materials

- Vibration isolation table (Newport; VIS3030-PG2-325A)

2.2 Treadmill System

Materials

- Two ½" diameter posts (12" length) (Thorlabs TR12)
- Two right-angle mounts for ½" diameter rods (Thorlabs RA90)
- Two ½" diameter brass or steel rods (6" length) (custom machined for holding headplate; schematic available at http://github.com/blinklab) (Fig. 1f)
- Two 2-56 screws for attaching headplate to brass rods
- One textured foam cylinder 6" diameter (Amazon (Exervo) TeraNova) (Fig. 1k)
- One ¼" diameter aluminum rod at least 3" longer than the cylinder (McMaster-Carr 9061K33)
- Two low-friction ball bearings (McMaster-Carr 3759T37)

- One bearing mount (custom machined from aluminum or 3D printed; schematic available at http://github.com/blinklab)
- One bearing and rotary encoder mount (custom machined from aluminum or 3D printed, schematic available at http://github.com/blinklab) (optional: use a second bearing mount if not using an encoder)
- One rotary encoder compatible with a ¼" diameter rod (e.g., Karlsson Robotics E6C2-CWZ3E 1024 count/revolution quadrature encoder) (optional) (Fig. 1l)
- One tube for connecting axle and rotary encoder (optional)

Construction

1. Cut the foam cylinder to 4–6" long, and drill a hole through the middle of the cylinder that is slightly less than ¼" diameter so that the ¼" rod can fit snugly. Take care not to deform the cylinder, and be certain that the hole runs directly through the center of the cylinder (*see* **Note 1.1**).

2. Cut the ¼" diameter aluminum rod to be 3–4" longer than your foam wheel. Insert the rod into the hole that you drilled.

3. Gently press the bearings into the custom aluminum mounts (*see* **Note 1.2**).

4. If using a rotary encoder, attach the rotary encoder to the aluminum encoder mount. Cut a ½" length of tubing and slide it onto the axle of the rotary encoder. If not using a rotary encoder, use a second custom bearing mount in place of the encoder mount.

5. Insert each end of the axle into one of the bearings. Make sure that one end of the axle nearly makes contact with the rotary encoder. Slide the tubing from the encoder partially onto the axle so that the axle and rotary encoder spin together. Set the treadmill, encoder, and mount aside.

6. Screw two 12" posts into about the center of your vibration isolation table or breadboard so that the posts can accommodate the treadmill and mounts. Make sure that the posts are screwed very tightly into the breadboard.

7. Slide the mounts/treadmill onto the ½" posts so that the treadmill can spin freely. The wheel is spinning freely enough if it rocks back and forth after a test spin.

8. Place one right-angle mount on each post, about 1 ½" above the surface of the treadmill, so that the headplate bars can be passed through the mounts and meet above the treadmill. Adjust the right-angle mounts until the headplate bars can hold a headplate without straining the metal. Rotate the headplate bars so that they are at a known angle relative to the surface of the vibration isolation table. Also make sure that this angle will hold the mouse's head in a reasonable position.

Maintenance

1. Thoroughly clean the surface of the wheel with 70% EtOH or a laboratory cleaner of your choice ASAP after removing the mouse from the wheel. This will help prevent too much urine from soaking into the foam.

2. Clean the area around the foam wheel. Urine and feces are sprayed off of the wheel when the mouse runs. It can be helpful to lay a paper towel or other disposable materials underneath the wheel to collect the majority of the mouse's waste.

3. When you apply cleaner to the wheel, moisten the paper towel or other disposable wipes away from the wheel. Doing otherwise can cause the cleaning solution to splatter on the equipment around the wheel.

2.3   Camera System

Materials

- One high-speed camera (e.g., Allied Vision Technologies GE680 gigabit Ethernet camera) (Fig. 1c)
- One lens for camera (Tamron TAM-23FM25-L)
- One extension tube for lens (1st Vision LE-EX-10 or LE-EX-5)
- One Ethernet card with support for jumbo packets (if using gigabit camera) (e.g., Intel PRO/1000 GT PCI)
- One Cat6 Ethernet cable (if using gigabit camera)
- One IR light source and power supply (Bosch EX12LED or similar) (Fig. 1d)
- One ½" diameter post (12" length) (Thorlabs TR12)
- Two knuckle joint (Panavise 851-00)
- One adapter to couple the knuckle joint and high-speed camera (custom-made, aluminum)
- One ½" diameter post (2–4" length) (Thorlabs TR2-TR4)
- Two right-angle mounts for ½" diameter posts (Thorlabs RA90)
- One BNC to SMB cable for triggering camera from data acquisition system (Fairview Microwave FMC0816315-72 is compatible with GE680)

Setup

1. Install high-speed camera drivers and software according to manufacturer instructions.

2. Attach the camera to a 2–4" long ½" diameter post using the knuckle joint and the custom camera mount.

3. Slide the 2–4" long post into one of the right-angle mounts.

4. Screw a 12" long ½" diameter post into the breadboard.

5. Slide the right-angle mount connected to the post and camera onto the 12" post. Check that the camera can be positioned roughly to point toward the future location of the mouse's face. If not, repeat until you get a good approximate location.

6. Attach the infrared light to a knuckle joint, and attach the knuckle joint to a 2–4" post. Slide this post into a right-angle connector.

7. Slide the IR light + post + connector onto the 12" post you screwed in before so that the light can be easily directed to the same point as the camera's focal plane.

8. Connect the camera to the gigabit Ethernet card in the PC with the Cat6 Ethernet cable.

9. Connect the camera power supply.

10. Open Vimba Viewer or your camera's video streaming software.

11. Stream video from your camera. Check that you can focus on the near headplate bar, which is approximately the depth where the mouse's face will be. Check that the image is not too bright or too dark. Adjust the aperture/focus of your camera lens and the position of your camera until you find something that you like. Depending on the space available in your setup, you may find that you need a 10 mm or 5 mm extension lens to achieve the right focus and field of view. Experiment with the different extension tubes until you can to find a good video image.

12. Once the camera is set up in approximately the right location, test again with a mouse in the rig. Use the grayscale images shown in Fig. 2 as a guide.

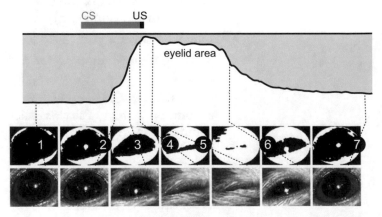

Fig. 2 Eyelid movement detection algorithm. Illustration of algorithm used to measure eyelid closure using FEC. Key time points during a paired CS-US trial are indicated by numbered lines connecting video frames to the corresponding points in the eyelid traces. Used with permission from [12]

2.4 Surgery   *Materials*

- Stereomicroscope with fiber-optic illumination (e.g., Zeiss Stemi 2000)

- Sterile physiological saline

- Sterile 1 mL syringes with needle

- Canned compressed air

- 7.5% povidone-iodine

- 70% ethanol

- C&B Metabond Quick Cement System (Parkell Inc. S396, S398, S371, S379, and S387)

- Dental acrylic (e.g., Lang Jet 1406, 1220)

- Depilatory cream

- Ophthalmic ointment or artificial tears

- Sterile nylon sutures (LOOK 18" Nylon Suture with C-17 needle, 5-0; ref. no. 917B)

- Antibiotic ointment

- Sterile gauze and cotton balls torn and rolled into 2–3 mm balls

- GELFOAM (Fisher Scientific NC04090659)

- Kwik-Sil (WPI Kwik-Sil)

- Instruments
  - Precision jeweler screwdriver compatible with skull screws
  - Small scissors (e.g., Fine Science Tools (FST) 14094-11)
  - Small spatula (e.g., WPI 504022)
  - Curved forceps (e.g., Dumont #7 from FST 11271-30)
  - Curved serrated forceps (e.g., Dumont #7b from FST 11271-30) (file off one to two teeth to grip screws better)
  - Fine-tipped forceps with angle (e.g., Dumont #5/45 from FST 11251-35)
  - Dental drill (e.g., Osada EXL-M40)
  - 0.5 mm burr drill bits compatible with dental drill

- 1/mouse, headplates (1" × 5/16" × 1/32") (custom machined from titanium or stainless steel; schematic available at http://github.com/blinklab)

- Custom headplate adapter (Fig. 3)

- 3D printed custom recording chamber (Fig. 5g; can be 3D printed by NeuroNexus Technologies or similar; schematic available at http://github.com/blinklab) or plastic ring that can serve as a recording chamber

- Skull Screws (e.g., Antrin Miniature Specialties, Inc. AMS120-1B, 000-120×1/16 SL BIND MS SS)

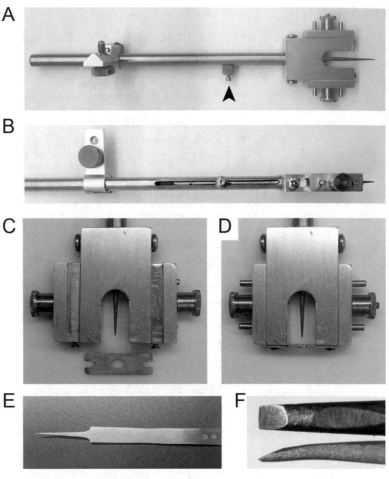

Fig. 3 Surgery tools. (**a–d**) A custom-made head plate holder that can be mounted on the stereotaxic manipulator arm. The stylus is retractable and can be held by the stopper screw (*arrow head*). Zoomed-in view of the holder before (**c**) and after (**d**) holding the headplate. The headplate is held between the horizontally moving screws. (**e, f**) A hand made dura cleaning tool. The tool was made by splitting the arms of FST forceps and sharpening the tip (**f**) by fine grit polishing paper

- Small animal stereotaxic surgical rig (Kopf 930)
- Universal stereotaxic adapter (e.g., Kopf Model 1772)
- Stereotaxic adapter for implanting cannulae (e.g., Kopf model 1766-AP)
- Silver wire (optional)

2.5 Eyeblink Conditioning

Materials

- Data acquisition and stimulus delivery system. For example:
  - TDT RZ5
  - TDT Medusa preamp with high impedance headstage such as RA16

- TDT P05e PC interface card
- TDT zBus ZB1PS Caddy
- Multi-Field Magnetic Speakers (TDT MF1) (Fig. 1a)
- Stereo power amp for speakers (TDT SA1)
- Windows PC with specs good enough to run the software and handle the video frames in near real time (e.g., 3.4 GHz Core i7 CPU and 16 GB RAM)
- BNC to LED cable with ~330 Ohm resistor for LED (Fig. 1b)
- One pressure injector (Applied Scientific Instrumentation MPPI-3)
- One 23G blunt needle (Fig. 1e)
- Tubing for air puff delivery (Clippard URH1-0402-CLT) (Fig. 1e)
- One plastic 1/16" barbed to luer fitting to couple the blunt needle and air puff tubing (e.g., WPI 13160) (Fig. 1e)
- Sound pressure meter (optional)
- MATLAB (or other data processing software)
- Neuroblinks software from http://github.com/blinklab (for use with MATLAB and TDT, may be substituted with other behavioral training software)

- Nitrogen gas with regulator
- 1/stimulus location, ½ diameter post (6–12" length, depending on your preferred setup)
- 1/stimulus, third hand (e.g., alligator clip and stem from SUNYEE Helping Hand MZ101B)
- 1/stimulus, right-angle mounts that can accommodate the third hands (OptoSigma CCHN-12.7-6 is compatible with SUNYEE Helping Hand)
- Air pressure measurement tool (e.g., PCE Instruments, PCE-P50; or construct your own based on a barbed pressure sensor like Honeywell SSCDANN150PG2A3)
- Tubing that will fit the pressure sensor barb and the luer adapter in the US delivery tubing (e.g., Clippard URH1-0804-CLT)

Construction

1. For each stimulus, screw a ½" diameter post somewhere around the front of the treadmill.

2. Attach the stimulus to the post using a right-angle mount, third hand, or other appropriate connector. Position the stimulus so that the mouse can reasonably be expected to perceive it.

3. Connect the various stimuli that you are using to the stimulus triggering/data acquisition hardware. Interface with the appropriate software. If you are using the TDT RZ5 system, you can use the Neuroblinks software from this laboratory listed above.

2.6 Electrophysiology

Materials

Apparatus

- Stereotaxic micromanipulator (e.g., Narishige SMM-100 with custom shaft (EDMS16-054b) and modified MA-2 micromanipulator stand (SDMS16-033a)) (Fig. 1j)
- One-axis micromanipulator (e.g., Narishige MMO-220A one-axis oil hydraulic micromanipulator modified to remove the ball joint) (Fig. 1i)
- Handheld microscope camera (Dino-Lite AD-4013 MTL) (Fig. 1h)
- Adapter for microscope camera (Dino-Lite HD-P1)
- Three-axis manipulator for microscope camera (e.g. Narishige UMC3) (Fig. 1g)
- 10 mm OD post (if you use the UMC3)
- 7 mm OD post (for holding the camera adapter in the micromanipulator)
- Four ½" diameter 12" posts (Thorlabs)
- Two ½" diameter 6" posts (Thorlabs)
- Three right-angle mounts for ½" posts (Thorlabs RA90)
- 1/stimulus, third hand (e.g. alligator clip and stem from SUNYEE Helping Hand MZ101B)
- 1/stimulus, right-angle mounts that can accommodate the third hands (OptoSigma CCHN-12.7-6 is compatible with SUNYEE Helping Hand)
- Solid core copper ground wires and panel

Supplies

- Custom dura scraper (see Fig. 3e, f)
- Sterile saline
- Small cotton balls
- 7.5% povidone-iodine
- Electrodes (see **Note 6**)
- Optrodes (see below for construction instructions)
- Implantable optical fibers (e.g., Thorlabs or custom-made)

Construction of the Recording Apparatus

1. Attach your materials to the vibration isolation table in appropriate positions relative to the treadmill using the Thorlabs posts and RA90 right-angle connectors. Make sure everything is very stable and screwed tightly down onto the breadboard. See Fig. 1 for example arrangement.

2. Install the software to run the Dino-Lite camera. Position the camera so that it can focus at the height of the mouse's craniotomy and so that it will not block your micromanipulator from moving freely around the craniotomy.

3. You will need to use a mouse with a headplate and craniotomy to make sure that the positions of all your devices are appropriate for your purposes.

Eliminating Electrical Noise

- Depending on the specific features of your setup, you may have more or less ambient electrical noise. To counteract noise coming from outside of your recording rig, it can be helpful to place a Faraday cage around the rig or to work in an electrically shielded room. However, most of the noise you encounter will probably originate from the equipment you are using for your experiment. Here are some tips for ways to deal with this kind of noise.

- Ground everything to a single point.

- Use a stainless steel or copper braided mesh to shield all of the power cables going to your rig (e.g., McMaster 5537K14). Ground these.

- Wrap other sources of noise in aluminum foil and ground these.

- Make sure that all of your ground points are sitting at the same voltage to eliminate potential ground loops.

- Electrostatic discharge from the mouse can be another noise source. Place the equipment out of the mouse's reach. If impossible (e.g., air puff needle, reward tubing, etc.), the object can be electrically connected to the mouse, or both the object and the mouse can be connected to the ground.

- Vibration is another potential noise source. The wires and clips between the headstage/preamp and the mouse and the ground wire should be stabilized.

2.7 Optogenetics

Materials

- Implantable Fiber Optic Cannulae (can be purchased from multiple vendors, including Thorlabs and Doric Lenses or constructed following the *Thorlabs Guide to Connectorization and Polishing of Optical Fibers; see* **Notes 2** and **3**)

- Laser or LED with driver for supplying fixed wavelength illumination (e.g., Blue Sky Research or Thorlabs)
- 100 or 200 μm optical fiber with silica cladding (if not purchasing preassembled, e.g., Thorlabs FG105LCA or FP200URT)
- 1.25 mm Zirconia ferrules with ID matching optical fiber (if not purchasing preassembled, e.g., Kientec FAZI-LC-230)
- Patch fiber with same diameter as implantable fiber for transmitting light between laser and implanted fiber (e.g., Thorlabs FG200UCC; or can be made by hand)
- Butane torch
- Stainless steel hypodermic tubing (e.g., 23XX and 26RW gauge) to protect the optical fiber and provide rigidity (optional, if making long optical fibers for acute use (see below))
- Light meter (Thorlabs PM100D and S140C)

Pulling Optical Fibers for Use in Acute Preparations

Connectorized optical fibers that have a silica cladding can be heated and pulled to produce finer tip sizes that allow the fiber to more easily pass through the dura and provide more precise targeting of cell populations [7, 21] (Fig. 4). Once pulled, they

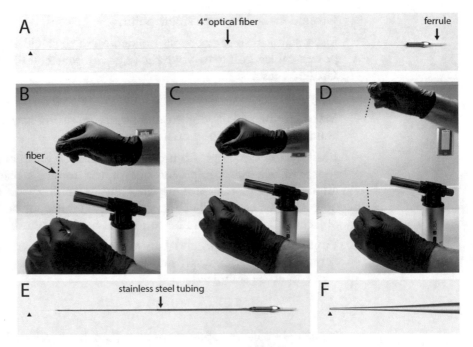

Fig. 4 Pulling optical fibers for acute photostimulation. Start with an approximately 4" connectorized fiber (**a**). Gently heat the fiber around its midpoint over a butane flame, and pull the two ends in opposite directions to stretch the glass and reduce the diameter (**b–d**). Secure the shaft of the pulled fiber in stainless steel tubing using cyanoacrylate glue (**e**). Cleave the fiber tip to desired diameter (**f**). Black arrowheads indicate location of fiber tip

can be glued to an electrode to make a custom "optrode." Fine control over the pull can be achieved using a pipette puller capable of melting quartz glass, such as a Sutter P-2000. However, with practice, optical fibers can be manually pulled by heating them over a small butane torch. To manually pull an optical fiber, start with a connectorized optical fiber that extends at least four inches from the unpolished end of the ferrule (Fig. 4a). Hold the center of the fiber vertically in front of the flame, gripping the ends with forceps or pliers in each hand, with the connectorized end on top (Fig. 4b). Gently pull the ends apart until the glass becomes fluid and starts to give (Fig. 4c). Then simultaneously remove the fiber from the flame while rapidly pulling the ends apart (Fig. 4d). If done properly, the fiber will taper to a submicron diameter, and then the two ends will break away from each other. At this point the tip is usually too small to be useful, so use a diamond or ruby scribe to cleave the connectorized fiber along its taper to the desired diameter. We typically use fibers with ~20 μm tip diameters (Fig. 4f). The cleaved end can be sharpened by holding it at an acute angle against the rotating surface of a pipette beveling device or hard drive platter covered with fine grit polishing paper. A conical tip shape can be produced by gently rotating the fiber about its long axis as it contacts the polishing paper, but bear in mind that this will remove the cladding at the tip of the fiber and allow light to leak out along the cone. Glue the optical fiber inside appropriately sized hypodermic tubing to protect it and make it more rigid, leaving at least 5 mm of the tip exposed (Fig. 4e).

Testing Optical Fiber Light Transmission

1. Connect light sensor (e.g., Thorlabs S140C) to the meter (e.g., Thorlabs PM100D), and power up the system.

2. Connect a patch fiber to the laser and point at the light sensor. Turn on the laser.

3. Move the light-emitting end of the patch fiber so that the light sensor reaches its maximum reading. Record this value. Turn off the laser.

4. Connect your optical fiber to the patch fiber. Repeat step 3, this time with the optical fiber instead of the patch fiber.

3 Methods

3.1 Surgical Procedures

Two separate surgical procedures are required to prepare mice for head-fixed single-unit extracellular recording: (1) implantation of a headplate for head fixation and (2) drilling of a craniotomy and implantation of a recording chamber. These procedures can be performed during the same surgery or separated by several weeks, depending on your experimental needs and institutional protocol.

3.1.1 Anesthesia and Preoperative Care

Inhalation anesthesia is induced by 5% isoflurane (% by volume in O2; SurgiVet) and maintained with 1.0–2.0% isoflurane throughout the surgery. Mice are placed in a stereotaxic apparatus (*see* **Note 4.1**), and a nonsteroidal anti-inflammatory drug (e.g., meloxicam) is given subcutaneously to reduce swelling and provide analgesia. During surgery, core body temperature is maintained by placing the mouse on a heating pad (DC temperature controller, FHC) to prevent hypothermia. Both eyes are covered with "artificial tears" ointment to keep them moist. Fur over the surgical site is removed by electric clippers and depilatory cream. The skin is then cleaned by three alternating scrubs of povidone-iodine surgical scrub followed by 70% alcohol.

3.1.2 Stereotaxic Implant of Headplate and Ground Screw

Surgery is performed under a stereomicroscope with fiber-optic illumination. A midline incision is made to expose the skull, and the underlying periosteum is cleared with cotton swabs so that bregma and lambda can be clearly seen. Two layered muscles attached to occipital bone (the splenius capitis and trapezius, Fig. 5e, f) are detached at their insertion point, reflected back, and covered by sterile Kimwipes or cotton balls to obtain a clear surgical field. To position the mouse in the stereotaxic plane (as defined by Paxinos and Franklin [22]), the midline suture should be aligned parallel to the anterior/posterior axis of the stereotaxic manipulator arm. The head of the mouse should be rotated around the ear bar axis (pitch) so that the dorsal/ventral difference between bregma and lambda is less than 50 μm. A shallow outline is then drilled to mark the craniotomy location that targets EBC-related areas (2–3 mm diameter craniotomy centered 6.5 mm posterior and 2.0 mm lateral from bregma, Fig. 5a). This area includes several lobules of the cerebellar cortex, including paravermal IV/V and VI, and hemispheric simplex and crus I, as well as the inferior colliculus (Fig. 5h). If an optical fiber or guide cannula is to be implanted, also mark their target locations using the drill. Next, two screw holes are drilled on either side of the midline just posterior of bregma (being careful to not puncture the dura, Fig. 5a), and two #000-120 self-tapping stainless steel screws are screwed into the holes (Fig. 5b), using the curved forceps with filed teeth to hold the screw. A third hole is drilled into the interparietal bone contralateral to the craniotomy site, and a #000-120 stainless steel screw is screwed into this hole (Fig. 5a, b). The third screw serves as both an anchor for the cement to provide more stability for the implant and as a reference for the extracellular recording. A short length of bare silver wire can be connected on/around the third

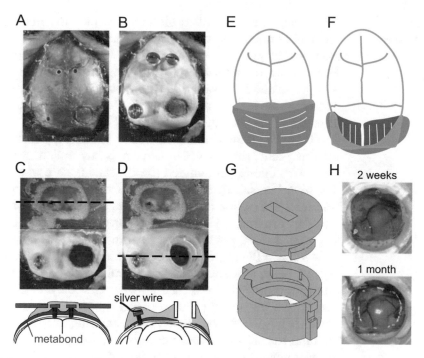

Fig. 5 Headplate and craniotomy surgery. (**a–d**) View of mouse's skull after drilling screw holes and marking the outline for craniotomy (**a**), after implanting screws and covering the skull with dental cement (**b**), after securing the headplate and craniotomy (**c**), and after securing the recording chamber (**d**). Coronal schematic view is shown at the level of dashed line in (**c**) and (**d**). (**e**) The first layer of the occipital muscle (the splenius capitis) is cut in the midline and detached from the occipital bone. (**f**) The second layer of the occipital muscle (the splenius trapezius) is seen under the folded first layer. The second layer is also detached from the occipital bone (not shown) so that the skull surrounding the craniotomy can be covered by dental cement. (**g**) Technical drawing of 3D printed recording chamber and interlocking lid. Lid is locked in place by inserting it into grooves of chamber and turning 45° clockwise using a small flathead screwdriver. (**h**) Healthy dura 2 weeks (*top*) and 1 month (*bottom*) after the craniotomy surgery

screw and covered with conductive epoxy to enable easy access to the reference screw (Fig. 5d). Next, the screws and most of the dorsal surface of the skull are covered in cement, leaving only the future location of the craniotomy clear (C&B Metabond, Fig. 5b). A custom-made stereotaxic adapter (Fig. 3a–d) is used to hold the headplate (Fig. 3c) in the stereotaxic plane such that the headplate can be held securely in place over bregma, while dental cement is applied (transparent dental acrylic or Metabond, Fig. 5c) to a very dry bone surface (*see* **Notes 4.2** and **4.3**) using a small spatula (e.g., WPI 504022). When the dental cement is almost completely cured, a small indentation is placed over bregma using the stereotaxic adapter's retractable stylus. The indentation will be used as a reference mark for guiding recording tracks based on stereotaxic position. Once the cement has fully cured and is hard to the touch, the headplate adapter can be removed.

3.1.3 Craniotomy and Recording Chamber

The craniotomy is drilled along the previously marked outline. To avoid puncturing the dura, it is crucial to make the craniotomy very slowly, gradually removing thin layers of bone and stopping periodically to irrigate the tissue with sterile saline. The saline should be blotted away with small cotton balls and any bone dust blown away with compressed air. Once the bone has been cut through along the entire perimeter of the craniotomy, the remaining bone flap should feel loose when gently pushed with forceps. At this point, the bone flap can be gently pulled away from the dura with fine-tipped, angled forceps, taking care to avoid rupturing blood vessels near the surface of the dura. Wetting the bone and dura with saline for a few minutes before attempting to lift off the flap can make it easier to remove and reduce trauma. The exposed dura is covered with a thin layer of silicone elastomer (Kwik-Sil, WPI) to protect it between recording sessions (see Sect. 3.1.6). Next, a plastic recording chamber (Fig. 5d, g) is placed on the craniotomy site and secured by dental acrylic. This chamber serves two purposes: (1) it protects the dura between recording sessions by preventing debris from entering the craniotomy and holding the silicone elastomer in place, and (2) it serves as a bath to maintain saline on top of the dura during recording sessions.

3.1.4 Implantation of Chronic Optical Fibers and Guide Cannulae

The principles underlying the implantation of chronic optical fibers or guide cannulae for drug delivery are essentially the same. After preparing and leveling the skull as described in Sect. 3.1.2, a small craniotomy is drilled over the stereotactic coordinates of the target. The implant is secured in the appropriate stereotaxic attachment (e.g., Kopf 1766-AP for guide cannulae or Kopf 1772 holding a pin vise for optical fibers). Then, a small incision is made in the dura using a sharp tool or sterile needle so that the implant can easily enter the brain instead of deforming the tissue. The implant is lowered to the desired depth. After verifying that the skull is very dry (*see* **Note 4.3**), a small spatula is used to apply Metabond around the implant, leaving enough of the implant free of Metabond to attach dummy cannulae or fiber-optic patch cables.

3.1.5 Completion of Surgery and Postoperative Care

Any remaining wound from the midline incision should be sutured and a topical antibiotic applied to the wound margins. In the first week following surgery, mice should be monitored twice daily. Topical antibiotics should be applied to the wound to prevent infection, and analgesics should be delivered according to your laboratory's approved animal protocol. Careful attention should be paid to the general level of activity of the mouse, including grooming behavior, vocalization, respiration, urination, defecation, eating, and drinking. In addition, the mice should be monitored for any signs of ataxia (one of the cardinal symptoms of damage to the cerebellum). Mice that develop ataxia should be euthanized.

3.1.6 Maintaining the Craniotomy and Health of the Dura

Covering the craniotomy with a silicone elastomer dramatically improves the health of the dura and prevents scar tissue buildup and reossification. Despite this, it is occasionally necessary to remove excess scar tissue or bone from the dura before performing a recording experiment. Once the original silicone plug has been removed and experiments have begun, we use an antiseptic (povidone-iodine) to clean the craniotomy daily before and after each recording session to keep it in good condition.

To access the craniotomy for cleaning, the chamber lid is opened, and povidone-iodine is applied into the chamber on the surface of silicone elastomer. The povidone-iodine is left in place for at least 1 min and is then removed with small sterile cotton balls (2–3 mm, torn from larger cotton balls) before opening the silicone. The silicone is removed by gripping it along its edge with curved forceps and gently lifting it away. After removing the silicone, povidone-iodine is applied to the dura and left to sit for a few minutes. Excess povidone-iodine solution is then removed with small sterile cotton balls, and the chamber is rinsed with sterile saline. At this point the dura can be examined to determine if the brain is acceptably visible or if there is excess bone or scar tissue buildup that needs to be removed before recording. If the dura is transparent and all blood vessels and cerebellar lobules are visible, the dura is usually thin enough to skip the dura cleaning step. If the dura is thicker, the excess granulation tissue can bend or otherwise damage the tip of the electrode. The granulation tissue is carefully removed by scraping off thin layers with a handmade tool (made from FST forceps, Fig. 3e, f) being careful not to rupture the dura. Stop when the dura and vasculature are visible. Saline is then applied into the chamber to keep the dura moist. A local anesthetic can be added to the saline solution (1% lidocaine) to reduce pain when the mouse wakes up.

After the experiment, the dura should be cleaned with povidone-iodine again, followed by a sterile saline rinse. Last, a thin layer of silicone elastomer (Kwik-Sil, WPI) is applied and the chamber closed with a disinfected lid. Using these procedures, a healthy dura can be maintained for more than 1 month (Fig. 5h).

It is also beneficial to trim the fur on the dorsal surface of the neck on a weekly basis to prevent fur from entering the chamber when the mouse hunches its shoulders on the treadmill. Fur entering the chamber can provide a route for infection and can be a source of electrical artifacts during recording.

3.2 Eyeblink Conditioning

Eyeblink conditioning can be carried out using stimulus delivery and data collection hardware and software like those described in *Materials*. It is essential that stimulus delivery be performed with at least 1 ms precision and that stimulus delivery, behavioral data, and electrophysiological data be timestamped according to the same clock.

To reduce stress during training sessions, the mouse is habituated to the conditioning apparatus. During this phase of the experiment, the animal is head-fixed (*see* **Note 5.1**) atop the treadmill via the headplate and bar system for 1 h/day until the mouse can walk comfortably on the wheel and maintains a normal posture (*see* **Note 5.2**). This process normally takes 2–3 days. All features of the apparatus and environment should be identical between the habituation sessions and normal training sessions, with the exclusion of CS and US presentation (*see* **Note 5.3**).

Following habituation, EBC sessions can begin. The mouse is once again head-fixed atop the treadmill in the conditioning apparatus. The CS may be any neutral stimulus that your system can deliver (e.g., LED illumination, a pure tone, etc.). The relative timing of the CS and US presentations (interstimulus interval, ISI) can be altered to engage different brain structures in the task. To train mice using a delay EBC paradigm that recruits the cerebellum [12, 16, 17], the US is presented while the CS is still on. Using a light CS, mice can effectively learn to make CRs in delay EBC with ISIs between approximately 200 and 500 ms [11]. To train mice using a trace EBC paradigm that recruits the cerebellum and other brain areas, including dorsal medial prefrontal cortex, hippocampus, and amygdala [23], the US is presented 200–500 ms after the CS terminates. During both delay and trace conditioning, trials are typically delivered at 10–25 s intertrial intervals (*see* **Note 5.4**). Mice are typically trained for at least 100 trials/day (*see* **Note 5.5**). EBC can be carried out using 100% CS + US trials. However, it can be informative to include some (e.g., 20%) CS-only trials (*see* **Note 5.6**). These trials allow the experimenter to examine the CR without interference from the UR. This is important, as the peak of the CR occurs after the US is delivered [11], and CR peak time is an important factor for assessing cerebellum dependency (see Sect. 3.8.2 for more details). When designing the EBC paradigm, the experimenter should consider whether it is important to use CS-only trials and whether delivering CS-only trials will confound the research question.

Depending on task parameters such as ITI, puff strength, and ISI, mice should start showing CRs within 2–3 days of training [12]. If a mouse does not start showing CRs within this period, verify that there are no problems with the conditioning apparatus or stimulus delivery parameters. If there are no clear problems and the animal has not acquired any conditioned responses within 5 days of training, it is best to start working with another mouse (*see* **Note 5.7**).

US intensity affects the rate of learning and the degree to which the learning depends on the cerebellum. Using a stronger US during task learning may recruit the amygdala [10, 24], and using a weaker US delays or fails to support learning of the eyeblink CR. If the US intensity decreases after CR acquisition, the CR may be partially or completely extinguished [25]. For consistent US intensities during training, the US delivery needle must be the same distance from the cornea each day, as the puff pressure at the cornea decreases with the square of the distance. The experimenter should also measure air puff strength (e.g., PCE-P50) to be sure that US intensity is appropriate (*see* **Note 5.8**). This laboratory has successfully used an air puff US with peak pressures from 6.5 to 8.5 PSI and positioned the US delivery needle ~3 mm from the cornea for cerebellum-dependent eyeblink conditioning (*see* **Notes 5.9** and **5.10**).

3.3 Acute Single-Unit Recording and Stimulation during Behavior

Acute single-unit recording or stimulation experiments can be performed daily during eyeblink conditioning sessions. Once the mouse is prepared for single-unit recording as described above, it is placed on the treadmill, and its head is fixed in place by the headplate. The session can begin after allowing approximately 10 min for the mouse to recover from anesthesia. The electrode (*see* **Note 6**) is lowered into the recording chamber under visual inspection with a stereomicroscope or Dino-Lite digital microscope. The anteroposterior and mediolateral locations of recording tracks are determined relative to the mark placed above bregma during the headplate surgery. The surface of the dura serves as the dorsal-ventral reference position.

The general strategy we use is to make recording tracks at evenly spaced intervals (100–200 μm) along the anteroposterior and mediolateral axes in a grid pattern. This ensures good coverage of the target region and is especially valuable during initial mapping experiments (see below). By taking care to place the mouse in the headplate holder consistently from day to day, the same region can be reliably targeted daily based on stereotaxic coordinates or visible surface landmarks. Figure 8 shows recording data during EBC for an example Purkinje cell and an example DCN cell.

3.4 Identifying Cerebellar Layers and Cell Types

The cerebellar cortex consists of three distinct layers that can be identified based on their characteristic activity profiles [26, 27] (Fig. 6): (1) molecular layer, (2) Purkinje cell layer, and (3) granular layer. These layers are comprised of different cell types and processes, giving each a distinct electrophysiological signature. When recording in the cerebellar cortex, it can be useful to listen to the auditory features of the recorded data. Listening to the activity can be perplexing at first, but over time the experimenter develops an ear for the different characteristic sounds of the cerebellar cortical layers.

The molecular layer contains the Purkinje cell dendrites, granule cell axons (parallel fibers), and molecular layer interneurons. When the electrode tip is in the molecular layer, recordings contain almost exclusively prominent complex spikes from the Purkinje cell dendrites and otherwise quiet background activity [27, 28] (Fig. 6a).

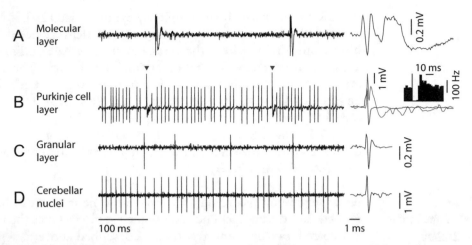

Fig. 6 Neural activity patterns in each layer of cerebellar cortex and deep cerebellar nuclei. (**a**) In molecular layer, complex spike from Purkinje cell dendrite can be isolated. Zoomed-in view of complex spike waveform (*right*) shows long tail with spikelets, which can be detected as a low-frequency sound in audio monitor. (**b**) In Purkinje cell layer, simple spike and complex spike (*red arrowhead and red trace*) can be simultaneously isolated. Simple spike firing (y-axis of the inset histogram is firing rate) pauses for 10–30 ms after each complex spike (*red line*). (**c**) In granular layer, a low firing rate cell (putative Golgi cell) can be isolated. Complex spikes should not be detectable in this layer. (**d**) In deep cerebellar nuclei, cells with high spontaneous firing rate are often isolated (putative excitatory projection neuron). The cerebellar nuclei should be separated from the cerebellar cortex by white matter (which lacks prominent spiking activity and sounds quiet on the audio monitor)

Complex spikes are the postsynaptic response of Purkinje cells to inputs from the inferior olive and have a distinct discharge rate of approximately 1 Hz [29]. Complex spikes make a characteristic popping sound (like popcorn kernels popping). Other action potentials arising from molecular layer interneurons can occasionally be found, but the signals are typically smaller than complex spikes and more difficult to isolate.

The Purkinje cell layer contains densely packed Purkinje cell bodies that each emit high- frequency simple spikes and low-frequency complex spikes, giving recordings in this layer a high background activity (Fig. 6b). Complex spikes are recognizable by their characteristic low-frequency (~1 Hz) discharge and the presence of calcium spikelets [29, 30] (Fig. 6b, *right*), which contribute to the "popping" sound. Simple spikes are characterized from other action potentials by at least three criteria: (1) the mean spontaneous firing rate of simple spikes should range from 50 to 100 Hz [29], (2) simple spikes should pause briefly after complex spikes (10–30 ms) [29, 31] (Fig. 6b, *inset*), and (3) simple spike amplitude should be roughly comparable with complex spike amplitude (though the most prominent direction—positive or negative—of the simple and complex spikes can sometimes vary considerably). If these simple spike criteria are not met, then the recording may be missing Purkinje cell action potentials, or the recording may contain multiple units.

The granular layer contains the tiny, densely packed granule cells, mossy fiber terminals, and cell bodies of Golgi cells and other granular layer interneurons [26]. The granular layer cells receive much of their synaptic input in glomeruli that are innervated by mossy fiber terminals (called rosettes). The mossy fiber rosettes are often larger than granule cell bodies and may contribute to a characteristic "hashy" sounding background activity in the granular layer. Interneurons, such as Golgi cells, can often be isolated in the granular layer or heard in the background (Fig. 6c). Criteria have been developed in anesthetized animals to identify some of the cerebellar cortical interneurons based on their spike patterns, but it remains to be seen whether these criteria translate to the awake mouse [32, 33].

The cerebellar cortex surrounds the deep cerebellar nuclei, which are mostly identified based on depth and absence of the characteristic signatures of the cortex (Fig. 6d).

3.5 Using Sensory Stimulation to Map Climbing Fiber Receptive Fields

The cerebellar cortex is organized into microzones, which are longitudinal bands in which Purkinje cells are innervated by climbing fibers with similar sensory receptive fields [34]. The microzones can be mapped by recording climbing fiber responses to stimulation of different points on the body surface to determine the receptive field [35, 36, 37]. Climbing fiber responses can be recorded as either single-unit complex spikes or as evoked potentials within a local field potential (low-pass filtered, 300 Hz cutoff; Fig. 6a, b and 7a), and in our experience, there is a strong correlation

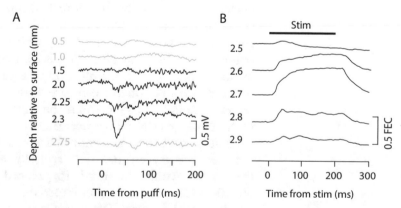

Fig. 7 Functional mapping of cerebellar cortex and nuclei. (**a**) LFP recordings at different depths through cerebellar cortex during periocular stimulation (air puff). Region with periocular climbing fiber receptive fields is indicated by large negative deflection in LFP signal following air puff to eye. (**b**) Eyelid traces during microstimulation at different depths through DCN. The eyeblink region of DCN is indicated by the location at which microstimulation at currents less than approximately 10 μA produces large and reliable eyelid closure that is sustained for the duration of the stimulation (indicated by *black line at top*). Used with permission from [12]

between the climbing fiber receptive fields identified by these two methods. It is best to begin mapping using stereotactic coordinates, but you should ultimately rely on recorded responses to sensory stimulation to determine whether the electrode tip is in a microzone of interest (e.g., Fig. 7a). In practice, local field potential recording is more useful for coarse mapping, and single-unit recording is more useful for finer-scale mapping. However, the source of the local field potential is still unclear, and the signal can be contaminated by non-climbing fiber sources, such as those arising from the mossy fiber pathway [36], so more care must be taken in interpreting the local field potential recordings.

As an additional precaution, stimulation trials should only be started when the animal is not moving (without locomotion, face grooming, licking, etc.) because the local field potential is easily contaminated by movement artifacts.

3.6 Using Microstimulation to Map DCN Motor Outputs

Unlike in the cerebellar cortex, there are no accepted criteria for identifying the different neuron classes of the cerebellar nuclei based on extracellular recordings. These neuronal classes include large excitatory projection neurons and smaller inhibitory neurons that make local synaptic contacts and/or project to the inferior olive or cerebellar cortex [38]. However, due to their substantially larger size, the excitatory projection neurons are more likely to be isolated in an extracellular recording, and we often assume these are the majority of the neurons that we record.

Because the excitatory projection neurons of the DCN project to premotor structures, including various brainstem and midbrain nuclei, the DCN can be functionally mapped by using microstimulation to elicit movements [39]. This is particularly true for the anterior interposed nuclei (AIP), which is where the essential region for eyeblink conditioning is located [12, 17]. The movements evoked by microstimulation in a particular location within the AIP give a reasonable estimate of the underlying somatotopic representation and a clue about the climbing fiber receptive fields of the Purkinje cells that project there [40].

Microstimulation is performed through platinum/iridium or tungsten microelectrodes that typically have lower impedances than what would be used for recording (in the range of 50–100 KOhms), but higher impedance recording electrodes can also be used if desired. We typically use a monopolar stimulation configuration in which the reference lead is attached to the same implanted reference screw used for recording, but bipolar stimulating electrodes can be used as well and may help contain the current spread, particularly at higher stimulation intensities. The stimulating electrode should be attached to the positive lead of a stimulus isolation unit capable of delivering currents in the range of 1–50 μA through the electrode. The stimulator should be configured to deliver biphasic pulses (pulse width 200 μs for each phase) at 200–500 Hz for train durations up to 500 ms. We have found that 200 ms pulse trains at 500 Hz reliably and reproducibly evoke movements that are easy to detect with high-speed videography (e.g., Fig. 7b) but that longer train durations can make the movements more clear to the naked eye. Microstimulation mapping begins when the electrode is positioned approximately 500 µm above the expected depth of the DCN, typically in the overlying cerebellar cortex or white matter. Stimulation at this depth with currents around 15–20 µA can evoke a variety of movements, but these movements are often non-specific and probably result from antidromic activation of motor-related mossy fibers projecting to the cerebellar cortex [41]. As the electrode is advanced toward the DCN, the ability to evoke movements with microstimulation intensities below 20 µA will briefly diminish and then will increase as the electrode enters the DCN. Within the DCN, microstimulation intensities as low as 1–2 µA are often sufficient to evoke motor twitches, but discrete movements usually require currents between 5 and 10 µA. Once the threshold current is found for evoking movement at a particular depth within the DCN, the electrode should be advanced 50–100 µm and the threshold movement tested again (Fig. 7b). This is continued through the entire depth of the DCN at multiple rostral-caudal and mediolateral locations such that a somatotopic map of evoked movements can be compiled. The essential location for eyeblink conditioning within AIP is identified based on the ability to evoke discrete eyelid closure with low currents (5–10 µA) [12]. Once this "hot spot" is identified, it can be targeted for single-unit recording.

3.7 Acute Optogenetics and Electrophysiological Recording During Behavior

Photostimulation during extracellular recording is a powerful technique for identifying genetically defined cell classes [42], relating optogenetically induced changes in neural activity with coincident changes in behavior [7, 43–45], or simply confirming that an optogenetic manipulation has the desired effect on neurons within the circuit [46]. We perform combined photostimulation and recording in awake mice using custom-made "optrodes" that consist of a length of optical fiber that is connectorized to a ferrule at one end and pulled to a fine tip at the other (Fig. 4). The fiber is then reinforced by placing it inside hypodermic tubing and attached to a tungsten electrode using cyanoacrylate glue. Once assembled an optrode can typically be used for multiple penetrations across multiple days until the electrode impedance drops too much to maintain single-unit isolation. When this happens, the electrode can be removed by gently heating the assembly with a heat gun and separating it from the optical fiber. The optical fiber can then be attached to a new electrode, and this can be repeated until the optical fiber becomes too damaged to transmit light. See Materials for a description of how to measure light transmission through the optical fiber.

During an experiment, the optrode is lowered into the brain using a micromanipulator, as with the normal single-unit extracellular recording procedure described above. The only additional requirement is that the optical fiber must be coupled to a fixed wavelength light source, such as a laser or LED. Be aware that photostimulation can sometimes induce artifacts in the recording due to the photoelectric effect [47]. This can be partially mitigated by ensuring that bright light coming from the fiber optic does not directly illuminate the electrode surface.

3.8   Data Analysis

A full description of data analysis is outside the scope of this chapter. However, in this section, we provide some guidelines for basic analysis of neural and behavioral data that are relevant for the cerebellum and EBC.

3.8.1   Spike Sorting

The best way to ensure that you are only picking up the activity of a single neuron is to take your time positioning the electrode during the experiment so that the spikes of a single neuron appear several times larger in amplitude than spikes from the surrounding neurons. If you have achieved "single-unit isolation," the spikes can be detected and stored during the experiment using a time-amplitude window discriminator (either hardware or software based, depending on the data acquisition system). However, it is sometimes possible to use post hoc ("offline") spike sorting algorithms based on principal component analy+sis (PCA) or direct measurements of spike features to separate the signals from multiple units that were picked up on the same electrode [48, 49]. Offline spike sorting can also be used to separate simple and complex spikes in a single-unit Purkinje cell recording [30].

There are several steps involved in sorting spikes, some of which depend on the particular properties of the cell type. Here we focus on sorting simple and complex spikes of a Purkinje cell. In an extracellular recording, simple spikes typically appear as brief negative and/or positive deflections in the extracellular field potential that last for only 200–300 µs (Fig. 6b). On the other hand, complex spikes begin with a similar brief deflection in the field potential that is then followed by a train of "spikelets" or low-frequency oscillations in the field potential that can last for up to several ms (Fig. 6b, *right*). Based on these differences in features, it is often possible to separate simple and complex spikes (Figs. 6b and 8a). First, the simple and complex spikes are extracted from the background activity generated by nearby neurons by setting an amplitude threshold. Next, exemplar simple and complex spikes are examined to determine promising criteria for distinguishing them. Often, a distinguishing feature is the deflection in field potential that occurs in complex spikes just following the rapid deflection that is shared by both simple and complex spikes. Using template matching, PCA, or other automatic methods for this period is often enough to obtain good spike sorting. Unfortunately,

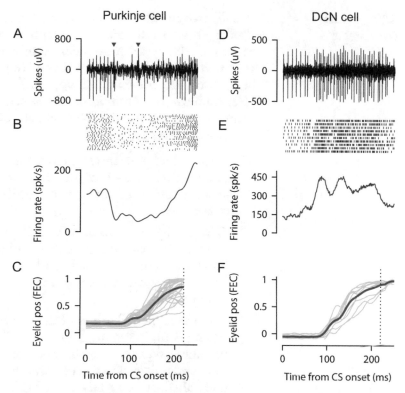

Fig. 8 Example Purkinje (**a–c**) and DCN (**d–f**) cell recorded in eyeblink hot spots of cerebellar cortex and anterior interpositus during EBC. (**a, d**) Extracellular signal during a single trial; (**b, e**) raster of spike times (*top*) and peristimulus time histogram (*bottom*) for multiple paired trials; (**c, f**) eyelid positions for the same trials (*light thin traces*). Average eyelid position trace indicated by dark thick trace. Used with permission from [12]

some complex spike profiles are difficult to sort using automatic methods, particularly when the initial phases completely overlap with those of simple spikes and the only distinguishing characteristic is the presence of spikelets. This is because the timing of spikelets is not the same from spike to spike, and automatic spike sorting algorithms are not good at dealing with time-jittered components. But for the trained eye, spikelets are obvious and easy to pick out, so complex spikes with these characteristics can be manually sorted.

Regardless of the method used, detected complex spikes should always be checked by eye, using the presence of spikelets or low- frequency oscillations to positively identify complex spikes. Once simple and complex spikes are sorted, it is essential to verify that they were emitted from the same neuron by generating a complex spike-triggered histogram of simple spike times. Presence of a 10–30 ms pause in simple spikes following each complex spike is the gold standard for showing that the simple and complex spikes were generated by the same Purkinje cell and is confirmation that the Purkinje cell was well isolated (Fig. 6b, *inset*).

Eyelid movements are monitored using a high-speed camera (200 frames/s) and near real-time processing of the video stream to extract the eyelid movement trace over time. You can write your own software for this, or if you are incorporating MATLAB into your experimental setup, you can use our software (Neuroblinks, see *Materials*). If you choose to write your own software, you can use the following algorithm to measure eyelid movements (Fig. 2).

1. Select a region of interest (ROI) in your video, and discard information outside of the ROI in each frame. Select as small a region of interest as possible while including the whole eye.

2. Convert the image to binary, setting the threshold for black/ whiteness so that the pupil and iris are black but the surrounding fur is white. You may need to adjust the lighting in the chamber to eliminate any shadows that prevent you from collecting clear images.

3. Sum the numerical values of the binary image (the number of white pixels, indicating fur). Normalize the values such that the summed pixels range between zero (eyelid fully open) and one (eyelid fully closed). This transforms the pixel values into units of "fraction eyelid closure" (FEC).

We typically extract the eyelid trace in near real time during the experiment to control the task, but the time resolution of these data is not usually good enough for further analysis. Instead, after the session, we re-extract and renormalize the eyelid traces from the saved video frames. After the eyelid trace is normalized (i.e., between 0 and 1 FEC) based on a trial in which the mouse produces a full blink (for instance, a US alone trial), data analysis can begin. Learning can be measured in terms of the percentage of trials on which a CR occurred (%CR), amplitude of eyelid closure during a CR, CR timing, CR velocity, etc. There are multiple ways to operationally define a CR. Our laboratory uses a criterion of a 10% eyelid closure. It is possible to analyze eyelid movements in terms of absolute eyelid closure. However, mice sometimes squint during EBC, which can be confused for a small CR if eyelid position is not measured relative to baseline.

When quantifying CRs, it is critical to consider the timing of the eyelid movement, as this parameter reflects the degree of the movement's dependence on the cerebellum (recall from Sect. 3.2 that US intensity during training affects the brain structures recruited during EBC). If an eyelid response reaches its peak amplitude before the end of the CS-US interval, the response is poorly timed and may not be cerebellum dependent [10–12, 50, 51]. Additionally, tone CSs elicit unlearned, cerebellum-independent startles that are reflected in eyelid movements peaking 15–30 ms after CS onset [10].

It can also be informative to examine movement velocity [11, 52]. To calculate CR velocity, take the derivative of the eyelid position trace across all time bins in a given trial.

4   Notes

Note 1. Treadmill

Note 1.1. Make the cylinders as perfectly as possible. If the axle is not centered in the foam, the treadmill will wobble when rotated, which can reduce recording stability and stress the mouse.

Note 1.2. Do not force the bearings into the mounts, as this will deform them and make it more difficult for the mouse to move the treadmill.

Note 2. Optical Fibers

Note 2.1. Make sure to measure the light transmission through your fibers using the patch fiber that you intend to use during your experiment, as differences in the patch fiber connectorization quality will affect light transmission.

Note 2.2. You can measure light transmission both before and after an experiment, even if the optical fiber has been chronically implanted. If the fiber has been chronically implanted, carefully extract the fiber from the brain after euthanizing your animal. Soak the extracted implant in 70% EtOH for a few hours to disinfect it. You may also rinse the extracted implant a few times in distilled water. After you have allowed the implant to dry, you can then measure the degree of light transmission through it as you did before implantation. This will allow you to know whether the fiber may have been damaged during your experiment.

Note 3. Constructing Optical Fibers

Note 3.1. In addition to the advice given in *Thorlabs Guide to Connectorization and Polishing of Optical Fibers*, it can be helpful to use a pin vise to protect the fibers while polishing.

Note 3.2. It is a good idea to perform the most difficult fiber construction steps as early as possible while making each fiber. This way, you will waste less time building fibers that you ultimately break.

Note 3.3. Instead of cleaving the fibers by taping them to a desk and manually scoring them, you can use Thorlabs posts and an RA90 to support a micromanipulator. Then, hold a fiber and ferrule in the micromanipulator's electrode holder using a pin vise. Use another RA90 to hold a ruby scribe at a right angle to the fiber. You can then use the micromanipulator to carefully score the fiber for more precise cleaving.

Note 4. Surgery

Note 4.1. Securing the mouse in ear bars stabilizes the skull, and skull stability is important when stereotactically implanting in multiple locations on the skull. Using non-rupture ear bars for mice (e.g., Kopf 822) will facilitate pitch adjustments for achieving a stereotactic plane.

Leveling the skull will be the easiest/fastest if the mouse is secured in ear bars such that:

- The mouse's head pitch can be easily and smoothly adjusted.
- The mouse's head yaw cannot be adjusted with gentle pressure.
- The mouse's nose is straight.
- The interaural distance is ~8 mm. If you are obtaining interaural distance measurements much smaller than this, you run the risk of damaging the ear canals (which could affect conditioning using tone stimuli) or compressing the airway.

Note 4.2. In order to increase the available surface area for dental cement to bind to, the surgeon can use Dumont #7 forceps or any other suitable surgical instrument to score the skull. This should be done carefully, however, as the mouse skull can be easily punctured, especially in the more anterior regions.

Note 4.3. The skull should be very clean and thoroughly dry when dental cement is applied. Moisture trapped between the dental cement and skull will (1) prevent the dental cement from bonding to the bone and (2) facilitate infection and bone decay by making pockets for bacteria to colonize. This second point is particularly relevant around the margins of the dental cement. To aid in drying the skull, the surgeon can use canned air. It can also be helpful to wipe the skull with a cotton swab moistened by hydrogen peroxide or 70% EtOH, although this should only be done as long as the skull is intact (i.e., you have not made any holes in the skull) in order to avoid damaging the brain. The surgeon should monitor the margins of the dental cement while it cures, drying any interstitial fluid or other seepages from the incision before it can contact the curing dental cement.

Note 5. Eyeblink Conditioning

Note 5.1. If you observe the mouse's eye moving up/down/ side to side substantially while he is head-fixed, the headplate may be loose. Remove the animal from the rig, and ascertain in a surgical setting whether the headplate can be secured to the skull better or whether the animal must be euthanized.

Note 5.2. Make sure that the mouse fits comfortably on the treadmill (i.e., the mouse reaches the surface of the wheel without trouble, the mouse is not squished against the wheel, the wheel spins freely).

Note 5.3. Making changes from the habituation conditions can dishabituate and frighten the animals. This includes playing masking white noise (or not) at the same volume level from day to day, positions of the different CS and US delivery mechanisms, etc. In addition, keep the positions of all parts of the conditioning apparatus as consistent as possible throughout your experiment to make sure that features of the stimuli do not differ between trials or sessions with the same mouse.

Note 5.4. Trials should not be delivered if the animal is too squinty (e.g., above 20% baseline eyelid closure), as this will make it difficult to detect CRs and can decrease the perceived strength of the US.

Note 5.5. The Neuroblinks EBC conditioning software used in this laboratory requires the experimenter to deliver one US-only trial at the beginning of each session. This calibration trial is used to extract the fraction eyelid closure in real time during the session (see Fig. 2). If delivering this calibration trial would confound the experimental question being addressed, the open source Neuroblinks software should be modified accordingly to suit the needs of the experiment.

Note 5.6. Including too many CS-alone trials during acquisition will delay/prevent learning. We would not recommend delivering more than 25% CS-alone trials. However, there is some evidence that rabbits can learn to make CRs during EBC with 75% CS-alone trials [53].

Note 5.7. Damage to the cerebellum during surgery or due to insult on the craniotomy can render animals incapable of learning the task, so be impeccable with your operative and post-op care.

Note 5.8. The time course of the air puff delivered will differ from the time course of the trigger sent to the pressure injector. The onset of the air puff will be delayed, with the duration of this delay depending on the length of tubing between the pressure injector and the puff needle. In addition, air puff pressure will peak after the time that the trigger ends. For example, in our system air pressure begins to rise 12 ms after trigger onset, and peak pressure for a 30 ms trigger is ~50 ms after trigger onset. You should test your system to ascertain the delay to puff peak and onset in order to better inform your experimental design.

Note 5.9. The US must be strong enough to elicit full eyelid closure during reflex blinks, but not so strong that it damages the cornea. As a rule of thumb, the puff should produce a reflex blink with about 100 ms of complete eyelid closure.

Note 5.10. Test the pressure injection system for the US regularly.

Note 6. Choosing an Electrode

Note 6.1. Tungsten electrodes are standard for a wide variety of purposes. These electrodes can be manufactured with thin outside diameters (e.g., 80 μm; FHC) and are rigid enough that they can penetrate the dura and remain straight even after many recording sessions. These electrodes can also be used to make electrolytic lesions at recording sites.

Note 6.2. Platinum/iridium (Pt/Ir) electrodes are not as rigid as tungsten for a given diameter, but they don't get damaged as much as tungsten by electrical stimulation or electrolytic lesioning, so they can be used longer. They are also available for purchase in thin outside diameters (e.g., 80 μm; Alpha-Omega).

Note 6.3. Pulled glass electrodes are useful for marking recording sites by dye injection (e.g. 2% pontamine sky blue in 3 M potassium acetate). These electrodes can also be constructed to have small tip sizes (tip size = 2 μm, around 5 MOhm), allowing for large signal-to-noise ratio recordings of single Purkinje cells and excellent resolution of the calcium waves of the complex spike. Alternately, glass electrodes can be constructed to have larger tip sizes (tip size = 5–8 μm, around 2 MOhm), which makes it easier to locate Purkinje cells.

Note 6.4. Electrodes with larger tip sizes (impedance <1.0 MOhm) are ideal for recording climbing fiber local field potentials. The impedance of electrodes used for recording simple and complex spikes from Purkinje cells or action potentials from the deep cerebellar nuclei is somewhat a matter of personal preference. Larger tip sizes (lower impedance) allow the experimenter to detect electrical activity from further away but can make it more difficult to achieve single-unit instead of multiunit isolation. Smaller tip sizes (higher impedance) allow the experimenter to more easily isolate single-units, but it can be difficult to find cells using these electrodes.

Note 6.5. For microstimulation and electrolytic lesioning, be sure to select an electrode that can pass the desired current. Use Ohm's law to determine the maximum electrode resistance (impedance) that can be used given the supply voltage of your stimulus isolation unit and the desired current.

References

1. Humphrey DR, Schmidt EM (1990) Extracellular single-unit recording methods. In: Boulton AA, Baker GB, Vanderwolf CH (eds) Neurophysiol. Tech. Appl. to Neural Syst. Humana Press, Totowa, NJ, pp 1–64

2. Bryant JL, Roy S, Heck DH (2009) A technique for stereotaxic recordings of neuronal activity in awake, head-restrained mice. J Neurosci Methods 178:75–79. https://doi.org/10.1016/j.jneumeth.2008.11.014

3. Schonewille M, Khosrovani S, Winkelman BHJ et al (2006) Purkinje cells in awake behaving animals operate at the upstate membrane potential. Nat Neurosci 9:459–61; author reply 461. https://doi.org/10.1038/nn0406-459

4. Goossens HHLM, Hoebeek FE, Van Alphen AM et al (2004) Simple spike and complex spike activity of floccular Purkinje cells during the optokinetic reflex in mice lacking cerebellar long-term depression. Eur J Neurosci 19:687–697. https://doi.org/10.1111/j.1460-9568.2003.03173.x

5. Cheron G, Gall D, Servais L et al (2004) Inactivation of calcium-binding protein genes

induces 160 Hz oscillations in the cerebellar cortex of alert mice. J Neurosci 24:434–441. https://doi.org/10.1523/JNEUROSCI.3197-03.2004

6. White JJ, Lin T, Brown AM et al (2016) An optimized surgical approach for obtaining stable extracellular single-unit recordings from the cerebellum of head-fixed behaving mice. J Neurosci Methods 262:21–31. https://doi.org/10.1016/j.jneumeth.2016.01.010

7. Heiney SA, Kim J, Augustine GJ, Medina JF (2014) Precise control of movement kinematics by optogenetic inhibition of Purkinje cell activity. J Neurosci 34:2321–2330. https://doi.org/10.1523/JNEUROSCI.4547-13.2014

8. Paré WP, Glavin GB (1986) Restraint stress in biomedical research: a review. Neurosci Biobehav Rev 10:339–370

9. Li S, Fan Y-X, Wang W, Tang Y-Y (2012) Effects of acute restraint stress on different components of memory as assessed by object-recognition and object-location tasks in mice. Behav Brain Res 227:199–207. https://doi.org/10.1016/j.bbr.2011.10.007

10. Boele H-J, Koekkoek SKE, De Zeeuw CI (2010) Cerebellar and extracerebellar involvement in mouse eyeblink conditioning: the ACDC model. Front Cell Neurosci 3:19. https://doi.org/10.3389/neuro.03.019.2009

11. Chettih SN, McDougle SD, Ruffolo LI, Medina JF (2011) Adaptive timing of motor output in the mouse: the role of movement oscillations in eyelid conditioning. Front Integr Neurosci 5:72. https://doi.org/10.3389/fnint.2011.00072

12. Heiney SA, Wohl MP, Chettih SN et al (2014) Cerebellar-dependent expression of motor learning during Eyeblink conditioning in head-fixed mice. J Neurosci 34:14845–14853. https://doi.org/10.1523/JNEUROSCI.2820-14.2014

13. Hilgard ER, Marquis DG (1935) Acquisition, extinction, and retention of conditioned lid responses to light in dogs. J Comp Psychol 19:29–58. https://doi.org/10.1037/h0057836

14. Schneiderman N, Fuentes I, Gormezano I (1962) Acquisition and extinction of the classically conditioned eyelid response in the albino rabbit. Science 136:650–652

15. Chen L, Bao S, Lockard JM et al (1996) Impaired classical eyeblink conditioning in cerebellar-lesioned and Purkinje cell degeneration (pcd) mutant mice. J Neurosci 16:2829–2838

16. Kim JJ, Thompson RF (1997) Cerebellar circuits and synaptic mechanisms involved in classical eyeblink conditioning. Trends Neurosci 20:177–181

17. McCormick DA, Thompson RF (1984) Cerebellum: essential involvement in the classically conditioned eyelid response. Science 223:296–299

18. Schonewille M, Gao Z, Boele H-J et al (2011) Reevaluating the role of LTD in cerebellar motor learning. Neuron 70:43–50. https://doi.org/10.1016/j.neuron.2011.02.044

19. Aiba A, Kano M, Chen C et al (1994) Deficient cerebellar long-term depression and impaired motor learning in mGluR1 mutant mice. Cell 79:377–388

20. Shibuki K, Gomi H, Chen L et al (1996) Deficient cerebellar long-term depression, impaired eyeblink conditioning, and normal motor coordination in GFAP mutant mice. Neuron 16:587–599. https://doi.org/10.1016/S0896-6273(00)80078-1

21. LeChasseur Y, Dufour S, Lavertu G et al (2011) A microprobe for parallel optical and electrical recordings from single neurons in vivo. Nat Methods 8:319–325. https://doi.org/10.1038/nmeth.1572

22. Paxinos G, Franklin KBJ (2013) Paxinos and Franklin's the mouse brain in stereotaxic coordinates. Elsevier Academic Press, London

23. Siegel JJ, Taylor W, Gray R et al (2015) Trace eyeblink conditioning in mice is dependent upon the dorsal medial prefrontal cortex, cerebellum, and amygdala: behavioral characterization and functional circuitry. eNeuro 2:1–29. https://doi.org/10.1523/ENEURO.0051-14.2015

24. Sakamoto T, Endo S (2010) Amygdala, deep cerebellar nuclei and red nucleus contribute to delay eyeblink conditioning in C57BL/6 mice. Eur J Neurosci 32:1537–1551. https://doi.org/10.1111/j.1460-9568.2010.07406.x

25. Kehoe EJ, White NE (2002) Extinction revisited: similarities between extinction and reductions in US intensity in classical conditioning of the rabbit's nictitating membrane response. Anim Learn Behav 30:96–111. https://doi.org/10.3758/BF03192912

26. Eccles J, Ito M, Szentágothai J (1967) The cerebellum as a neuronal machine. Springer, Heidelberg

27. Miles FA, Fuller JH, Braitman DJ, Dow BM (1980) Long-term adaptive changes in primate vestibuloocular reflex. III. Electrophysiological observations in flocculus of normal monkeys. J Neurophysiol 43:1437–1476

28. Lisberger SG, Fuchs AF (1978) Role of primate flocculus during rapid behavioral modification of vestibuloocular reflex. I. Purkinje cell activity during visually guided horizontal smooth-pursuit eye movements and passive head rotation. J Neurophysiol 41:733–763

29. Thach WT (1968) Discharge of Purkinje and cerebellar nuclear neurons during rapidly alternating arm movements in the monkey. J Neurophysiol 31:785–797

30. Ohmae S, Medina JF (2015) Climbing fibers encode a temporal-difference prediction error during cerebellar learning in mice. Nat Neurosci 18:1798–1803. https://doi.org/10.1038/nn.4167

31. Granit R, Phillips CG (1956) Excitatory and inhibitory processes acting upon individual Purkinje cells of the cerebellum in cats. J Physiol 133:520–547

32. Van Dijck G, Van Hulle MM, Heiney SA et al (2013) Probabilistic identification of cerebellar cortical neurones across species. PLoS One 8:e57669. https://doi.org/10.1371/journal.pone.0057669

33. Ruigrok TJH, Hensbroek RA, Simpson JI (2011) Spontaneous activity signatures of morphologically identified interneurons in the vestibulocerebellum. J Neurosci 31:712–724. https://doi.org/10.1523/JNEUROSCI.1959-10.2011

34. Apps R, Hawkes R (2009) Cerebellar cortical organization: a one-map hypothesis. Nat Rev Neurosci 10:670–681. https://doi.org/10.1038/nrn2698

35. Garwicz M, Jorntell H, Ekerot CF (1998) Cutaneous receptive fields and topography of mossy fibres and climbing fibres projecting to cat cerebellar C3 zone. J Physiol 512:277–293

36. Mostofi A, Holtzman T, Grout AS et al (2010) Electrophysiological localization of eyeblink-related microzones in rabbit cerebellar cortex. J Neurosci 30:8920–8934. https://doi.org/10.1523/JNEUROSCI.6117-09.2010

37. Jörntell H, Ekerot CF, Garwicz M, Luo XL (2000) Functional organization of climbing fibre projection to the cerebellar anterior lobe of the rat. J Physiol 522(Pt 2):297–309

38. Uusisaari MY, De Schutter E (2011) The mysterious microcircuitry of the cerebellar nuclei. J Physiol 589:3441–3457. https://doi.org/10.1113/jphysiol.2010.201582

39. Dow RS, Moruzzi G (1958) The physiology and pathology of the cerebellum. University of Minnesota Press, Minneapolis, MN

40. Ekerot CF, Jörntell H, Garwicz M (1995) Functional relation between cortico-nuclear input and movements evoked on microstimulation in cerebellar nucleus interpositus anterior in the cat. Exp Brain Res 106:365–376. https://doi.org/10.1007/BF00231060

41. Hesslow G (1994) Correspondence between climbing fibre input and motor output in eyeblink-related areas in cat cerebellar cortex. J Physiol 476:229–244

42. Kravitz AV, Owen SF, Kreitzer AC (2013) Optogenetic identification of striatal projection neuron subtypes during in vivo recordings. Brain Res 1511:21–32. https://doi.org/10.1016/j.brainres.2012.11.018

43. Smear M, Shusterman R, O'Connor R et al (2011) Perception of sniff phase in mouse olfaction. Nature 479:397–400. https://doi.org/10.1038/nature10521

44. Lee S-H, Kwan AC, Zhang S et al (2012) Activation of specific interneurons improves V1 feature selectivity and visual perception. Nature 488:379–383. https://doi.org/10.1038/nature11312

45. Lee KH, Mathews PJ, Reeves AMB et al (2015) Circuit mechanisms underlying motor memory formation in the cerebellum. Neuron:1–12. https://doi.org/10.1016/j.neuron.2015.03.010

46. Zhao S, Ting JT, Atallah H et al (2011) Cell type–specific channelrhodopsin-2 transgenic mice for optogenetic dissection of neural circuitry function. Nat Methods 8:745–752. https://doi.org/10.1038/nMeth.1668

47. Kozai TDY, Vazquez AL (2015) Photoelectric artefact from optogenetics and imaging on microelectrodes and bioelectronics: new challenges and opportunities. J Mater Chem B Mater Biol Med 3:4965–4978. https://doi.org/10.1039/C5TB00108K

48. Lewicki MS (1998) A review of methods for spike sorting: the detection and classification of neural action potentials. Network 9:R53–R78. https://doi.org/10.1088/0954-898X/9/4/001

49. Rey HG, Pedreira C, Quian Quiroga R (2015) Past, present and future of spike sorting techniques. Brain Res Bull 119:106–117. https://doi.org/10.1016/j.brainresbull.2015.04.007

50. Mauk MD, Ruiz BP (1992) Learning-dependent timing of Pavlovian eyelid responses: differential conditioning using multiple interstimulus intervals. Behav Neurosci 106:666–681

51. Domingo JA, Gruart A, Delgado-García JM (1997) Quantal organization of reflex and conditioned eyelid responses. J Neurophysiol 78:2518–2530

52. Gruart A, Blazquez PM, Delgado-García JM (1995) Kinematics of spontaneous, reflex, and conditioned eyelid movements in the alert cat. J Neurophysiol 74:226–248

53. Powell DA, Churchwell J, Burriss L (2005) Medial prefrontal lesions and Pavlovian eye-blink and heart rate conditioning: effects of partial reinforcement on delay and trace conditioning in rabbits (Oryctolagus Cuniculus). Behav Neurosci 119:180–189. https://doi.org/10.1037/0735-7044.119.1.180

Chapter 4

Multielectrode Arrays for Recording Complex Spike Activity

Eric J. Lang

Abstract

The multielectrode technique described here was developed for simultaneously recording complex spike activity from arrays of Purkinje cells in anesthetized rodents. The technique involves placing a platform on the surface of the brain and implanting microelectrodes through this platform, which then serves to hold the individual electrodes in place and to protect the brain surface. With this technique, stable recordings of complex spike activity for over 24 h have been achieved. Typical arrays consist of ~40 electrodes that are arranged as a 4 × 10 matrix with electrodes spaced by ~250 μm; however, denser and larger arrays are possible.

The technique was designed for recording complex spike activity in anesthetized animals. However, it has also been successfully used to record Purkinje cell simple spikes and activity from other neurons. Moreover, it is possible to use this approach with awake, head-fixed animals performing behavioral tasks.

Key words Multielectrode recording, Purkinje cell, Complex spikes, Synchrony, Cerebellum

1 Introduction

The multielectrode technique described here is based on the one originally developed by Llinás and colleagues to record complex spikes from arrays of Purkinje cells in order to test hypotheses about their patterns of synchronous activity [1–3]. One of its strengths is its flexibility in allowing both implantation of high-density arrays, with which a significant percentage of the Purkinje cells in a local area (20–25%) can be recorded [3], and multiple arrays for simultaneous recording from widely separated regions of the cerebellar cortex [4–6].

The characteristics of the cerebellar cortex, and of complex spikes in particular, allow this technique to be particularly successful when employed for recording complex spike activity. Specifically, because complex spikes can be recorded over a wide range of depths, from as shallow as 50 μm below the brain surface down to the Purkinje cell layer (about 250 μm in rats), the single-unit isolation

Roy V. Sillitoe (ed.), *Extracellular Recording Approaches*, Neuromethods, vol. 134,
https://doi.org/10.1007/978-1-4939-7549-5_4, © Springer Science+Business Media, LLC 2018

process is easy, rapid, and highly successful. Moreover, the relatively superficial location of Purkinje cells in the cerebellar cortex and their high susceptibility to damage require an approach that minimizes tissue damage, which the use of glass microelectrodes with ~1–2-μm-diameter tips allows.

Although this technique was primarily designed for recording of complex spikes in anesthetized animals, its use has been expanded to include recording from awake, head-fixed animals [7, 8] and recording of Purkinje cell simple spike activity [9] and deep cerebellar nuclear activity [10, 11]. However, the yield in these cases is significantly lower than for recording complex spikes in anesthetized animals.

2   Materials

2.1   Electrode Holders

Rubber.
Brass strips.
Tungsten rods.
Acrylic sheet and rods.
Double-sided scotch tape.
Female 15-pin D-Sub connector.
Acrylic rod.
Epoxy.

2.2   Microelectrodes

Solution: NaCl, glycerin.
Capillary tubes.
Narishige electrode puller.
Electrode storage jars.

2.3   Electrode Platform

Dow Corning 3140 RTV coating silicon rubber.
Tungsten rods.
Electron microscope grids.
Epoxy.
Vibratome.
Acrylic rectangular bar.

2.4   Surgical Procedures

Stereotaxic frame.
Standard surgical tools, including forceps, vessel clips, clamps, mini-screwdriver, scalpel, suture (sizes 5–0, 3–0), cyanoacrylate glue, Gelfoam, cotton-tipped applicators, cauterizer, Rongeurs.
Tracheal tube.
3-axis micromanipulator.
Polyethylene tubing.

2.5   Recording Procedures

Temperature controller and heating pad.
Syringe pump.
3-axis micromanipulator.

Heating coil.
Surgical scope.
Multichannel amplifier system.
A cable with D-Sub 37 connectors (one male, one female).

3   Methods

3.1   Microelectrode Holder Assembly

The microelectrode holders are used for housing groups of 15 final stage electrodes prior to their implantation into the cerebellum (Fig. 1b). A set of holders that is sufficient to house enough electrodes for an experiment needs to be constructed before beginning experiments. Each holder is assembled from an acrylic rod, a 15-pin D-Sub connector (841-17DAFR-B15S, Amphenol, Wallingford, CT), a rectangular brass strip, and a silicon rubber block that contains a series of 15 closely spaced holes running from top to bottom.

The rubber blocks are made using a mold. The mold is constructed from an acrylic sheet (that forms the floor of the mold) and rectangular acrylic rods (that form the long and short walls of the mold). Two sets of two long rods and one set of two short rods

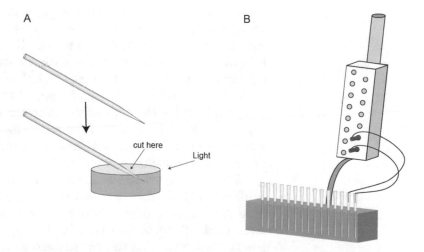

Fig. 1 Electrode assembly. (**a**) A pulled capillary tube is placed in a small petri dish filled with electrode solution. The dish is illuminated by a small light from the side. The pipette is oriented perpendicular to the light path in order to check for air bubbles. The tip is then cut using a small pair of scissors while submerged. (**b**) An electrode holder loaded with glass tips. The holder consists of a silicon rubber block attached, via a brass metal strip, to the base of a plastic rod to which a 15-pin D-Sub connector is also attached. The rubber block is loaded with glass tips. Wires are inserted into the glass tips from *right* to *left* with the corresponding socket pins inserted into the D-Sub connector from *bottom* to *top* in the *right column* followed by the *left column*. During an experiment, the electrodes are used in the reverse order

are needed. For a mold that allows four rubber block holders to be made at one time, the dimensions of the long rods are 7 cm (L), 2 cm (W), and 0.4 cm (H). The dimensions of the floor sheet are not critical but must be large enough to accommodate the arrangement of the rods. Using acrylic cement (SciGrip 4, SciGrip, Durham, NC), one set of the long rods is glued to the floor sheet, such that the two rods run parallel to each other with a gap of 1.5 cm between them. Each end of the channel running between the long rods is then capped by cementing in place a short rectangular acrylic rod that is 0.8 cm in height and whose width (1.5 cm) matches that of the channel (the short rods should be inserted into the channel slightly, so that their sides overlap with those of the long rods). The short rods should be glued to both the floor sheet and to the sides of the long rods, sealing the channel. Once the cement has dried, the mold can be used to make the rubber blocks that will hold the electrodes.

To form the rubber blocks, a piece of double-stick Scotch tape (3 M; St Paul, MN) is applied to the top surface of each of the long walls of the mold. The floor and side walls of the channel are then coated with a thin layer of Vaseline. Next, two sets of 15 pieces of bare tungsten rod (diameter, 0.004 in, # 719000, A-M Systems, Carlsborg, WA) are cut (length 2.5 cm). Each piece of tungsten rod is placed such that it spans the gap between the two long side walls, runs perpendicular to the long axis of the channel, and extends onto the tape on each side. The 15 rods in each set should be spaced at 1-mm intervals, and a somewhat larger gap should be left between the two sets of rods. Once all of the tungsten rods are in place, a second piece of tape is applied to the top of each long wall, sandwiching the ends of the tungsten rods. Vaseline is then applied to one side of each of the remaining long wall pieces (the side that will line the channel), and those pieces are placed on top of the tape, completing the mold.

Silicon (Sylgard 184 Silicone Elastomer Kit, Dow Corning, Midland, MI) is then poured into the mold and allowed to harden for several days. There may be some leakage, and if so, additional silicon should be added after the initial silicon has partially set, such that the entire channel is filled. Once the silicon has gelled, the upper long walls are removed, and the silicon block is taken out from the mold with the tungsten rods in place. The tape is then removed, and tungsten rods are individually pulled out from the block using forceps. Using a razor blade, the silicon block is cut into two pieces so that each piece contains 15 holes. Each of these pieces is then cut lengthwise down its center with the razor blade. The resulting smooth cut surface should be used as the top surface when the block is assembled into the electrode holder.

Once the rubber blocks have been completed, the entire electrode holder can be assembled. First, a 15-pin D-Sub connector and a rectangular brass strip (4–5 cm by 0.5 cm) are attached to an ~15-cm-long cylindrical acrylic rod using epoxy (diameter ~0.5 cm). The D-Sub connector should be affixed so that its long axis aligns with the acrylic rod, its back is to the rod, and its bottom edge coincides with the bottom end of the rod (Fig. 1b). The brass strip is attached at that end and should extend down ~3 cm from it. To complete the holder assembly, the rubber block is epoxied near the free lower end of the brass strip, as shown in Fig. 1b. The brass strip can subsequently be curved to create an ~45° angle between the holes in the silicon block and the acrylic rod.

Fully assembled holders are stored in plastic cups with lids. A hole is cut in the lid, so that the acrylic rod can slide through it but that is small enough that it firmly grips the rod in place. The hole should be cut somewhat off-center because this better allows the holder to fit inside the cup when it is loaded with electrodes. When storing electrodes, the cup should contain a small amount of electrode solution so that the rubber block is partially submerged in the solution without touching the bottom of the cup. Cups containing the solution and electrodes should be stored in a refrigerator.

3.2 Micromanipulator and Heating Coil System

A 3-axis, joystick-controlled micromanipulator (SM 3.25, Marzhauser, Wetzlar, Germany) is used to retrieve electrodes from the holder and implant them into the cerebellum. The manipulator is mounted on a magnetic base via a pair of metal poles, which are connected at right angles to allow flexible positioning of the manipulator.

A cylindrical metal or acrylic rod is held by the manipulator and on its free end is attached a 16-gauge syringe needle whose canal has been plugged and whose Luer lock attachment has been cut off (the needle should be attached to the rod at the cut end, leaving the beveled end free).

A heating coil is wrapped around the lower half of the needle, starting just above the beveled end. The heating coil is connected to a voltage source that is controlled by a foot pedal switch. The coil is heated to melt sticky wax (00625, Kerr, Emeryville, CA) to fill the space between the coil and needle, creating a reservoir of wax that will be used during the experiment to allow for temporary attachment of the electrodes to the syringe needle.

3.3 Microelectrode Assembly

Recordings are obtained with glass microelectrodes assembled from D-Sub connector pins, platinum-iridium wires, and pulled capillary glass filled with microelectrode solution. Large stocks of pulled capillary glass filled with solution, and of connecting wires attached to their pins, should be prepared ahead of the experiment.

The final assembly of the microelectrodes in the holder is usually done the day before (or day of) the experiment.

Connecting wires. Pieces of teflon-coated platinum-iridium (90/10%) wire (0.014 in diameter, California Fine Wire, Grover Beach, CA) are cut to a length of 7 cm. The ends of each piece are briefly flamed to remove the insulation from the terminal 2–3 mm. Finally, one end of the wire is soldered into a male D-Sub pin (Multicomp SPC15395, Newark element14, Chicago, Il). These wires can be reused several times if they do not become twisted and the insulation remains intact. However, before each usage, the free end of the wire should be burned clean, and the integrity of the wire and its insulation should be tested with an ohmmeter by dragging the wire through a small drop of saline and verifying that the resistance is high throughout the wire, except at its end.

Microelectrode tips. Capillary tubes (1.5-mm OD, 0.86 ID) are pulled and filled with electrode solution (a 50–50 solution of 2 M NaCl solution and glycerin). The capillary tubes are pulled in a vertical electrode puller (PE-22, Narishige). The puller allows for adjustment of the intensity of the heat for melting the glass and of two magnets to supplement the gravitational force. It is important to not use the magnets because doing so results in electrodes that are too flexible to penetrate the recording platform. Note that in the case of the Narishige puller, one of the magnets can be shut off using the control knob; the other must be disabled by clipping its limit switch to the closed position. The heat intensity should be adjusted so that the tapered portion of the capillary tube is ~7 cm.

Once the electrode is pulled, the tips are viewed under a microscope and broken against the flat end of a thick-walled capillary tube to a diameter of 2–3 μm. The electrodes are then placed in an electrode holder jar (E215, WPI, Sarasota, FL) with their tapered ends facing down, and a small drop of electrode solution is added to the top end of each capillary tube. The drop only needs to be large enough to fill the tapered portion of the capillary tube, as the assembled microelectrode uses only this portion. Note that a small amount of water should be added to the jar in order to maintain a high humidity to prevent the electrodes from drying, but the electrodes themselves should remain completely above the water level. At this stage, the capillary tubes may be stored in a refrigerator for days to weeks before use. In fact, because the electrode solution is viscous, it often takes several days for the air bubbles trapped in the tapered end of the capillary tube to rise, and so it is highly recommended to prepare the capillary tubes at least several days in advance of their use.

Final assembly of microelectrodes. The final stage of the electrode assembly should be done the day before the experiment but can be done on the day of the experiment if needed. In this stage, the tapered ends of the capillary tubes are cut and mounted in an electrode holder, and then the free end of a wire connector is inserted into them.

To cut the ends of the capillary tubes, a small petri dish is filled with electrode solution and a small light (e.g., a desktop reading lamp) is placed to the side of the dish (Fig. 1a). The tapered end of the capillary tube is submerged, and the presence of air bubbles is checked for (they appear as regions of high reflectance of the light). If the end is free of air bubbles, it is cut while submerged using a small pair of scissors. Once the desired number of tips has been cut from the capillary tubes, they are mounted on an electrode holder.

To mount the electrodes, the electrode holder assembly should be positioned such that the holes in its rubber block are visible under a stereomicroscope. A glass tip is then gently grabbed toward its larger end with a fine forceps (#5) and removed from the petri dish. Once out of the solution in the dish, the forceps are allowed to open, and the glass tip will remain attached to, and should extend from, one of the tips of the forceps. The glass tip is inserted into one of the holes of the electrode holder. The tip will descend vertically into the hole and can then be detached from the forceps by moving the forceps horizontally. The electrode should end up standing vertically. If not, it should be removed and reinserted. This process is repeated until all of the tips have been inserted into the electrode holder.

Next, the free end of a connecting wire is inserted into each of the microelectrode tips (Fig. 1b). To do this, the pin of a connecting wire is first inserted into one hole of the D-Sub connector of the electrode holder. Then, a fine forceps is used to grab the wire gently near its free end and insert that end into a glass tip that has been mounted on the electrode holder. The wire should be inserted far enough that all of the uninsulated end is inside the tip. The wire is then bent where it emerges from the pin so that it makes close to a right angle with the pin and the length of the wire forms a smooth arc. The process is repeated until wires have been inserted into all of the electrode tips.

The order in which the wires are inserted is critical. The wires must be inserted into the electrodes in the reverse order in which the electrodes will be used. Similarly, the placement of the pin into the D-Sub connector must be done systematically (e.g., from top to bottom of each column of holes) such that the wire attached to the first electrode to be used lies on top of all of the other wires, the second wire lies on top of the remaining wires, etc. Figure 1b shows the ordered loading of electrodes in progress.

Once the holder is fully loaded, it is stored in its plastic cup with the rubber block halfway submerged in electrode solution. The cup should be placed in a refrigerator until used.

3.4 Assembly of Electrode Platforms

The electrode platform serves to protect the brain surface, holds the electrodes in place for the duration of the experiment, and acts as a guide for the placement and spacing of the electrodes. The platforms need to be prepared in advance of the experiment.

The first step in assembling the platform is to cement a tungsten rod frame to an electron microscope grid (GN100, Ted Pella, Tustin, CA) with 5-min epoxy. A fine-tipped tool is needed to apply the epoxy in building the frame of the platform (the capillary tubes from which the electrode tips were cut are good for this purpose).

Two types of tungsten rod pieces, which are cut from rod stock (0.008 in, 716,500, A-M Systems, Everett, WA), are needed for the frame (Fig. 2a). The first is a U-shaped piece. This is formed by bending a tungsten rod around a cylindrical object of the appropriate diameter and then cutting the resulting U-shaped piece from the main rod with wire cutters, such that the legs of the piece are ~10 mm in length, run parallel to each other, and are separated by a distance just shy of the diameter of the grid. Second, two short straight pieces of tungsten rod that are ~2/3 the diameter of the grid should be cut from the main rod.

To assemble the tungsten frame onto the grid, the grid is first placed on a flat piece of acrylic. A small amount of epoxy is mixed. When the epoxy becomes somewhat viscous, a drop is applied to the midpoint of each leg of the U-shaped rod, and (with the aid of a microscope) the rod is laid on top of the grid such that its legs lie on top of the perimeter of the grid and run parallel to the rows of the grid, and the epoxy drops connect the grid to each leg. Once this epoxy is dry, further drops are applied to the perimeter of the grid on the two sides that run between the legs of the U-shaped piece. A short tungsten rod piece is placed on the epoxy drop on each of these sides. Following this, a circular wall of epoxy is built up along the entire perimeter of the grid that also embeds the

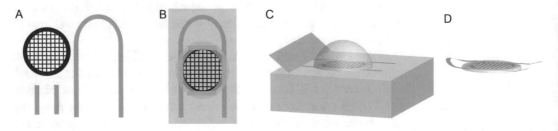

Fig. 2 Assembly of the recording platform. (**a**) Parts needed for assembling the platform: electron microscope grid and pieces of tungsten rod. (**b**) The grid should be placed on a flat acrylic block (*blue rectangle*), and the rods should be epoxied to the edges of the grid as indicated by *pink-colored ring*. (**c**) A silicon rubber drop is applied to grid and rods as shown. When the rubber is dry, a razor blade, either attached to a vibratome or manually, is used to shave the rubber drop using the tungsten rods as guides. (**d**) Completed platform after shaving of rubber drop and bending of the ends of the rods

tungsten rods around the grid (the loop and the free ends of the U-shaped piece remain free of epoxy) (Fig. 2b). The wall should be built to a height just higher than that of the tungsten rods.

The major technical difficulty in this first stage is having the epoxy run under the grid and fill its holes. This must be avoided, as it is hard to detect in the completed electrode platform and will prevent implantation of the electrodes. To avoid this situation, the epoxy should only be applied when it is relatively viscous.

After the epoxy has completely hardened, the platform is removed from the acrylic sheet using a razor blade, placed (grid side down) on a small rectangular block of acrylic, and a drop of silicon rubber (3140RTV, Dow Corning, Midland, MI) is applied such that the entire grid and the surrounding epoxy ring are covered (Fig. 2c). Air bubbles may form in the first 1–2 h after application, so the platform should be monitored, and any bubbles that form should be removed with a fine-tipped tool. The silicon is allowed to set (1–2 days).

Next, the rubber is shaved using a vibratome or manually with a razor blade (Fig. 2c). The acrylic block is placed such that the blade will cut along the top of the U-shaped piece. The rubber should be cut to a uniform thickness of ~250 μm. Any silicon rubber that remains outside of the epoxy ring can be trimmed with a scalpel or razor blade. Finally, the platform is detached from the acrylic block using a razor blade and flipped so that the rubber is face down, and both the legs and the curved portion of the U-shaped piece are bent upward about 30–45° without bending the grid (Fig. 2d). This can be done by using two razor blades: one to press down on the rod at the bending points and one to pull up from under the rod at a more distal point.

3.5 General Surgical Procedures and Placement of the Recording Platform

The animal is anesthetized, and if desired, a tracheal tube for ventilation and a femoral vein catheter for infusion of anesthesia and/or drugs are inserted. The animal is then placed on a heating pad controlled by a rectal thermometer, and its head is fixed in a stereotaxic frame (1430, Kopf, Tujunga, CA).

To expose the cerebellum, the skin overlying the dorsal surface of the skull is incised at the midline, and small vessel clamps are used to retract it. The tissue and muscles connected to the interparietal and occipital plates of the skull are removed using a scalpel or scraping tool. The tendon attaching the muscles at the junction of the interparietal and occipital plates is cut first, and then the muscle is peeled off of the bone to the foramen magnum by using gentle traction and a scraping tool. Laterally, toward the base of the occipital plate, several blood vessels exit the skull that should be cauterized to prevent bleeding. The muscles are then separated from the membrane spanning the base of the skull and first vertebra, by using traction and a scalpel. Using a 26-gauge syringe needle, a small hole is made in the dura in the space caudal to the base of the skull so that clear cerebrospinal fluid flows from the hole.

Next, a self-tapping metal screw (size 0) is inserted through the parietal or frontal plates to act both as an anchor for attaching the recording platform and as a ground for the recordings. For grounding purposes, a wire is soldered onto the screw prior to its insertion.

A craniotomy is then performed. The extent of the craniotomy is determined by the desired recording location, but our standard procedure is to remove all of the bone from approximately the middle of the interparietal plate to the foramen magnum. This will allow access to most of the lobules of the posterior lobe. The craniotomy is performed in two steps. First, a channel is drilled that outlines the desired region and creates an island of bone. Once the bony island is freely moveable, it can be removed by grabbing it with a Rongeurs and slowly but steadily peeling it away from the brain and dural surface. Sometimes, the caudal edge of the bony island is firmly attached to the dura at the foramen magnum. If so, a syringe needle or a microsurgical scissor can be used to cut this attachment. Bleeding from the dura often occurs but can be stopped by placing a small piece of Gelfoam at each bleeding point. Ideally, the Gelfoam pieces should be presoaked in Ringers or in 0.9% saline solution. While waiting for any bleeding points to clot, the dura should be kept moist by tenting it with saline or Ringers soaked pieces of Gelfoam or gauze.

While the dura is still intact, the access for the recording platform should be visually checked. In making this assessment, it is important to take into account that the brain surface is deep to the dura. If there is not enough clearance for the recording platform, greater access can be achieved in two ways. First, the silicon rubber and epoxy along the edges of the platform can be further trimmed and its tungsten rod legs can be angled further. Second, additional bone can be carefully removed using a Rongeurs.

Next, the dura is removed using a fine forceps and microscissors. Starting at the previously made hole in the dura, the forceps are used to raise the dura off of the brain surface and hold it, while microscissors are used to make two incisions that each run anteriorly from the hole for a distance sufficient to allow the dura to be peeled off of the recording location. Note that often there are one or two points where blood vessels run from the brain surface to the dura and care must be taken not to rupture or cut these vessels.

Once the dura is removed, the recording platform should be immediately but gently placed onto the recording area so that its silicon rubber surface is apposed to the brain and the grid surface faces outward. Ideally, the recording area should not be exposed for more than a few seconds following dural removal. The recording platform should be placed so that it is oriented with its rows running parallel to the longitudinal axis of the folium on which it has been placed, thus approximately aligning the columns of the array with the zebrin stripes and bands of synchronous complex spike activity. The underside of the platform should be flush with the brain surface. If platform's position is not correct, it can be raised off and repositioned (though one should try to avoid doing this). Although not always used, a pipette tip attached to the 3-axis micromanipulator can be used to press gently on the top of the platform to make the underside flush with the brain surface.

Once the platform is correctly positioned, small pieces of Gelfoam soaked in Ringers are used to cover any remaining exposed areas of the cerebellar cortex. Finally, dental cement (repair liquid, Durabase, Reliance Dental Manufacturing Co., Worth, IL) is used to fix the platform to the skull by connecting the exposed portions of the U-shaped tungsten rod to the skull and ground screw and to build an inclined wall surrounding the recording platform.

3.6 Electrode Implantation and Recording Procedures

To implant the electrodes, the head is positioned so that the recording platform is horizontal. Next, the female connector of the D-Sub 37-pin cable, which connects to the recording system, is positioned ~2 cm caudal to the caudal edge of the recording platform, the needle attached to the micromanipulator is positioned above the platform, and a filled electrode holder is positioned just rostral to the platform with the surface of the rubber block horizontal (parallel to the platform surface). The heating coil of the manipulator is then primed with sticky wax.

A surgical scope (MC-M900, Seiler, Saint Louis, MO) is used during the rest of implantation procedure as needed. To begin implanting electrodes, a socket pin is transferred from the electrode holder to the D-Sub connector. The bend in the wire may need to be adjusted so that the wire makes a smooth, roughly horizontal arc from the glass tip (still in the holder) to the pin (in the connector). To achieve this, it may also be necessary to bend the wire at its exit from the glass tip. Next, the needle is positioned just behind the glass tip whose wire was transferred, and the heating coil is activated to melt just enough wax to form a bridge to the glass tip and to seal the open top of the glass tip. The wax is allowed to harden (~30 s), and then the manipulator is used to raise the electrode out of the holder and move it to the platform surface above one of the squares in the grid.

Insertion of the electrode through the grid should be done while observing the electrode signal on an oscilloscope and audio speaker to detect the moment the tip reaches the cerebellar surface. The micrometer is then set to zero, and the electrode is slowly lowered through the molecular layer of the cerebellum until single-unit activity is well isolated (typically ~100 μm for complex spikes alone or ~250 μm for simple and complex spike activity). The heating coil is then activated to melt the wax bridge and thus release the electrode. The electrode should stand either vertically or slant caudally. This orientation assures that there will be access to implant additional electrodes. If the electrode slants rostrally, the wire may be additionally bent or the D-Sub 37 connector can be slightly repositioned until the proper orientation is achieved.

The procedures described in the previous two paragraphs are repeated until the electrode array is fully implanted. Two things should be noted. First, the order of implantation should be systematic and is dependent on the direction of curvature of the wire connecting the electrode to the D-Sub 37 connector. The standard way is for the wire to curve from the array out toward the investigator and then back toward the connector. In this case, the holes of the grid are most easily implanted by starting with the caudalmost row of holes and proceeding from nearest to farthest hole in each row.

Once the array is completed, the signals on each electrode are verified and characterized. Activity can then be recorded using a multichannel amplifier system (Multi channel Systems MCS GMbH, Reutlingen, Germany). A SC2 × 32 connector box is used to interface between the D-Sub cable and the main amplifier system. Typically, both the continuous analog signals and the spike cutouts detected by a voltage threshold are recorded for off-line spike sorting and analysis. For details on the software and multichannel system hardware, see manuals at http://www.multi-channelsystems.com/.

4   Notes

1. Implantation-related issues: loss of activity upon release of electrode. Isolation of complex spike activity is relatively simple in an excellent preparation; however, maintaining the isolated activity varies with the number of implanted electrodes. Specifically, stability increases with each additional row of electrodes that is implanted. Thus, it is not uncommon to lose the activity upon release of the first several electrodes to be implanted. To counter this, one can usually reattach the electrode to the manipulator and try to reposition the electrode. If this is not possible, a second electrode can often be implanted to the same location. However, given that stability increases with each implanted electrode, it is often best just to continue implanting additional locations in the array.

2. Implantation-related issues: electrode implantation access. Access for implantation of electrodes after the first row can become an issue if the electrodes are not correctly oriented. If the spacing of the D-Sub connector and grid is optimal, the electrodes will lean slightly caudal (assuming the first row to be implanted is the caudalmost row of the array) or at worst

remain upright. In these cases, access for implanting subsequent rows generally will not become a problem. However, if some electrodes are angled rostrally or with very large arrays, access may become a problem. Several steps can be taken to mitigate the problem. In the case of individual misoriented electrodes that remain unattached to the other electrodes, the solution is to reattach the electrode and remove it. If removal is not possible, the D-Sub connector may be gently moved backward to shift the electrode angles as a whole. Other maneuvers include rotating the head angle in the stereotaxic or changing the angle of the manipulator. Of course, these manipulations entail the risk of losing the activity of previously implanted electrodes and must therefore be weighed against the benefit of increasing the array size.

3. Implantation-related issues: sticky wax levels. The amount of wax used to connect the electrode to the syringe needle of the micromanipulator must be enough to connect the electrode to the manipulator and to form a seal on the top of the glass tip once the electrode is released. This seal prevents the electrode solution from evaporating and helps prevent electrical cross talk between electrodes. However, if too much wax is used, access for implantation of subsequent electrodes will be progressively hindered. Additionally, large wax droplets can slide down the electrode and seal large areas of the grid and/or burn the brain surface.

References

1. Sasaki K, Bower JM, Llinás R (1989) Multiple Purkinje cell recording in rodent cerebellar cortex. Eur J Neurosci 1:572–586
2. Llinás R, Sasaki K (1989) The functional organization of the olivo-cerebellar system as examined by multiple Purkinje cell recordings. Eur J Neurosci 1:587–602
3. Fukuda M, Yamamoto T, Llinás R (2001) The isochronic band hypothesis and climbing fibre regulation of motricity: an experimental study. Eur J Neurosci 13:315–326
4. Lang EJ, Sugihara I, Llinás R (1996) GABAergic modulation of complex spike activity by the cerebellar nucleoolivary pathway in rat. J Neurophysiol 76:255–275
5. De Zeeuw CI, Lang EJ, Sugihara I, Ruigrok TJH, Eisenman LM, Mugnaini E et al (1996) Morphological correlates of bilateral synchrony in the rat cerebellar cortex. J Neurosci 16:3412–3426
6. Yamamoto T, Fukuda M, Llinás R (2001) Bilaterally synchronous complex spike Purkinje cell activity in the mammalian cerebellum. Eur J Neurosci 13:327–339
7. Welsh JP, Lang EJ, Sugihara I, Llinás R (1995) Dynamic organization of motor control within the olivocerebellar system. Nature 374:453–457
8. Lang EJ, Sugihara I, Welsh JP, Llinás R (1999) Patterns of spontaneous Purkinje cell complex spike activity in the awake rat. J Neurosci 19:2728–2739
9. Marshall SP, Lang EJ (2009) Local changes in the excitability of the cerebellar cortex produce spatially restricted changes in complex spike synchrony. J Neurosci 29:14352–14362
10. Blenkinsop TA, Lang EJ (2011) Synaptic action of the olivocerebellar system on cerebellar nuclear spike activity. J Neurosci 31:14708–14720
11. Lang EJ, Blenkinsop TA (2011) Control of cerebellar nuclear cells: a direct role for complex spikes? Cerebellum 10:694–701

Chapter 5

Large-Scale Tetrode Recording in the Rodent Hippocampus

Xiang Mou and Daoyun Ji

Abstract

Tetrode recording is a large-scale single-unit recording technique that has greatly advanced our understanding of neural circuits involved in spatial navigation and spatial memory processing, particularly those in the brain area hippocampus. The technique can simultaneously record spiking activities of hundreds of hippocampal neurons in a freely moving rat or mouse during spatial navigation tasks, making it possible to analyze large-scale neural activity patterns rather than activities of individual neurons. In this chapter, we first introduce design principles of large-scale tetrode recording and then describe methods and procedures of using this technique to record hippocampal neurons in freely moving rodents. Using replay of spatial memory patterns as an example, we also demonstrate how neural activity patterns of multiple neurons are analyzed. Finally, we discuss the strength and weakness of tetrode recording and address a few key technical issues.

Key words Tetrode, Hyperdrive, Hippocampus, Learning and memory, Place cells, Replay

1 Introduction

The use of extracellular recording to study neuronal activity has been employed over a century. By recording spiking activity of a single neuron (single unit) using single electrodes in vivo, great success has been made in relating single neuron' activities to sensory stimuli in a variety of settings. However, it is believed that neurons in the brain utilize a population code to represent information—a specific sensory stimulus or a particular thought in mind at any given moment is encoded by a particular spiking activity pattern of an ensemble of neurons [1]. To understand information processing in the brain, it is necessary to simultaneously record spiking activities of a large number of single neurons. Ideally, such recording needs to be performed in freely moving animals so that recorded activity patterns reflect actual neural computations underlying natural behavior. Tetrode recording is an extracellular recording technique that aims to simultaneously record a large number of single units in freely behaving animals, in particular, rats and mice. First employed over three decades ago in

Roy V. Sillitoe (ed.), *Extracellular Recording Approaches*, Neuromethods, vol. 134,
https://doi.org/10.1007/978-1-4939-7549-5_5, © Springer Science+Business Media, LLC 2018

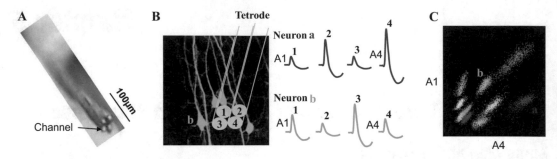

Fig. 1 Single-unit separation by tetrodes. (**a**) Tip of a tetrode with four channels. *Arrow:* one of the channels. (**b**) Schematics to show single-unit separation by signal triangulation. *Left:* a tetrode surrounded by a group of neurons. *Right:* an example spike simultaneously recorded by four channels of the tetrode for the neuron **a**, which is close to the channel four, and **b**, which is close to the channel three. Polarity of spike waveforms in the plots is reversed. A1, A4: spike amplitudes at channel one and four, respectively. (**c**) Spikes form separate clusters in the 2D space of amplitudes A1 vs. A4. Each cluster is a single unit presumably containing spikes from a single neuron like **a** or **b**. Clusters not separated in the space of A1_A4 can be separated in other 2D amplitude spaces

the study of hippocampal place cells in spatial navigation tasks [2, 3], it has been instrumental in elucidating ensemble codes of space and spatial memories.

A tetrode is an electrode made of four thin wires tightly twisted together, each of which is an independent recording channel (Fig. 1a). This special arrangement of four channels makes it possible to record and separate multiple single units from a single tetrode, due to the fact that spike amplitude is inversely related to the distance between the electrode and the cell that emits the spike. Different channels of a tetrode record the same spike with different amplitudes (Fig. 1b). When a large number of spikes are recorded, spikes emitted from the same neuron form a cluster in the parameter space made of spike amplitudes of the four channels. When a tetrode is carefully positioned in the extracellular space close to multiple cell bodies, such clusters become separable from each other and from other smaller background spikes (Fig. 1c). These clusters are single units corresponding to spikes from individual neurons, which can be verified by simultaneous intracellular and tetrode recordings [4].

This strategy of single-unit separation by electrical signal triangulation was first introduced by McNaughton et al. in 1983 [5]. They employed the two-channel "stereotrode", which considerably improves the discrimination of single units. However, it is the extension from two-channel stereotrodes to four-channel "tetrodes" that fully realizes the separation potential of this strategy and greatly increases the yield of number of simultaneously recorded cells by a single electrode, especially in the hippocampal pyramidal cell layer where pyramidal cell bodies are densely packed [2, 3, 6]. In principle, the yield depends on number of channels and the diameter of each channel in relation to the density of cells in a given brain area. In the hippocampus, although not rigorously tested, the tetrode with four channels is a proven design. The diameter of a channel should be close to the size of a cell, for example, 10–20 μm of hippocampal pyramidal cells. In our experience, a tetrode made of 13 μm diameter wires can simultaneously record up to ~15 single units from the CA1 pyramidal cell layer in the hippocampus.

Besides unit separation, another major development to increase the yield of simultaneously recorded cells is to increase the total number of channels used in a single experiment. Nowadays recording with 128 channels or 256 channels, which are divided into 24 or 48 tetrodes, is not uncommon. In addition, to reduce noise, most of the recording systems utilize a noise cancellation strategy that uses a few reference channels (see Discussion below). Although a tetrode can record a good number of single units, in practice, not every tetrode will be optimally positioned. In our experiments, simultaneous recording of ~100 or ~150 single units using 12 tetrodes or 24 tetrodes is not hard to achieve. Using 40 tetrodes, simultaneous recording of >250 neurons has been reported from a single rat [7].

Tetrode recording, as a chronic recording technique, was designed to record from freely moving rodents performing spatial tasks, which requires behavioral training lasting days to weeks. This requirement raises an important consideration: Tetrodes need to be placed close to the target neurons at the right time when the animal's behavior is ready. The traditional way of placing electrodes to a brain area in a surgery is not effective for this purpose. This is because the brain is dynamic during and after the surgery, due to inflammatory responses to the surgery and the recovery process afterward. Even if electrodes are perfectly placed at the CA1 pyramidal layer at the time of surgery, they can move several hundreds of micrometers or more a few days later. To overcome this issue, a mechanical "drive" that can move tetrodes to the target area after surgery is necessary. For recordings with multiple tetrodes, each tetrode needs to be independently movable because they may encounter target neurons at different depths. Therefore, an essential part of large-scale tetrode recording is the design of a "hyperdrive" that contains a large number of drives, each for a tetrode. When considering how many tetrodes to use, a factor is then the weight and size of the hyperdrive that an animal can handle without causing interference with behavior. A typical (~300 g) adult rat can wear a hyperdrive of up to ~30 g, which can hold 24–48 tetrode drives. A mouse can hold ~5 g with 6–12 drives.

After introducing basic principles of large-scale tetrode recording, in the following sections we demonstrate how tetrode recording is implemented, using one of our previous experiments as an example. In this example, we recorded multiple single units from the hippocampal CA1 area, while rats shuttled back and forth on a simple linear track for food rewards, located at both ends of the track. Many CA1 neurons, known as place cells, reliably fire spikes when animals are at specific locations on the track [8]. As an animal actively run through the track from one end to the other, place cells fire one after another with a distinguished pattern, which is believed to encode the animal's spatial memory of the track [3, 9]. Our question is whether such patterns were also reactivated (replayed) when animals stopped and rested at each end of the track. If so, the replay may be how the spatial memory is retrieved or consolidated [10, 11]. The replay phenomenon has been established by many studies [12–19]. Here we use it to demonstrate the procedure of tetrode recording. We first provide a step-by-step description of materials and methods in the experiment. Then, as an example of analyzing neuronal patterns of multiple neurons, we describe how replay is detected. Finally, we address several technical issues in the application of tetrode recording.

2   Materials and Methods

In this section, we describe how to manufacture various experimental devices, followed by a step-by-step experimental schedule. Our focus is on the materials and methods uniquely important for tetrode recording, including those related to hyperdrive assembly, tetrode fabrication, surgical implantation of hyperdrive, and tetrode placement.

2.1 Hyperdrive Assembly, Tetrode Fabrication, Tetrode Loading

We make a hyperdrive in three steps. First, we design the hyperdrive body using a computer-aided design software, such as Solidworks, and obtain the product by three-dimensional (3D) printing (Fig. 2a). For a 24-tetrode hyperdrive, the design sketches out a solid piece with 26 holes to hold 24 tetrodes plus two reference electrodes. The design can be printed out by a 3D printer with Accura-55 as printing material, due to its appropriate hardness and elasticity. The second step is to make 26 drives one by one. Each drive is made of a small (1.2 mm diameter) screw with 250 μm thread intervals (Fig. 2b). The top 2 mm of the screw head is milled into a shape of half moon, for easy turning after implantation. The length of the screw in our experiment is ~15 mm for hippocampal recording, but should be determined by the depth of targeted area in the brain. Below the half-mooned screw head is the neck of the screw of ~3 mm, where threads are carefully removed and smoothed. The screw is then held together with a

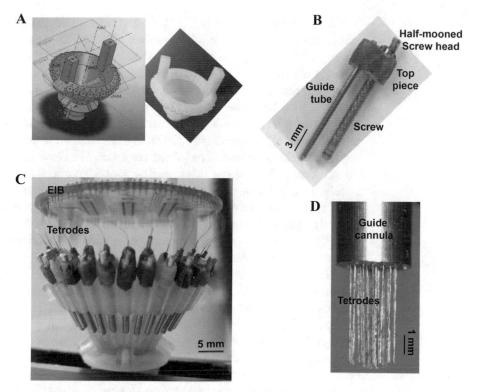

Fig. 2 Tetrode hyperdrive. (**a**) Solidworks sketch of a drive body (*left*) and its 3D–printed product (*right*). (**b**) A tetrode drive with a screw and a guide tube bound with a top piece. (**c**) An assembled 24-tetrode hyperdrive with tetrodes loaded to guide tubes and connected to an EIB. (**d**) Loaded tetrodes sticking out of a guide cannula at the bottom of the hyperdrive

small (23 ga) guide tube at the screw neck by a (top) piece of dental acrylic (Fig. 2b). The third step is to assemble all the drives and an electrical interface board (EIB, Neuralynx, Bozeman, MT) together to the hyperdrive body (Fig. 2c). Each drive is held by and can slide up and down along a 30-ga metal tubing. The 30-ga tubings are bundled within a guide cannula at the bottom and then fixed to the drive body by dental acrylic. The EIB is connected by two screws into the pre-designed posts in the drive body. For our 24-tetrode hyperdrive, an EIB with 96 channels plus 3 reference channels and 3 ground channels, is used. Now it becomes clear that the design of hyperdrive body needs to consider the number of tetrode drives, type of EIB, and estimated final size and weight of a fully assembled hyperdrive.

We next make tetrodes and load them into the hyperdrive. Each tetrode is made of a strand of ~40-cm-long insulated nichrome wire with a 13 μm diameter (Sandvik, Fair Lawn, NJ). The wire is folded twice with one cut at the open end to make four strands of ~10-cm equal length that are bundled together. The four-wire bundle is grasped with a clip at open ends and suspended on a metal rod. Once the rod holds the bundle in place and tension is evenly applied on all four wires, no wire should diverge from the others, and the clip should hang down freely. Then the clip is placed directly above a wire twister (Neuralynx, Bozeman, MT). An appropriate number of twisting rotations (~60 forward rotations followed by ~20 backward rotations) are applied to the bundle to ensure enough stiffness without breaking. A high-powered hot air blower (Steinel, Bloomington, MN) is then used to melt the coating so that the four wires are joined together. We blow the tetrode up and down from three sides, each about 10 s and ~2 cm away. Special care is taken to heat the part that will be loaded into the drive and implanted to the brain, but do not heat too much to short the wires. The heating should spare the top part of the tetrode such that the wires on the top can be separated and individually connected later to the EIB.

Tetrodes are then loaded into the hyperdrive one by one. A polyimide tubing (35 ga, MinVasive Components, Trenton, GA) is first inserted into the guide tube of each tetrode drive and glued at its top. A tetrode is then loaded and glued to the polyimide. The open ends of the four wires on the top are each connected to the pre-designated holes on the EIB (Fig. 2c). Each wire is secured in a hole by pushing a gold pin down from the top. Insulation at the open end is naturally cracked during this process, and each wire is thus electrically connected to the hole. For each reference electrode, we just connect one channel of a tetrode to one of the reference channels on the EIB and leave other three wires open. After all tetrodes are loaded, they are lowered out of the guide cannula by turning drive screws one by one and then undergo the steps of final cut and gold plating. The final cut is to trim each tetrode to the appropriate length for its targeted brain area. For the rat hippocampal CA1, we cut tetrodes to a length of ~4 mm from the bottom of the guide cannula to tetrode tips (Fig. 2d). The cut needs to be made under a scope and using a pair of sharp, fine scissors (sharp tip, 22 mm cutting edge; Fine Science Tools, Foster City, CA). The four wires of a tetrode need to be all exposed at their tips (Fig. 1a). We then plate each channel of a tetrode in a 1–5% gold chloride solution and measure its impedance at 1 kHz by an impedance meter (Model IMP-2, Bak Electronics, Umatilla, FL), which usually starts at 1–5 $m\Omega$ after appropriate final cut. We then pass DC current (0.1–0.5 mA) to its four channels one by one, each for 1–5 s, using a current isolator (Model A365, World Precision Instruments, Sarasota, FL). The tetrode is washed in distill water and the impedance is measured again. This step is repeated a few times, with decreasing current or plating time, to reach a desirable, stable impedance of 200–300 $k\Omega$. Finally, electrical connections between any two channels of a tetrode are checked to avoid shorts among channels. All tetrodes are retrieved back inside the guide cannula after plating.

Finally, we attach a protection cone, which is also 3D–printed, to hold the entire hyperdrive inside. The final assembled 24-tetrode hyperdrive is ~42 mm in diameter and ~47 mm in height and weighs ~17 g. Other hyperdrives of different sizes can be similarly designed, assembled, and loaded. For example, our hyperdrive for mouse recording contains eight tetrodes with a diameter of ~25 mm, a height of ~27 mm, and a weight of ~5 g [20].

2.2   Surgery

The loaded hyperdrive is mounted to an animal's skull in a surgery. The procedure is similar to other survival surgeries involving intracranial implants. Here we briefly describe the surgery, emphasizing on steps unique to tetrode recording. A rat is fixed in a stereotaxic under anesthesia and an incision is made to the scalp. The skull is exposed and cleaned. Using the Bregma and Lamda as markers, the recording site is identified on the skull surface and marked. For recording CA1 neurons, we use the coordinates of anteroposterior −3.8 mm and mediolateral 2.5 mm from the Bregma. We then secure anchoring screws to the skull. Due to the size of the hyperdrive, this step is critical for later stable, long-term chronic recordings. For our 24-tetrode hyperdrive, we put down ~10 anchoring screws (1.59 mm O.D., 4 mm length; Stoelting, Wood Dale, IL). The screws need to be further away from the recording site and should be close to the bone ridge as much as possible to cover a sufficiently large area of the skull. A hole is drilled through the skull for each screw without touching the brain. The screw is tightly turned into the hole.

The next step is to make an exposure at the recording site for the tetrodes. A hole is slowly drilled with all bone residues on top of the brain carefully removed. Since tetrodes are relatively soft and thus cannot penetrate the dura matter in rats, a durotomy is necessary. We use a fine (27 ga) needle to gently poke and slash through the dura and then drag it to the edge of the hole to make an exposure. To implant a hyperdrive, we lower the hyperdrive until its guide cannula is inserted into the hole but stops right above the brain. Silicone gel is applied around the guide cannula to completely cover up the hole. At this time, we also connect a grounding wire from the EIB to one of the anchoring screws (usually the one above the cerebellum). Then, the hyperdrive is secured to the anchoring screws and the skull by applying dental acrylic. Finally, the skin is sutured and a protective cap is added on top of the cone. The surgery in rats usually takes 3–6 h. The surgery for implanting a hyperdrive to a mouse is similar, but with two key differences. First, we still put in a similar number of anchoring screws, but the screws are much smaller (~0.8 mm diameter, ~1 mm length). Second, the mouse dura matter is penetrable by tetrodes and thus a durotomy is not necessary. A typical mouse surgery lasts for 2–3 h.

2.3 Tetrode Placement, Behavioral Testing, and Data Acquisition

After the animal recovers from surgery, tetrodes are moved down to the target brain area one by one. Tetrode placement needs to be slow in order to achieve long-term recording stability (see Discussion). Moreover, ideally all tetrodes need to be placed to their targets at the same time in order to boost number of simultaneously recorded cells. Our strategy for recording hippocampal CA1 neurons is a two-stage process. The first stage is to move all tetrodes and references to the white matter above the hippocampus within the first week, starting either on the day of surgery after the animal completely wakes up from anesthesia or 1 day after. In the first couple of days, we make sure that all tetrodes get into the cortex. In the following 3–5 days, tetrodes and references are moved down one by one, passing through different cortical layers. At the end of the first week, all tetrodes and references should reach and stabilize roughly within the white matter, without entering the hippocampal CA1 below.

In this stage, a key issue is how to tell which area (different cortical layers, white matter, CA1) a tetrode is inside the brain. The depth of a tetrode, which is calculated from number of turns made to a screw, is only a rough guide. The precise location of a tetrode at a given time needs to be judged from the electrical signals it records. This is a skill that can only develop through experience by repeated practice, but a data acquisition system that displays signals appropriately is important. Several commercial data acquisition systems are available, including Plexon (Dallas, TX) and Neuralynx (Bozeman, MT). The Digital Lynx system we use from Neuralynx directly digitizes unamplified, unfiltered raw signals from each channel and then processes the signals online for various ways of visualization. The signals recorded by a reference electrode are subtracted from those recorded by each tetrode channel. The resulted differential signals can be displayed as ongoing continuous traces in a broad or any specified frequency band. In addition, a threshold (50–70 μV) can be set to detect spikes on high-passed (e.g., >600 Hz) signals, and spike amplitudes are displayed to visualize spike clusters online. The electrical signals of any channel can also be monitored by an audio amplifier (e.g., a Grass speaker; Grass Instruments, West Warwick, RI). As a tetrode passes through different brain areas, different patterns of activities can be distinguished in the visual and audio monitors. The patterns are most distinguishable when the animal is in slow-wave sleep, during which cortical LFP (local field potential) shows delta waves (1–4 Hz) and slow (<1 Hz) waves with different shapes in different cortical layers (Fig. 3a) [21]. At the same time, high-frequency sharp-wave ripples (100–250 Hz) occur frequently in the hippocampal CA1 (Fig. 3b) [22]. The firing patterns associated with these LFP events are also distinguishable in the audio monitor, which is especially useful for identifying the white matter when

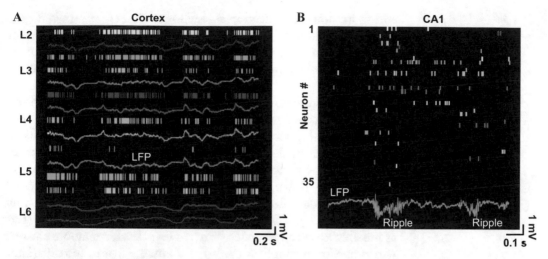

Fig. 3 Electrophysiological landmarks in the cortex and hippocampus. (**a**) Spiking and LFP activities recorded in different layers (L2–L6) of the rat primary visual cortex during slow-wave sleep. Ticks in a row show all spikes recorded by a tetrode and each trace is the LFP recorded by a tetrode. Tetrodes are ordered top-down from superficial to deep layers. (**b**) Spiking and LFP activities recorded in the hippocampal CA1 in slow-wave sleep. Ticks in a row represent spikes from a single neuron. Two sharp-wave ripple events are marked. The polarity of LFP traces in both panels is reversed

a tetrode is being turned and passing through cortical layers, as ongoing spiking sounds suddenly drop and enter a quiet territory.

Our daily procedure for tetrode placement at this stage starts with placing the animal on a small, flat plate, which is on top of a tall (>30 cm) flower pot. We then open the protective cap, plug in recording cables to the EIB, and start monitoring LFPs and spike clusters of each tetrode. The animal actively explores the plate for a while, especially during the first a few days, but eventually gets bored and falls sleep if left alone. After watching/listening the signals while the animal in sleep, we draft a plan of how much to move for each tetrode. When we decide to move down a tetrode for a certain distance, we gently but firmly hold the cone and turn its associated screw with the corresponding number of turns. For the turning, we use a custom-built screw driver, which fits tightly with the half-moon screw head of each drive. We can turn a screw by a minimum of a 1/8th of a turn. With the type of screws we use, this corresponds to a resolution of ~25 μm. If a tetrode is determined to move down too fast or too close to the target area, it can be pulled back up and its location is reevaluated. It is our experience that once a tetrode misses its recording site, pulling back and reentering the recording site yields much less number of single units. Therefore, a rule of thumb is never to pass through the recording area, and any pulling back needs to be done before reaching the target. After turning tetrodes, it is important to estimate and write down the location of each tetrode before returning the animal to its home cage at the end of the day. This practice is key to estimate how much tetrodes move overnight, which can be surprisingly substantial during the first week when the brain is dynamic.

The second stage of tetrode placement is to move down tetrodes to their final recording sites and hunt for single units (clusters). For recording rat hippocampal CA1 neurons, this step still needs to move tetrodes 0.2–1 mm from the white matter and takes 1–3 weeks. The principle is to move relatively fast at the beginning and then slow down, as slow as 25 μm/day. The CA1 pyramidal cell layer is identified by the largest/loudest ripples among all depths (Fig. 3b). The goal is to obtain as many separable clusters as possible across all tetrodes. When a few clearly separable clusters with a maximum amplitude of ~1 mV emerge from a tetrode (later offline sorting usually identifies more clusters than the number seen at the time of recording), it is at or close to its optimal recording position. The cluster patterns and maximum amplitudes are carefully monitored for 2 or more days without any tetrode movement to check for stability. If they are fairly similar across days, data acquisition may start.

2.4 Data Acquisition During Behavioral Tasks

In most applications of tetrode recording, data are acquired when an animal is performing a behavioral task. In our example of simple linear track running, the animal can learn the task quickly. In this case, we train the animal on the track for ~20 min every day with recording cables plugged, during the second stage of tetrode placement until a desirable level of task performance is achieved. This step is necessary to ensure the animal will be used to running/moving while wearing the hyperdrive and recording cables. A custom-made weight-counter-balancing system or a commutator is used to accelerate this step. Recording starts when a target performance is achieved and when tetrodes are at appropriate positions. If appropriate surgical and tetrode placement procedures are followed (see Discussion), our experience is that recording from hippocampal neurons can last for weeks to months.

Neural data can be acquired in two different modes. The first is the continuous mode that records voltage values at every time point of all channels in a broad band (0.1–9 kHz) with a high sampling frequency (e.g., 32 kHz). In our experiment, the recorded signals are already differential, i.e., after subtraction of a specified reference signal. The acquired data can be filtered at different frequency bands offline later to obtain LFP (0.1–2 kHz) or spike (> 600 Hz) data. The advantage of this recording mode is that every data point is saved and no data are lost. The concern is that a large amount of data are generated and storage space may be a problem if the recording lasts too long. Typically, an hour's recording of 128 channels can yield ~30 GB of data in our Neuralynx system, which is not unmanageable in this digital era. If data storage is a concern, for example, when longtime overnight recording is needed, data can be acquired in a second mode: Data are preprocessed online so that LFP and spike data are stored separately. In this case, for each tetrode we typically select only one channel of each tetrode for LFP data and store the data using a lower sampling rate of ~2 kHz, because LFP data across the four channels of a tetrode are essentially identical. Spikes are identified online by a preset threshold (50–70 µV), and we only store 1-ms-long waveforms of identified spikes at 32 kHz sampling rate.

Besides neural data, another key type of data in tetrode experiments is the animal's behavior. For spatial navigation tasks, we attach two color diodes (red, green) to the hyperdrive to monitor the animal's position and head direction at any given time. An overhead video camera (Sony SSC-DC374) is used to capture video frames at 33 Hz. The video data are recorded using the same timestamps as in neural data. The resolution of the camera and the size of the behavioral apparatus determine the resolution of position recording, which is ~0.3 cm per pixel in our experiments.

2.5 Histology

After data acquisition, the recording site of every tetrode and reference is identified by electrical lesion and standard histology. Briefly, the animal is euthanized. A 30 µA current is passed for 10 s on each tetrode/reference to generate a small lesion at its tip. The brain is then fixed using formaldehyde and sectioned at 50–100 µm. Brain sections are stained with 0.2% cresyl violet. Lesion sites are identified, and tetrode recording sites are determined by matching lesion sites with tetrode depths and their relative positions.

2.6 Summary of Experimental Procedure

We have described major steps in tetrode recording and addressed related issues. Here we summarize these steps for recording rat CA1 neurons during track running.

1. Hyperdrive construction (1–3 days)
2. Tetrode preparation, loading, final cut, and plating (1–3 days)
3. Surgery (3–6 h)
4. Tetrode placement: down to the white matter (5–7 days)
5. Pre-training and tetrode placement: to CA1 (7–21 days)
6. Data acquisition (7–60 days)
7. Euthanasia, lesion, and histology (1–3 days)

3   Data Analysis

The acquired data are sorted offline into three types of data: single-unit spike data, LFPs, and position data, which are analyzed for a particular question under consideration. Here we use the replay of place cell patterns on the linear track as an example to demonstrate basic steps of data analysis.

3.1   Single-Unit Sorting (Clustering)

The first step is to sort all spikes from a tetrode to individual single units. Unlike the traditional waveform matching method, tetrode data are better sorted using clustering. Four waveforms are recorded for every spike by a tetrode, each by a channel. The parameters that represent the most characteristic features of these waveforms are extracted. When parameters of all spikes recorded by a tetrode are plotted, individual clusters emerge, and spike sorting is a process of "cutting" these clusters (Fig. 1c). There are two main considerations here. One is what parameters to use for clustering. A natural choice is to apply principal component analysis and identify the components that contribute most to spike waveform variations. This is used by MClust, which is freely available (http://redishlab. neuroscience.umn.edu/MClust/MClust.html). However, one disadvantage of this approach is that physical meanings of the extracted parameters are lost. Alternatively, we can use the amplitudes across four channels as principal parameters for clustering, which, after all, are what tetrodes are designed for. Additional parameters, such as spike width, can be added. These parameters have clear physical meanings and sometimes help us reject false clusters generated by noise or overcutting, a mistake that produces "good-looking" clusters by eliminating too many spikes. For example, CA1 pyramidal cells usually burst and generate complex spikes with decreasing amplitudes. As a result, a cluster of CA1 pyramidal cell appears oval in amplitude spaces, whereas an interneuron cluster tends to be more round. Pyramidal cells in the CA3 area of hippocampus burst even more, which produce clusters of baseball-bat shape with a long tail. This type of information helps us reject clusters with unnatural shapes in the amplitude space. The second consideration is how to cluster. Clustering seems a task that computers do well. Although several algorithms are available including MClust and others [23, 24], to the best of our knowledge, none of current algorithms are satisfactory. Our own experience is that these algorithms tend to produce too many clusters (overcutting) and manual verification or resorting is necessary. Therefore, we take a semi-manual approach that uses the software xclust (M. Wilson, MIT; https://github.com/wilsonlab/mwsoft64/tree/master/src/xclust). The software displays spikes in the projected 2D planes of the high-dimensional parameter space and allows the experimenter to manually draw cluster boundaries. The software then automatically assign spikes to individual clusters. It is sometimes beneficial to rank the quality of each cluster, i.e., how far it is separated from the rest of the spikes, which can be approximately quantified by the so-called L-ratio [25]. Once spikes are clustered, we now save timestamps (peak times) of the sorted spikes in individual clusters for further analysis.

3.2 Mapping of Place Cell Patterns

The power of tetrode recording is its unique capability of linking multineuronal activity patterns to specific natural behavior. To illustrate an example of this, we next describe our analysis to address the question of whether CA1 place cell activity patterns associated with track running are replayed when the animal rests. Here we start with constructing the pattern during track running. We sorted a total of 53 single units, each presumably from a neuron, from the spike data recorded by a 12-tetrode hyperdrive on one particular recording day. Among them, 19 neurons are place cells that were active when the animal ran the trajectory from left to right on the track (not all neurons active on this trajectory) [26]. We examine where each place cell fired on the track, by plotting a firing rate curve: its firing rate along each location of the trajectory (Fig. 4a). We divide the 2.8-m long track, considered a 1D space, into 2-cm spatial bins. Each spike occurring during each running lap along the left-to-right trajectory is assigned to a spatial bin according to the animal's location at the time of the spike. The firing rate of a cell at a bin is the number of spikes divided by the total time the animal spending at the bin among all laps. We compute firing rate curves of all 19 place cells on this trajectory. The cells are then sorted according to the order of their peak firing locations on the track (Fig. 4b). It is clear that, when the animal ran from left to right, these cells fired spikes one by one in a sequence, which is considered a neural code of the animal's spatial memory of the track (on one direction) [27]. In this way, we have mapped out the activity pattern of a group of simultaneously recorded place cells as the animal ran through a trajectory on the track, which we call a template pattern.

3.3 Replay of Place Cell Patterns

We then analyze whether the template pattern during running was replayed during resting on the track. As shown in Fig. 4c, we find that when brief ripple events (50–400 ms) occurred during resting, the same group of 19 neurons burst together in a sequence with an order that is highly similar to the template pattern. To formally identify that this is indeed a replay event, we quantify the similarity between the firing pattern within a ripple event and the template pattern by Bayesian decoding. To do so, we divide the time within a ripple event into individual 20-ms bins. For each time bin, assuming that its spiking activity encodes a location on the track, we ask what would be the probability for each track location, which can be

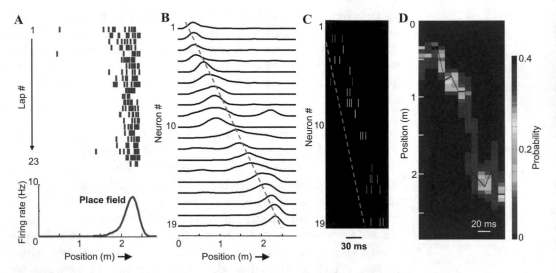

Fig. 4 Place cell patterns during running are replayed during resting on a linear track. (**a**) Spike raster of a CA1 place cell during every lap of running the track from *left* to *right* (*arrow*: *left-right* trajectory). *Bottom*: firing rate curve (rate vs. position) of the cell showing its place field. (**b**) Firing rate curves of all place cells active on the left-right trajectory. *Dashed line*: sequential place fields as the animal traveled through the trajectory. (**c**) Burst firing of the same group of cells within a ripple event when the animal was resting at one end of the track. Ticks in a row represent spikes from a cell. *Dashed line*: sequential firing of the cells. (**d**) Decoded trajectory for the firing pattern in (**c**). Color depicts the probability of the firing pattern at each time bin (*x*-axis) encoding a position along the track (*y*-axis). *Red line*: decoded trajectory. Rank-order correlation between positions in the decoded trajectory and time bins: $r = 0.92$; $p = 0.00032$, compared to shuffled correlation values

computed from the template pattern using the standard Bayesian decoding approach [28]. We consider the location with the maximum probability as the decoded position for the time bin. By linking the decoded positions in all time bins of a ripple event, the firing pattern within the event is transformed into a decoded trajectory (Fig. 4d), which can be considered a "mental" trajectory traveled by the animal within the brief ripple event. We then examine whether the decoded trajectory is similar to a physical trajectory that the animal actually traveled. On the linear track, "similar" means moving along the track as time goes by. Therefore, we define a similarity score as the rank-order correlation between decoded positions and time bins [16, 18]. To statistically determine whether a similarity score is significant, we randomly shuffle positions among time bins ("local" shuffle) 1000 times and recompute the similarity score for each shuffle. A ripple event is considered as a replay, if the percentage of shuffles (*P* value) producing the same or a greater similarity score than the actual one passes a threshold (e.g., 5%). The Bayesian approach takes advantage of the intuition that a firing pattern similar to part of the template implies that the animal might be mentally traveling through this part of the trajectory. Unlike earlier methods of identifying replays such as sequence matching [13, 19], this approach does not require precise quantification of place fields and is less prone to noise in the data [29].

However, replay events could arise from chance by randomly ordering cells. To examine the overall significance of replay, we analyze all ripple events in a track session. We identify all ripple events from filtered CA1 LFPs within the ripple band (100–250 Hz), as 50–400 ms time windows with peak values passing a threshold (e.g., 6 standard deviations of the filtered LFPs). Since not every place cell fires spikes within a ripple event, we designate those ripple events with at least five active place cells as candidate events. We determine how many of the candidates are replay events by the Bayesian method described above. We then perform a "global" shuffle procedure, where the identities of place cells in the template pattern are randomly permuted for 1000 times and number of replays is recounted for each globally shuffled template. The overall significance (P value) is then the percentage of shuffled templates that generate the same or more replays than the actual template.

In our data on the linear track, we identified 89 candidate events during resting and found that 42% of candidates are replays, which is highly significant ($P < 0.001$). The analysis thus shows that the template pattern associated with running the left-right trajectory was indeed replayed within brief periods of ripple events when the animal rested on the track. The template pattern associated with the other trajectory (right to left) was also replayed. We point out that the number of identified replay events ($N = 37$) does not necessarily reflect the actual number in the brain, because we only sampled 53 cells among tenths of thousands of CA1 cells. Obviously, the more cells we can record, the more sensitive the analysis is. Therefore, the power of patterned analysis depends on the large number of single units simultaneously recorded by a tetrode hyperdrive.

4   Discussion

Focusing on its ability to simultaneously record a large number of single units in freely moving rodents, we have described design principles and experimental procedures of large-scale tetrode recording, as well as an example of data analysis. However, tetrode recording is a fairly sophisticated technique and by no means is ideal for every application. Here we discuss several technical and application considerations important for a successful implementation.

**4.1 Strength
and Weakness**

There are other techniques for simultaneously recording neural activity patterns of a large number of neurons. Among them, silicone probe recording is similar to tetrode recording, but with one or more thin probes where a number (8–64) of channels are custom arranged vertically on the shank of each probe [30]. Optical imaging can also simultaneously monitor a large number of neurons, but record spike-associated intracellular calcium signals instead of actual spikes [31, 32]. Compared to these two techniques, tetrode recording offers unique strengths under certain experimental conditions, but is less powerful under others.

The strength of tetrode recording, compared to silicone probes, is that it is ideal for recording from a brain area with densely packed cells on a thin, tight layer, such as the CA1 and CA3 areas of the hippocampus. Each tetrode can be individually and precisely placed to the layer by a drive. This flexibility is lost in silicone probes, where all recording channels on a shank are bound together. Since recording sites are located at tetrode tips, all channels can record from neurons in the same layer. In contrast, channels are distributed vertically on a silicone probe, not all channels can be placed within a thin layer. However, silicone probes may be more efficient in brain areas with more loosely distributed cells such as the cortex. In the cortex, recording sites along a silicone probe can make contact with more neurons in different cortical layers. In addition, unlike ripples as a sign of CA1 single units, there are no clear neurophysiological markers for single units in the cortex. Therefore, recording in the cortex is more like fishing. It is harder to predict when single units appear on what channel. In this case, silicone probes presumably increases yields by making more random contacts with cortical neurons. Another difference between tetrode and silicone probe recording is that tetrodes are more economic and are easier to make and replace. Silicone probe recording is costly in chronic experiments, because there is no easy way to reuse silicone probes after being implanted to an animal.

Compared to optical imaging, tetrode recording offers superb temporal resolution and behavioral flexibility. Tetrodes can record individual spikes at sub-millisecond resolution. Current imaging techniques rely on calcium signals with a time scale of >100 ms, although faster voltage-sensitive imaging may substantially increase the temporal resolution [33]. Second, tetrode recording is especially powerful in experiments with freely moving animals. Optical recording using two-photon imaging, in combination with virtual reality tasks [34], has become a powerful technique. But two-photon imaging can only be applied to head-fixed animals, which severely limits the type of behavioral tasks to be studied. Alternatively, optical recording with head-mounted mini-scopes can be used in freely moving animals [31], but the issue of low temporal resolution remains, and spatial resolution may become an issue without using two-photon imaging. However, optical recording offers a major advantage: It can visualize individual neurons and their anatomical configuration. This ability allows more precise identification of recorded neurons and the brain area they reside, which sometimes are key to understand the function of neuronal assembles [35].

4.2 Noise Control

Like any other electrophysiological techniques, tetrode recording is prone to noise. Noise control is one of the most difficult, and annoying, steps in a tetrode experiment. Here we discuss several strategies to control noise. First, the relatively large diameter of wires we use and the plating reduce tetrode impedance, which produces less electrode-related noise, compared to traditional sharp electrodes. Second, as in any electrophysiology experiment, reducing or eliminating sources of noise is important. A strategy is, if possible, to record within a shielded Faraday cage and keep all noisy power sources outside the cage. In freely moving animals, one type of noise is caused by the animal moving through a nonuniform electromagnetic field in space, which produces slow, large swing in signal baselines. For this type of noise, a Faraday cage is particularly useful. However, noise can be controlled without a Faraday cage, if the following strategies are carefully considered.

Obviously, proper grounding is key to noise control. The animal needs to be grounded by a separate wire connecting a skull crew to the recording equipment. Second, any materials the animal can touch should be carefully arranged. They should be free of statics. The electrical contact between the animal and behavioral apparatuses should be evaluated. If an apparatus is made of metals, it should not be grounded. Otherwise, double grounding (one through apparatus and one through the animal skull) would generate a ground loop. In this case, the behavioral apparatus needs to maintain constant contact with the animal throughout the experiment. Any "on and off" contact with a metal object would create large noise. For this reason, isolated conductive objects should be avoided around animals. In particular, if liquid is present in an apparatus, like liquid milk often used as reward, it needs to be electrically connected to the metal apparatus. Alternatively, if direct liquid contact can be avoided, the apparatus can be grounded and then electrically isolated from the animal by using nonconductive, static-free plastic wraps.

However, one type of noise cannot be eliminated by shielding or grounding: the one associated with chewing and shaking. When animals consume solid food or just become bored, they grind their teeth. Animals sometimes also shake their heads. In these types of behavior, large-amplitude high-frequency noise occurs across all channels, presumably due to mechanical vibrations at various electrical contacts in or outside the hyperdrive. This type of noise can only be reduced by noise cancellation via referencing, which also greatly reduces other, conventional noises. This is why we use differential recording that subtracts signals recorded from a reference electrode from those of tetrode channels. Since noise appears across the reference and tetrode channels, the subtraction is effective at reducing all types of noise. Obviously, the closer the reference is to the tetrode, the more effective the subtraction is. Therefore, the reference should be local, i.e., near the tetrodes. However, the reference cannot be too close to subtract spike signals. In addition, referencing makes the signals at a tetrode channel contaminated by the signals at the reference. For this reason, references should be placed in a "quiet" area. Furthermore, if LFPs are important for an experiment, the reference needs to be placed in a different brain area such that local features of LFPs are preserved. For these considerations, we place reference electrodes in the nearby white matter for recording hippocampal or cortical neurons.

4.3 Recording Stability

The stability of recording is always important in data interpretation. It is essential that the same population of neurons are tracked throughout the course of an experiment. Recording stability can be monitored by observing changes in cluster patterns and spike amplitudes. One source of instability in chronic recording is the long-term biocompatibility of tetrodes, which can be improved by gold plating. Another source is that the implanted hyperdrive may become unstable, due to either loose anchoring screws or infections caused by a bad surgery. Therefore, a good surgery is critical for tetrode recording. However, in most cases, recording instability is caused by tetrode shifting inside the brain. Here we discuss a few issues related to tetrode shifting and strategies to address them.

The first issue is that the brain is dynamic after the surgery. It takes about 7–10 days for the brain to recover to its stable state. For this reason, it is risky to conduct recording within the first week or so after the surgery. Second, there is a dynamic response of brain tissue to tetrode movement. When a tetrode is moved down, the tissue is dragged and dip down together with the tetrode. It takes a long time for the tissue to relax back to its stable state. The exact time of this relaxation depends on the distance a tetrode travels, but can take hours or days judging from the stability of signals it records. Therefore, it is important to approach targeted neurons slowly and conduct actual recording after spike clusters appear stable for the desired duration (hours to days). For hippocampal CA1 neurons, since there is a clear electrophysiological landmark (ripples), approaching slowly is not that difficult, but requires patience and time. It is our experience that tetrode placement needs 2–4 weeks. Once tetrodes are appropriately placed at the CA1 pyramidal layer, CA1 neurons can be recorded for weeks to months, which makes it possible to study CA1 neurons in multiple behavioral tasks. For cortical recordings without a clear target, this long-term stability is harder to achieve. Yet it remains important to always move slowly, e.g., <250 μm/day. Otherwise, cells will disappear before an experiment is completed. In our practice, cortical clusters may stay for many hours to many days. Another issue is that moving one tetrode may disturb the tissue and affect the stability of other tetrodes. It is thus important to synchronize locations among all tetrodes in a recording area. They should all approach the target approximately at the same time. Since the target may occur at different depths for different tetrodes, for example, the CA1 pyramidal layer is curved along the mediolateral axis, tetrodes may need to move at different speed to maintain approximately the same distance from the target. Constant monitoring of LFP and spiking activities is essential for this purpose.

Ideally, data acquisition should take place only when recording is stable, and the animal's behavior is ready at the same time. Because stable recording requires careful monitoring and adjustments, it is not easy to predict when acceptable recording stability can be achieved. A related issue then is the timing of behavior training. For simple tasks, like running back and forth on a linear track for food rewards, we conduct behavioral training on the same days when tetrodes are slowly moved to the target. For more complex tasks that takes more time to learn, pre-training prior to surgery is necessary. The pre-training, often involving food or water deprivation if the task is reward-based, familiarizes animals with the type of reward and behavioral apparatus and enables them to achieve a desirable level of performance.

4.4   Conclusion

Tetrode recording is a powerful technique for studying neuronal ensembles in freely behaving animals. Its large recording capacity, superb temporal and spatial resolution, and excellent behavioral flexibility allow us to study what neural activity patterns underlie natural behavior, how they dynamically adapt in different learning environments, and how they are altered in models of neurological and psychiatric disorders. Although it takes time and patience to collect high-quality tetrode data, the data collected are rich in contents and permit the study of multiple questions in the same dataset. Future technological development, such as recordings with wireless devices and with more channels, will further enhance the capacity and applicability of tetrode recording.

References

1. Buzsaki G (2010) Neural syntax: cell assemblies, synapsembles, and readers. Neuron 68(3):362–385

2. O'Keefe J, Recce ML (1993) Phase relationship between hippocampal place units and the EEG theta rhythm. Hippocampus 3(3):317–330

3. Wilson MA, McNaughton BL (1993) Dynamics of the hippocampal ensemble code for space. Science 261(5124):1055–1058

4. Harris KD, Henze DA, Csicsvari J, Hirase H, Buzsaki G (2000) Accuracy of tetrode spike separation as determined by simultaneous intracellular and extracellular measurements. J Neurophysiol 84(1):401–414

5. McNaughton BL, O'Keefe J, Barnes CA (1983) The stereotrode: a new technique for simultaneous isolation of several single units in the central nervous system from multiple unit records. J Neurosci Methods 8(4):391–397

6. Gray CM, Maldonado PE, Wilson M, McNaughton B (1995) Tetrodes markedly improve the reliability and yield of multiple single-unit isolation from multi-unit recordings in cat striate cortex. J Neurosci Methods 63(1–2):43–54

7. Pfeiffer BE, Foster DJ (2013) Hippocampal place-cell sequences depict future paths to remembered goals. Nature 497(7447):74–79

8. O'Keefe J, Dostrovsky J (1971) The hippocampus as a spatial map. Preliminary evidence from unit activity in the freely-moving rat. Brain Res 34(1):171–175

9. O'Keefe J, Nadel L (1978) The hippocampus as a cognitive map. Oxford University Press, Oxford

10. Buzsaki G (1989) Two-stage model of memory trace formation: a role for "noisy" brain states. Neuroscience 31(3):551–570

11. Carr MF, Jadhav SP, Frank LM (2011) Hippocampal replay in the awake state: a potential substrate for memory consolidation and retrieval. Nat Neurosci 14(2):147–153

12. Wilson MA, McNaughton BL (1994) Reactivation of hippocampal ensemble memories during sleep. Science 265(5172):676–679

13. Lee AK, Wilson MA (2002) Memory of sequential experience in the hippocampus during slow wave sleep. Neuron 36(6):1183–1194

14. Foster DJ, Wilson MA (2006) Reverse replay of behavioural sequences in hippocampal place cells during the awake state. Nature 440(7084):680–683

15. Diba K, Buzsaki G (2007) Forward and reverse hippocampal place-cell sequences during ripples. Nat Neurosci 10(10):1241–1242

16. Karlsson MP, Frank LM (2009) Awake replay of remote experiences in the hippocampus. Nat Neurosci 12(7):913–918

17. Gupta AS, van der Meer MA, Touretzky DS, Redish AD (2010) Hippocampal replay is not a simple function of experience. Neuron 65(5):695–705

18. Davidson TJ, Kloosterman F, Wilson MA (2009) Hippocampal replay of extended experience. Neuron 63(4):497–507

19. Ji D, Wilson MA (2007) Coordinated memory replay in the visual cortex and hippocampus during sleep. Nat Neurosci 10(1):100–107

20. Cheng J, Ji D (2013) Rigid firing sequences undermine spatial memory codes in a neurodegenerative mouse model. elife 2:e00647

21. Steriade M, Amzica F (1998) Slow sleep oscillation, rhythmic K-complexes, and their paroxysmal developments. J Sleep Res 7(Suppl 1):30–35

22. Buzsaki G, Horvath Z, Urioste R, Hetke J, Wise K (1992) High-frequency network oscillation in the hippocampus. Science 256(5059):1025–1027

23. De Benedetti E, Lew SE, Zanutto BS (2010) A wavelet approach for on-line spike sorting in tetrode recordings. Conf Proc IEEE Eng Med Biol Soc 2010:6662–6665

24. Takahashi S, Anzai Y, Sakurai Y (2003) A new approach to spike sorting for multi-neuronal activities recorded with a tetrode—how ICA can be practical. Neurosci Res 46(3):265–272

25. Schmitzer-Torbert N, Jackson J, Henze D, Harris K, Redish AD (2005) Quantitative measures of cluster quality for use in extracellular recordings. Neuroscience 131(1):1–11

26. McNaughton BL, Barnes CA, O'Keefe J (1983) The contributions of position, direction, and velocity to single unit activity in the hippocampus of freely-moving rats. Exp Brain Res 52(1):41–49

27. Harris KD, Csicsvari J, Hirase H, Dragoi G, Buzsaki G (2003) Organization of cell assemblies in the hippocampus. Nature 424(6948):552–556

28. Zhang K, Ginzburg I, McNaughton BL, Sejnowski TJ (1998) Interpreting neuronal population activity by reconstruction: unified framework with application to hippocampal place cells. J Neurophysiol 79(2):1017–1044

29. Kloosterman F, Layton SP, Chen Z, Wilson MA (2014) Bayesian decoding using unsorted spikes in the rat hippocampus. J Neurophysiol 111(1):217–227

30. Buzsaki G (2004) Large-scale recording of neuronal ensembles. Nat Neurosci 7(5):446–451

31. Ziv Y et al (2013) Long-term dynamics of CA1 hippocampal place codes. Nat Neurosci 16(3):264–266

32. Ohki K, Reid RC (2014) In vivo two-photon calcium imaging in the visual system. Cold Spring Harb Protoc 2014(4):402–416

33. St-Pierre F et al (2014) High-fidelity optical reporting of neuronal electrical activity with an ultrafast fluorescent voltage sensor. Nat Neurosci 17(6):884–889

34. Dombeck DA, Harvey CD, Tian L, Looger LL, Tank DW (2010) Functional imaging of hippocampal place cells at cellular resolution during virtual navigation. Nat Neurosci 13(11):1433–1440

35. Malvache A, Reichinnek S, Villette V, Haimerl C, Cossart R (2016) Awake hippocampal reactivations project onto orthogonal neuronal assemblies. Science 353(6305):1280–1283

Chapter 6

A Guide to In Vivo Optogenetic Applications for Cerebellar Studies

Oscar H.J. Eelkman Rooda and Freek E. Hoebeek

Abstract

The mammalian cerebellum consists of a superficial cortex and centrally located output nuclei, which together with brainstem nuclei are organized in a modular fashion. Regardless of the function, these cerebellar modules consist of the same cell types, and their connectivity has been unraveled to some detail using electrical stimulation experiments. To unravel the highest level of detail, cell-specific stimulation experiments are warranted, which cannot be accomplished using electrical stimulation. To reach this unprecedented level of specificity, optogenetic applications are now being implemented in cerebellar studies. Due to the extensive knowledge about cell-specific markers in both the cerebellar cortex and the cerebellar nuclei, optogenetics can be applied cell specifically. Ideally the anatomical and electrophysiological characteristics of the cerebellum can be utilized for designing future optogenetic studies. In this chapter we review the opportunities and pitfalls for optogenetic studies in the cerebellum. We provide insights into the technical issues at hand and which solutions are currently available.

Key words In vivo, Extracellular, Cerebellar cortex, Cerebellar nuclei, Functional output, Learning, Locomotion

1 Introduction: Shining Light on Cerebellar Optogenetics

Since the beginning of electrical stimulation, the cerebellum has been probed for functional relevance. Early work from Moruzzi in the 1940–1950s indicated that motor responses and body posture could be precisely adapted by electrical stimulation of defined parts of the cerebellar surface [1]. It didn't take long before the therapeutic power of cerebellar stimulation was evaluated for brain disorders, like epilepsy. Although the antiepileptic effects of cerebellar stimulation were extensively probed in experimental animals and confirmed in initial clinical trials, subsequent double-blind studies revealed that the exact location of stimulation determined the efficacy (reviewed in [2]). The spatial precision of cerebellar stimulation has been of great importance for gaining insight in cerebellar functioning. Anatomical tracing studies in the late 1900s revealed that specific regions of

Roy V. Sillitoe (ed.), *Extracellular Recording Approaches*, Neuromethods, vol. 134, https://doi.org/10.1007/978-1-4939-7549-5_6, © Springer Science+Business Media, LLC 2018 the cerebellar cortex form functional modules together with downstream cerebellar nuclei and the inferior olive nuclei in the ventral brainstem [3, 4]. Especially in the cerebellar cortex, this anatomical differentiation has been shown to be very precise and results in the need of neuromodulation techniques with high spatial resolution. Although novel electrode designs brought new options for neuromodulation on a microscale, the recent development of optogenetics launched a new era of investigating brain functioning. Apart from the option of applying neurostimulation to a specific type of neuron rather than a volume of brain tissue, many light-sensitive ion channels have been described, which not only allow researchers to excite neurons but also to inhibit their action potential firing (extensively reviewed in [5] and other literature). Alike for other brain areas also for cerebellar research optogenetics provide numerous opportunities. This chapter combines reports of optogenetic applications in the cerebellar field with a technical guideline for questioning cerebellar interactions with up- and downstream targets using optogenetics.

2   Optogenetic Approaches in the Cerebellar Cortex

The foliated cerebellar cortex (CC) consists of three readily distinguishable layers of gray matter and a core of white matter, the latter of which is a combination of afferent mossy fibers and climbing fibers and efferent Purkinje cell axons. Adjacent to the white matter is the cytological diverse granular layer, which contains numerous granule cells, Golgi cells, Lugaro cells, and unipolar brush cells. Distal to the granular layer, a monolayer of Purkinje cells separates granule cells from the molecular layer, which contains molecular layer interneurons, i.e., stellate and basket cells. This outer layer also contains the dendritic trees of Purkinje cells and Golgi cells as well as the granule cell axons. In principle each of the cell types can be targeted using optogenetic tools (Fig. 1), which would have been impossible using electrical stimulation techniques (juxta-cellular stimulation of single neurons is not discussed in this chapter, but could be used to increase action potential firing in a single neuron [6]). The increasing knowledge of anatomical connections of cortical neurons and their concurrent activity [7–9] enables one to control the activity in functionally distinctive modules or microzones using specific (sub-) populations of cerebellar neurons [10].

2.1   How to Induce Opsin Expression

A common method to accomplish cell-specific expression of light-sensitive channels is the Cre-LoxP system, which is based upon the expression of Cre recombinase and its capacity to cause DNA

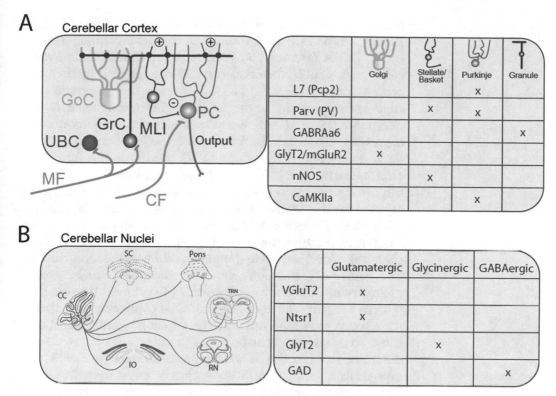

Fig. 1 Circuitry and cell-specific promoters of cerebellar neurons. (**a**) (*left*) Schematic and simplified illustration of cerebellocortical circuitry. (*right*) Table with promoters identifying promoters for different cortical cells. (**b**) (*left*) Schematic and simplified illustration of cerebellar nuclei output and its downstream targets. (*right*) Table with promoters identifying promoters for different CN neurons. Apart from the listed promoters, several nonselective promoters like synapsin or CAG have been utilized in the cerebellum. See main text for references and examples (mGluR2 is specific for a subset of Golgi cells) [115]. *UBC* unipolar brush cell, *GrC* granule cell, *GoC* Golgi cell, *MLI* molecular layer interneuron, *PC* Purkinje cell, *MF* mossy fiber, *CF* climbing fiber, *CC* cerebellar cortex, *SC* superior colliculus, *TRN* thalamic relay nuclei, *RN* red nucleus, *IO* inferior olive; *L7-PCP2* L7 Purkinje cell protein 2, *Parv* parvalbumin, *GABRA6* alpha6-subunit of GABAA receptor, *GlyT2* glycine transporter type II, *nNOS* neuronal nitric oxide synthase, *CaMKIIa* calcium-calmodulin-activated kinase type II alpha

synapsis and site-specific recombination of DNA strands at genetically engineered loxP sites [11]. For optogenetic constructs, the transgene can be preceded by a LoxP-flanked stop cassette and thus will only be expressed if the cell expresses Cre recombinase. This commonly used strategy can be applied, for instance, by crossbreeding a Cre mouse line [12] (Fig. 1) and any of the AI mouse lines [13, 14]. Of these, the AI-27 or AI-32 mouse lines result in cell-specific expression of the channelrhodopsin-2 (ChR2) construct, which encodes for light-sensitive ion channels that mediate neuronal depolarization [5], combined with a red or yellow fluorescent protein construct, respectively. To drive ChR2 expression in Purkinje cells, the most commonly used approach is to crossbreed with a transgenic mouse line in which the L7 (Purkinje cell protein 2) promoter for Purkinje cell-specific transgene expression [15] is linked to Cre recombinase [16] and thereby can be used for Purkinje cell-specific mutations (reviewed by [17]). For instance, L7Cre*Ai32 mutant mice show Purkinje cell-specific ChR2 expression and can be identified by their YFP-positive identity (Fig. 2a) [18]. Of note is that in several recent studies, the specificity of the L7/Pcp2-Cre mouse lines has been questioned [9, 19] and warrants a detailed evaluation of the expression. To avoid the potential a-specific expression of Cre lines, one option is to induce ChR2 transfection using in utero electroporation [20]. To direct the transfection specifically to cerebellar neurons, one option is to inject ChR2-encoding constructs into the fourth ventricle and electroporate the cerebellar progenitor cells in the rhombic lip or the subventricular zone near the fourth ventricle during embryonic days 10.5 and 12.5 [21] (Fig. 2b). Novel electrode designs should warrant acceptable success ratios [22], and numerous constructs that can be used for excitatory or inhibitory optogenetics [13] should provide ample options for cerebellar optogenetics. One consideration is that the transfection rates of in utero electroporation typically fall short of those reached by crossbreeding transgenic mouse lines—a difference that can also be advantageous if mosaic but cell-specific expression is the goal. Another option is to transfect the cerebellum by cerebrospinal fluid injections, which results in mosaic expression throughout the whole CNS [23, 24].

To reach a more spatially restricted expression of opsins, often viral vectors are used. From the pre-optogenetic era of gene therapy, it is known that various vectors can be used to transfect cerebellar cell types. Apart from lentiviral backbones (see, for instance, refs. [25–27]), cerebellar tissue can be transfected using adeno-associated viral (AAV) vectors [28–30] (Fig. 2c). As has been shown in other parts of the CNS [31], the various AAV serotypes can lead to widely varying transfection rates and even serotype-specific tropism. Apart from the serotypes, also the promoters can be utilized to tune the transfection in cerebellar tissue. In case the location of transfection rather than the cell-type specificity is most important, human synapsin (hSyn) is commonly used and results in neuron-specific staining. Another commonly used promoter is ubiquitous CMV and chicken β-actin fused to CMV enhancer (CAG). When packaged in a AAV1 vector, the CAG promoter showed an elevated specificity for Purkinje cells [30]. Another commonly used promoter is parvalbumin, which is endogenously expressed in Purkinje, stellate, and basket cells [32–34]. Apart from these generally used promoters, cerebellar scientists have also been exploring the use of cell-specific markers. For cerebellar granule cells, the α6 subunit of the GABAA receptors (GABRα6) is a selective marker [35]. Kim and colleagues produced an AAV1-GABRα6-GFP

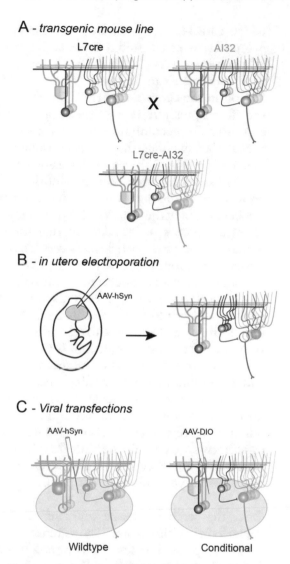

Fig. 2 Transfection techniques for (non)specific expression. (**a**) Cell-specific expression of ChR2 using a Purkinje cell-specific Cre mouse line (L7cre) and a transgenic mouse in which the ChR2-construct and the fluorescent reporter are preceded by a floxed stop codon (AI32). Crossbreeding of these two lines results in Purkinje cell-specific expression of ChR2 throughout the whole cerebellum. (**b**) In utero electroporation using a-specific promoters like hSyn can still result in a cell-specific transfection due to the spatial and temporal specificity of the viral presence. (**c**) (*left*) Viral expression using a-specific promoters will result in opsin expressing in multiple cell types. (*right*) Injecting a L7Cre mouse that expresses Cre recombinase Purkinje cells only with a viral vector that encodes the opsin construct preceded by a floxed stop codon will result in Purkinje cell-specific expression of ChR2, but only in the injected region

vector, which also resulted in selective expression of GFP molecules in granule cells [36]. In the same study, also an AAV1 vector was produced that packaged the αCaMKII promoter, which is selective for Purkinje cells in the cerebellar cortex. Finally, the Augustine lab and collaborators generated a transgenic mouse in which ChR2 is expressed under a neuronal nitric oxide synthase promoter, which resulted in selective expression in molecular layer interneurons [34, 37, 38]. These results indicate that the viral constructs can be readily tuned to drive cell-specific transfection in the cerebellar cortex. Another approach to reach cell-specific expression is by injecting a transgenic mouse that expresses Cre recombinase in a cell-specific manner with a viral vector that encodes the opsin construct preceded by a floxed stop codon (Fig. 2c) (see for instance refs. [39, 40]).

As has been pointed out previously [41], using the extensive knowledge of the anatomical connections within the cerebellum can help to study the functional connectivity of identified cell types. Great care needs to be taken so that the volume, location, and spread of the viral particles are restricted. Viral vectors can be injected using iontophoresis (see ref. [42] for a functional protocol), although it is more common to use air pressure. Pressure injections can be performed using injection systems that can be purchased from various vendors or can easily be manufactured using an injection syringe, a piece of tubing, and a glass injection pipette. Important for pressure injections is that the speed and volume are tightly controlled. Typically, the speed of injection is <50 nL/min and the volume for intracerebellar injections below 200 nL. Another measure taken to limit the spread of the viral vectors beyond the injected area is to leave the injection pipette in place for several minutes before retracting. To reach a stable opsin expression level, experiments are typically started after >3 weeks of incubation time, although the latency until stable levels have been reached could vary per serotype and promoter [31].

2.2 Combining Light and Recording in the Cerebellar Cortex

Once the cerebellar tissue expresses opsins, the next step is to apply light to the brain. The intensity of light needed ranges between opsins, but for the commonly used variants of ChR2, 1–5 mW/mm^2 is typically sufficient to evoke action potential firing [5, 43]. Although some experimental studies on ChR2 stimulation in the cerebellum aimed to record the spread of light in tissue and its effects on Purkinje cell action potential firing [44, 45], it is not clear whether the foliation and mixture of gray and white matter layers in the cerebellar cortex affect the spread of light. Although simplified calculation tools are readily available to the scientific community [46], these do not take into account that light attenuation coefficients [47] and other important characteristics vary widely in the brain [48]. Recently developed predictive models like the OptogenSIM platform by Liu et al. [48] provide the scientific community with readily accessible tools for more realistic estimations of how far the light of particular wavelengths will travel. This could be particularly informative for experiments that report the physiological and behavioral responses following

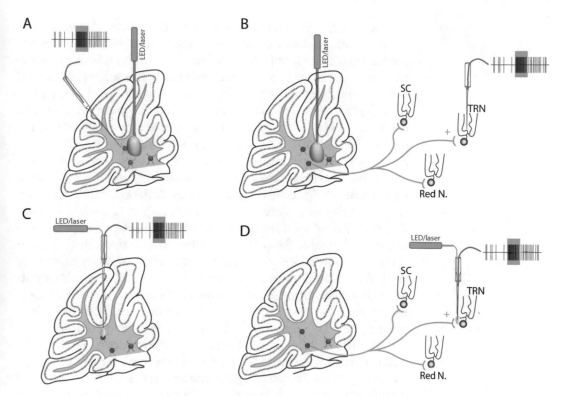

Fig. 3 Stimulation and recording methods. (**a**) Schematic for recording CN neurons (*left top*, schematic trace in which *black vertical lines* represent action potentials and the *blue box* the time of optical stimulation of ChR2) using a glass pipette (transparent tube including a wire) with a separate optic fiber (*vertical blue probe*) for light delivery. (**b**) Schematic for recording downstream targets with light delivery in CN. Note that this will also activate other pathways which might cause network effects. (**c**) Schematic for combined recording and light delivery through glass pipette. (**d**) Schematic for combined recording and light delivery at CN terminals in downstream nuclei to isolate the specific effect of cerebellar output on downstream targets without initiating network effect. *SC* superior colliculus, *TRN* thalamic relay nuclei, *Red N* red nucleus

optical stimulation of the cerebellum with large (200–400 μm diameter) optic fibers (Fig. 3a).

To limit the spread of light, small-diameter optic fibers (<100 μm diameter) can be purchased from several vendors. Another option is to strip optic fibers and pull them to sharp tip using heating and grinding (see for instance ref. [34]). Once the diameter is sufficiently limited, they can be attached directly to recording electrodes to apply optical stimulation directly to the recorded area (Fig. 3b). One important consideration for local optical stimulation is the fact that local optical stimulation causes stimulus artifacts evoked by photoelectrochemical effects that were first described by the French physicist Becquerel in 1839 and are considered as the main mechanism of stimulus artifact for conventional metal electrodes [49–51]. It is produced by photonic excitation of electrons at the electrode valence band that absorbs the photon energy; subsequently an electric potential is generated because these excited electrons leave their orbit [49, 51]. There are several ways to deal with this artifact. One of them is to place the recording electrode at a distance from the light source (Fig. 3b). If possible another solution would be to switch to low energy photons (green light instead of blue) which in combination with decreasing light intensity to such levels that the amplitude of the artifact falls below the background noise. However opsins require a minimum light intensity to be activated; a more ideal solution would be an artifact-free recording method. Several groups tried to eliminate the photoelectric artifact by developing fully transparent electrodes with increasing manufacturing complexity [52–54]. The use of tin-doped indium oxide and graphene as material does not seem favorable for (chronic) in vivo applications since indium tin oxide oxidizes over time and thereby alters the biocompatibility [55]. Still, a straightforward solution is to use the classic liquid-filled glass pipettes and insert an optic fiber into the taper (Fig. 3c, d). By placing a Pt or Ag/AgCl recording electrode at the back of the pipette, i.e., far from the light source, the charged ions move through the liquid instead of through the conduction band of the brain. This custom-made "optrode" returns high quality single cell recordings virtually free of photoelectric artifacts. Note that this fiber-in-glass approach is not suited for chronic applications since the glass pipettes are very thin and fragile and are subject to increasing tip resistances that limit the recording quality.

2.3 Optical Stimulation: Upper and Lower Limits

Ideally, optogenetic experiments are designed to encompass a recording electrode at the stimulation site. In this way the experimenters can tune their optical stimulation paradigm to the intended neuromodulation. Several landmark papers scrutinized the kinetics of opsin constructs and the neuronal responses [43, 56–61], which clearly indicate that optical stimulation cannot only be too low but also too high. In the case of ChR2, over-stimulation would result in a decrease rather than in increase of action potential firing. Such an inhibitory effect is typically mediated by a "depolarization block" caused by ChR2-mediated depolarization that inactivates sodium channels and subsequently inhibits action potential firing [62]. Obviously, such effects may lead to false assumptions, for instance, when correlating "prolonged action potential generation" to a behavioral readout, while in fact the action potential firing was only initially elevated and then silenced. In a similar fashion, the use of inhibitory opsins should be carefully considered. For instance, it was recently shown that when using chloride-conducting channelrhodopsins, the expected inhibition of neurotransmitter release was effectively reversed [63]. To prevent previously mentioned over-activation artifacts, the maximum pulse length and the power should be critically considered. It is possible to prevent over-activation effects by implementing a duty cycle. This will limit the amount of light exposure but can also dramatically change the output compared to a continuous pulse of light. It is easy to recognize that a 200 ms long 50 Hz train of short (3 ms) light pulses (17 ms pause) and a 200 ms long continuous long light pulse will have a completely different effect on the opsin-expressing neuron as well as on its downstream targets [62].

Sustained activation of any type of rhodopsin by light can also lead to heat induction. Heating depends on many properties such as power of light, duration, and the wavelength, but also on (and not limited to) intrinsic brain properties, temperature loss (due to skull removal), blood flow, etc. Commonly used light power densities for optical stimulation have been reported to marginally increase local temperatures near the stimulation site, leading to enhanced activity of neurons [64–66]. Recently it has been shown that prolonged light introduction of 10 mW can lead to temperature increases ranging between 1 °C and 4 °C across a large volume of tissue, which resulted in directly increased firing rates by up to 40% [67]. Another important issue is the diameter of optic fiber. It has been shown that the core peak temperature change for 62 μm fibers with the same light output (10 mW) is several °C higher compared to a 200 μm diameter fiber [67]. To control for the impact of such heating effects on neuronal firing and thereby on the study results, it is of upmost importance to include a set of control experiments in which the same optical stimulation is applied in ChR2-negative animals. In case prolonged activation of the opsins is an absolute necessity, a potential solution could be to use step-function opsins, which require only a short light pulse to activate long-lasting increases in action potential firing [5, 68].

2.4 Mediating Purkinje Cell Activity Using Opsins

It is of great importance to select the type of opsin and the optical stimulation paradigm that match the electrophysiological characteristics of the target cells. One of the main reasons for this is that opsins are membrane bound and thus the membrane surface area is related to the number of light-sensitive ion channels that can be incorporated. For instance, granule cells, which are the smallest cerebellar neurons, can fit less opsin channels than the larger Purkinje cells. The impact of activating light-sensitive ion channels is also related to the membrane properties—granule cells have a ~tenfold higher membrane resistance than Purkinje cells [69], and thus, the activation of a certain number of ion channels has a bigger impact on the membrane potential of granule cells than on Purkinje cells. Apart from the passive membrane properties, it is likely that also the intracellular constellation has an impact on the neuronal response to opsin activation. For instance, Purkinje cells typically do not express voltage-gated sodium channels in their dendritic tree; activation of a high density of sodium-permeable ion channels, i.e., ChR2 channels, in this membrane may induce pathophysiological processes. Indeed, in our hands light stimulation on freshly prepared in vitro slices containing Purkinje cells expressing ChR2(H134R) can induce excitotoxicity (unpublished observation, FEH).

Purkinje cells endogenously fire two types of action potentials (simple and complex spikes), which come about because of a delicately organized interaction between dendritic and somatic compartments [70, 71], and thus the use of optogenetics in Purkinje cells to manipulate action potential firing is precarious. The available data on direct manipulation of Purkinje cell activity [9, 18, 25, 27, 36, 40, 72, 73], which has been achieved using a myriad of opsins, should be evaluated critically. First and foremost, Purkinje cells are characterized by an intrinsic pacemaking activity [74] that can easily be disrupted by, for instance, abnormal levels of excitatory inputs [75, 76]. Moreover, opsins' actions may depend on the viral vector used: Jackman and colleagues recently showed that when ChR2(H134R) is expressed by AAV1, 5, or 8 serotypes, repetitive stimulation resulted artificially in paired-pulse depression, whereas this was not the case when the same construct was encoded by AAV9 or by crossbreeding of mutant mouse lines [77]. Another important consideration for selecting suitable optogenetic constructs is the fluorescent tag that is typically coupled to the opsin. Asrican and colleagues identified differences in light-evoked currents when tagging ChR2 with various fluorescent tags. Most notably, Purkinje cells that expressed ChR2-mCherry showed intracellular protein aggregates, and the membrane potential did not respond to optical stimulation, whereas Purkinje cells transfected with ChR2-YFP showed membrane-bound expression and responded with depolarizing currents upon optical stimulation [32] (but see refs. [39, 63] for mCherry-containing constructs without aggregates). These findings underline the importance of selecting the optimal optogenetic construct and inclusion of controls for the efficacy of optical stimulation.

Most of the studies that allow critical evaluation of optogenetics concern activating opsins, i.e., ChR2 and correlates. Yet, a critical evaluation of the impact of inhibitory opsins, i.e., proton pumps [78], chloride channels [56, 79, 80], and light-gated chloride pumps [81, 82], is equally important. For the most widely used halorhodopsin (eNpHR3.0) and archaerhodopsin (eArch3.0), it was shown that short illumination (millisecond range) of axonal terminals attenuates synaptic transmission, albeit that eArch3.0 is a more potent inhibitor than eNpHR3.0. Notably, such pulsed inhibition also resulted in potent rebound responses [63], which obviously is of importance for the interpretation of any readout.

For longer inhibition (minute range) the use of eArch3.0 resulted in counterintuitively elevated presynaptic calcium concentrations and hence increased spontaneous neurotransmitter release [63]. For future investigations that require long-term inhibition of neuronal firing rates, other options include chemogenetic approaches that utilize G-protein-coupled receptors that can be selectively activated by designer drugs (e.g., excitatory hM3Dq [83] and inhibitory hM4Di [84]), photoswitchable tethered ligands linked to metabotropic glutamate receptors [85], and vertebrate rhodopsins [58, 86]. By applying the inhibitory compounds to specific cell types, like molecular layer interneurons [34, 37, 38] or to Purkinje cells directly [40, 73], the net effect on the cerebellar output can be tuned to the specific research questions.

3 Optogenetic Manipulation of Cerebellar Nuclei Activity

Apart from the vestibulo-cerebellar Purkinje cell projections to the vestibular nuclei, all Purkinje cell axons terminate in the cerebellar nuclei (CN). Most, if not all [87], CN neurons receive Purkinje cell input, which is GABA-mediated. The medial, interposed, and lateral CN contain various types of neurons, up to six different types in the rodent lateral nucleus [88]. For several of these types of neurons, cell-specific markers have become available in the last years (see Fig. 1 for overview). Vesicular glutamate transporter type II (vGluT2) is expressed in the CN solely by the glutamatergic projection neurons [89], of which currently only a single type has been described albeit with varying projection patterns (see for instance refs. [90, 91]). Apart from using vGluT2-Cre transgenic mice, also the Ntsr1-Cre mouse line has been reported to selectively express in glutamatergic projection neurons of the CN [91]. The other cell types in the CN are inhibitory and release glycine, GABA, or both neurotransmitters from their terminals. By coupling Cre recombinase to the glutamic acid decarboxylase (GAD), GABAergic neurons can be identified, just as glycinergic neurons can be tagged selectively by coupling Cre to glycine transporter type II (GlyT2) [39]. Ankri and colleagues recently dissected the various types of CN neurons that can be targeted using the GAD-Cre and GlyT2-Cre mice. They convincingly showed that GlyT2-Cre labeled a single type of neuron (based upon somatic diameter and dendritic morphology), which also expresses GAD, and that the GAD-Cre separately labels a smaller type of neuron that is GlyT2-negative and projects to the inferior olive [39, 92–94]. Using the GlyT2-cre mice, it is in principle also possible to target the glycinergic projection neurons that selectively project to the brainstem's vestibular complex [95]. These cell-specific tactics provide researchers with a similar arsenal as previously mentioned for the cerebellar cortex. Importantly, also the same precautions for optical stimulation and recordings will have to be taken into account. Below we will review the currently available data important to design future optogenetic studies that can unravel the impact of afferents on CN neuronal activity and the impact of CN neurons on their downstream targets.

3.1 Manipulating Afferents to Cerebellar Nuclei Neurons

Neurons in the CN have been reported to receive a mixture of excitatory and inhibitory input from external inputs—for the current chapter, we disregard the intranuclear axon collaterals and local interneurons (for review see ref. [87]). The excitatory input may originate from the inferior olive and arrive as climbing fiber (CF) collaterals or from various reticular, pontine, or spinal regions as mossy fiber (MF) collaterals [96]. It was concluded from electrical stimulation experiments performed in vivo that CF inputs monosynaptically excited CN neurons [97–101]. Yet, a more recent study that used optogenetic approaches indicated that CF-evoked responses in CN could only be recorded in a few cells and, if present, these responses were relatively weak [102]. A potential confounder for some of the electrical stimulation studies is that not only inferior olive neurons and their axons were stimulated but also the neurons and axons passing nearby that may give rise to MF-evoked responses [98, 100, 102]. Electrical stimulation of isolated MF afferents to CN neurons is mostly limited to in vitro studies in which the CN-surrounding white matter is stimulated and the excitatory responses are pharmacologically isolated by blocking inhibitory currents (as reviewed by ref. [103]). These data on putative MF-CN transmission reveal that the excitatory input is subject to synaptic plasticity and mediates also inhibitory transmission and cellular excitability levels. Applying selective optogenetic stimulation of MF inputs will provide more insights in the differences between the MF and CF inputs that innervate CN neurons.

For the other major input to CN neurons, the Purkinje cells, it was long thought that this GABA-mediated input had a purely inhibitory effect on CN action potential firing patterns. This thought was fueled by the extreme convergence of Purkinje cell axons onto single CN neurons, the perisomatic location of these axon terminals on CN neuronal membrane, and the relatively slow time constants of the inhibitory postsynaptic potentials. However, the landmark paper by Person and Raman [104] revealed that synchronous Purkinje cell firing could elicit time-locked spiking in CN neurons, which can be used to design optogenetic paradigms for Purkinje cell stimulation. For instance, short pulses (1–2 ms pulse width) of 470 nm light on ChR2-expressing Purkinje cells could result in a short inhibition of CN spiking, followed by a well-timed action potential (see for instance refs. [100, 104]). This will probably occur once a sufficient number of Purkinje cells are synchronously entrained by the optical stimulation, i.e., when a sufficient level of Purkinje cell synchronicity is reached [104, 105]. By lengthening such potent optogenetic stimulation to tens or hundreds of milliseconds, it may occur that CN neurons are depolarized to such levels that once the stimulation stops, the membrane potential shows a rebound depolarization (see also refs. [98, 100, 105–107]). The accompanying increase in CN firing has indeed been shown in various optogenetic studies from several labs albeit ranging from a rather limited increases [44, 102] to more pronounced elevations that even accompanied behavioral responses [34, 40, 73]. These responses can be tuned quite precisely by varying the light intensity and/or location of optical Purkinje cell stimulation. It should be noted that the optical stimulation is merely enhancing the synchronicity of Purkinje cell firing for the duration of stimulation and that the local and downstream network will outlast the direct intervention.

3.2 Manipulating Cerebellar Nuclei Neuron Activity

The output neurons of the CN can be inhibitory or excitatory, and thus the effect on downstream targets can increase or decrease local action potential firing. In general it is thought that the excitatory projection neurons connect to the di-, mes-, met-, and myelencephalon and that the GABAergic and glycineric projection neurons project to the inferior olive and vestibular nuclei, respectively (Fig. 1b). So far, a limited number of optogenetic studies evaluated the impact of varying CN firing rates on these downstream targets. Nucleo-olivary neurons have been shown by electrical stimulation to provide a tonic and phasic inhibition to olivary neurons at the site of dendrodendritic gap junctions [108]. When selectively activating these neurons using a GAD-Cre mouse and a double-floxed ChR2(H134R) viral construct, Lefler et al. recently showed for the first time that this inhibitory input uncouples inferior olive neurons and thereby manipulates complex spike firing and Purkinje cell activity [93]. Moreover, it was recently shown for these nucleo-olivary neurons that optogenetic stimulation of Purkinje cells evoked postsynaptic responses that decayed tenfold slower than in excitatory projection neurons [94]. These two types of neurons can also we differentiated by the action potential half width as well as the dynamic range in action potential firing rates [94]. These apparent differences could be of importance for selecting optogenetic constructs to drive each of these neurons or their afferents.

So far, the studies which have been applying optogenetics directly in CN to study the impact of glutamatergic neurons all injected a hSyn promoter-driven ChR2 vector into the cerebellar nuclei. In an effort to stop generalized spike-and-wave discharges in epileptic mouse models, Kros et al. transfected interposed and lateral CN neurons, optically stimulated in these regions, and recorded CN action potential firing patterns and the electrocorticogram [109]. Apart from the fact that the generalized episodes were reliably stopped, the responses recorded in the CN showed a mixture of increased and decreased firing during optical stimulation. This latter finding is most likely due to the fact that using the hSyn promoter resulted in the co-transfection of both excitatory and inhibitory neurons, the latter of which synapses on the former and thereby could decrease the firing rate. Also Chen et al. transfected the lateral CN with the same a-specific promoter [110]. Although they did not extensively report on the CN activity during stimulation, the authors found that the electrophysiological responses in the striatum evoked by optical stimulation in the CN and those evoked by stimulation of the CN axon terminals in the centrolateral thalamic nucleus were in principle the same [110]. These results argue that although the transfection was most likely not specific for excitatory CN neurons, the impact of optical stimulation is mediated by the excitatory input in the downstream target nucleus. Although the local stimulation of axon terminals may be an option for many studies, it could still be advantageous to study an isolated subgroup of CN terminals projecting to a particular downstream target. One way to do so is to use a retrogradely transported viral vector and inject it in the downstream nucleus targeted by the CN. Several vectors for such retrograde expression have been generated; currently it seems that AAV-retro provides the best options [111].

Using the cell-specific stimulation options, recently there have been several studies that revealed the functional importance for CN axonal connections back to the cerebellar cortex. These axonal tracts had been described in various anatomical papers, but due to the anatomical specifications, it had not been possible to electrophysiologically isolate them sufficiently [112]. Using the hSyn promoter, Gao et al. transfected CN neurons with ChR2 and found that their fluorescently labeled, mossy fiber-like terminals evoked purely excitatory currents in granule and Golgi cells of the cerebellar cortex [113]. In another recent study, an inhibitory nucleocortical projection has been described from neurons that co-express GAD and GlyT2 and specifically target a subpopulation of cerebellar Golgi cells [39]. This latter study utilized GAD-Cre and GlyT2-Cre mice in combination with AAV vectors encoding for floxed-ChR2 constructs, which warranted cell-specific manipulations of inhibitory neurons that are GABAergic and glycinergic (and excluded GABAergic nucleo-olivary and glutamatergic projection neurons) [39]. Optogenetic approaches provided a similar level of unprecedented anatomical, physiological, and functional detail in a recent study of Purkinje cell axon collaterals to granule cells and neighboring Purkinje cells in the adult brain [8, 9].

4   Final Recommendations and Conclusions

Utilizing optogenetics will have a positive impact on cerebellar studies. Novel research questions about the functional impact of particular connections can now be adequately addressed. The information content of the results from optogenetic studies will be unprecedented, but will require a multitude of controls and sensitive analyses to filter out artifacts and identify the true impact of optogenetic manipulations. For instance, to recognize subtle changes in action potential firing patterns evoked by the activation of a subset of presynaptic axon terminals, a sufficient number of stimulation repeats has to be recorded to allow for Monte Carlo bootstrapping and subsequently identification of significant changes in firing patterns. Another aspect of optogenetics in cerebellar network physiology that deserves attention is the fact that various recurrent loops exist that can affect the electrophysiological responses in addition to the direct effects of the optical stimulation. These loops come in various shapes and sizes that not only include cortical neurons (MF-granule cell-Golgi cell-granule cell, Purkinje cell-granule cell-Purkinje cell, etc.) but also involve the CN neurons and inferior olive neurons (intra-CN, nucleocortical, and nucleo-olivary) and thus can have a delayed effect on cerebellar activity patterns. Moreover, the glutamatergic CN output can also return to the MF or CF sources by thalamocortical networks and midbrain nuclei or even via the sensory feedback triggered by cerebellar-evoked motor responses. Although most of these potential feedback loops will be characterized by a fixed latency, it is very likely that the evoked network effects can last for extended periods and need to be considered when setting interstimulus intervals. Likewise, for most optogenetic constructs, the interstimulus interval needs to be tailored to the desensitization and closed states [114]. Tuning these and other technical details will be decisive for the added value of optogenetics in cerebellar studies. Although the cerebellar field can ride on the constant wave of optogenetic innovations, this does not mean that optogenetic approaches can be bluntly copied from, for instance, the cerebral cortex field. By using the extensive anatomical knowledge of the cerebellar connectivity and cellular physiology, the most effective and functionally relevant optogenetic approaches can be achieved.

References

1. Moruzzi G (1950) Effects at different frequencies of cerebellar stimulation upon postural tonus and myotatic reflexes. Electroencephalogr Clin Neurophysiol 2:463–469

2. Kros L, Rooda OH, De Zeeuw CI, Hoebeek FE (2015) Controlling cerebellar output to treat refractory epilepsy. Trends Neurosci 38:787–799. https://doi.org/10.1016/j.tins.2015.10.002

3. Hawkes R et al (1993) Structural and molecular compartmentation in the cerebellum. Can J Neurol Sci 20:S29–S35

4. Pijpers A, Voogd J, Ruigrok TJ (2005) Topography of olivo-cortico-nuclear modules in the intermediate cerebellum of the rat. J Comp Neurol 492:193–213

5. Yizhar O, Fenno LE, Davidson TJ, Mogri M, Deisseroth K (2011) Optogenetics in neural systems. Neuron 71:9–34

6. Houweling AR, Brecht M (2008) Behavioural report of single neuron stimulation in somatosensory cortex. Nature 451:65–68. https://doi.org/10.1038/nature06447

7. Valera AM et al (2016) Stereotyped spatial patterns of functional synaptic connectivity in the cerebellar cortex. elife 5:e09862. https://doi.org/10.7554/eLife.09862

8. Guo C et al (2016) Purkinje cells directly inhibit granule cells in specialized regions of the cerebellar cortex. Neuron 91:1330–1341. https://doi.org/10.1016/j.neuron.2016.08.011

9. Witter L, Rudolph S, Pressler RT, Lahlaf SI, Regehr WG (2016) Purkinje cell collaterals enable output signals from the cerebellar cortex to feed back to purkinje cells and interneurons. Neuron 91:312–319

10. Apps R, Hawkes R (2009) Cerebellar cortical organization: a one-map hypothesis. Nat Rev Neurosci 10:670–681

11. Sauer B, Henderson N (1988) Site-specific DNA recombination in mammalian cells by the Cre recombinase of bacteriophage P1. PNAS 85:5166–5170

12. GENSAT (2016) Gensat website for Cre mice. http://www.gensat.org/cre.jsp

13. Madisen L et al (2012) A toolbox of Cre-dependent optogenetic transgenic mice for light-induced activation and silencing. Nat Neurosci 15:793–802. https://doi.org/10.1038/nn.3078

14. Madisen L et al (2015) Transgenic mice for intersectional targeting of neural sensors and effectors with high specificity and performance. Neuron 85:942–958. https://doi.org/10.1016/j.neuron.2015.02.022

15. Oberdick J, Smeyne RJ, Mann JR, Zackson S, Morgan JI (1990) A promoter that drives transgene expression in cerebellar Purkinje and retinal bipolar neurons. Science 248:223–226

16. Barski JJ, Dethleffsen K, Meyer M (2000) Cre recombinase expression in cerebellar Purkinje cells. Genesis 28:93–98

17. De Zeeuw CI et al (2011) Spatiotemporal firing patterns in the cerebellum. Nat Rev Neurosci 12:327–344. https://doi.org/10.1038/nrn3011

18. Nguyen-Vu TD et al (2013) Cerebellar Purkinje cell activity drives motor learning. Nat Neurosci 16:1734–1736

19. Mark MD et al (2011) Delayed postnatal loss of P/Q-type calcium channels recapitulates the absence epilepsy, dyskinesia, and ataxia phenotypes of genomic Cacna1a mutations. J Neurosci 31:4311–4326. https://doi.org/10.1523/JNEUROSCI.5342-10.2011

20. Petreanu L, Huber D, Sobczyk A, Svoboda K (2007) Channelrhodopsin-2-assisted circuit mapping of long-range callosal projections. Nat Neurosci 10:663–668. https://doi.org/10.1038/nn1891

21. Yamada M et al (2014) Specification of spatial identities of cerebellar neuron progenitors by ptf1a and atoh1 for proper production of GABAergic and glutamatergic neurons. J Neurosci 34:4786–4800. https://doi.org/10.1523/JNEUROSCI.2722-13.2014

22. Szczurkowska J et al (2016) Targeted in vivo genetic manipulation of the mouse or rat brain by in utero electroporation with a triple-electrode probe. Nat Protoc 11:399–412. https://doi.org/10.1038/nprot.2016.014

23. McLean JR et al (2014) Widespread neuron-specific transgene expression in brain and spinal cord following synapsin promoter-driven AAV9 neonatal intracerebroventricular injection. Neurosci Lett 576:73–78. https://doi.org/10.1016/j.neulet.2014.05.044

24. Lukashchuk V, Lewis KE, Coldicott I, Grierson AJ, Azzouz M (2016) AAV9-mediated central nervous system-targeted gene delivery via cisterna magna route in mice. Mol Ther Methods Clin Dev 3:15055. https://doi.org/10.1038/mtm.2015.55

25. Tsubota T, Ohashi Y, Tamura K, Miyashita Y (2012) Optogenetic inhibition of Purkinje cell activity reveals cerebellar control of blood pressure during postural alterations in anesthetized rats. Neuroscience 210:137–144

26. Hirai H (2008) Progress in transduction of cerebellar Purkinje cells in vivo using viral vectors. Cerebellum 7:273–278

27. Tsubota T, Ohashi Y, Tamura K, Sato A, Miyashita Y (2011) Optogenetic manipulation of cerebellar Purkinje cell activity in vivo. PLoS One 6(8):e22400

28. Kaemmerer WF et al (2000) In vivo transduction of cerebellar Purkinje cells using adeno-associated virus vectors. Mol Ther 2:446–457. https://doi.org/10.1006/mthe.2000.0134

29. Dodge JC et al (2005) Gene transfer of human acid sphingomyelinase corrects neuropathology and motor deficits in a mouse model of Niemann-pick type a disease. Proc Natl Acad Sci U S A 102:17822–17827. https://doi.org/10.1073/pnas.0509062102

30. Bosch MK, Nerbonne JM, Ornitz DM (2014) Dual transgene expression in murine cerebellar Purkinje neurons by viral transduction in vivo. PLoS One 9:c104062

31. Aschauer DF, Kreuz S, Rumpel S (2013) Analysis of transduction efficiency, tropism and axonal transport of AAV serotypes 1, 2, 5, 6, 8 and 9 in the mouse brain. PLoS One 8:e76310. https://doi.org/10.1371/journal.pone.0076310

32. Asrican B et al (2013) Next-generation transgenic mice for optogenetic analysis of neural circuits. Front Neural Circuits 7:160

33. Celio MR, Heizmann CW (1981) Calcium-binding protein parvalbumin as a neuronal marker. Nature 293:300–302

34. Heiney SA, Kim J, Augustine GJ, Medina JF (2014) Precise control of movement kinematics by optogenetic inhibition of purkinje cell activity. J Neurosci 34:2321–2330

35. Aller MI et al (2003) Cerebellar granule cell Cre recombinase expression. Genesis 36:97–103. https://doi.org/10.1002/gene.10204

36. Kim Y et al (2015) Selective transgene expression in cerebellar Purkinje cells and granule cells using adeno-associated viruses together with specific promoters. Brain Res 1620:1–16

37. Astorga G et al (2015) An excitatory GABA loop operating in vivo. Front Cell Neurosci 9:275. https://doi.org/10.3389/fncel.2015.00275

38. Kim J et al (2014) Optogenetic mapping of cerebellar inhibitory circuitry reveals spatially biased coordination of interneurons via electrical synapses. Cell Rep 7:1601–1613. https://doi.org/10.1016/j.celrep.2014.04.047

39. Ankri L et al (2015) A novel inhibitory nucleo-cortical circuit controls cerebellar golgi cell activity. elife 4

40. Witter L, Canto CB, Hoogland TM, de Gruijl JR, De Zeeuw CI (2013) Strength and timing of motor responses mediated by rebound firing in the cerebellar nuclei after Purkinje cell activation. Front Neural Circuits 7:133

41. Tsubota T, Ohashi Y, Tamura K (2013) Optogenetics in the cerebellum: Purkinje cell-specific approaches for understanding local cerebellar functions. Behav Brain Res 255:26–34. https://doi.org/10.1016/j.bbr.2013.04.019

42. Wang Q et al (2014) Systematic comparison of adeno-associated virus and biotinylated dextran amine reveals equivalent sensitivity between tracers and novel projection targets in the mouse brain. J Comp Neurol 522:1989–2012. https://doi.org/10.1002/cne.23567

43. Boyden ES, Zhang F, Bamberg E, Nagel G, Deisseroth K (2005) Millisecond-timescale, genetically targeted optical control of neural activity. Nat Neurosci 8:1263–1268

44. Chaumont J et al (2013) Clusters of cerebellar purkinje cells control their afferent climbing fiber discharge. PNAS 110:16223–16228

45. Kruse W et al (2014) Optogenetic modulation and multi-electrode analysis of cerebellar networks in vivo. PLoS One 9(8):e105589

46. Lab, s. u.-d. irradiance calculator tool (2016). http://web.stanford.edu/group/dlab/cgi-bin/graph/chart.php

47. Al-Juboori SI et al (2013) Light scattering properties vary across different regions of the adult mouse brain. PLoS One 8:e67626. https://doi.org/10.1371/journal.pone.0067626

48. Liu Y et al (2015) OptogenSIM: a 3D Monte Carlo simulation platform for light delivery design in optogenetics. Biomed Opt Express 6:4859–4870. https://doi.org/10.1364/BOE.6.004859

49. Brabec C, Dyakonov V, Parisi J, Sariciftci NS (2003) Organic photovoltaics: concepts and realization. Springer, Berlin

50. Gratzel M (2001) Photoelectrochemical cells. Nature 414:338–344

51. Williams R (1960) Becquerel photovoltaic effect in binary compounds. J Chem Phys 32:1505–1514

52. Ledochowitsch P et al (2015) Strategies for optical control and simultaneous electrical readout of extended cortical circuits. J Neurosci Methods 256:220–231

53. Park DW et al (2014) Graphene-based carbon layered electrode array technology for neural imaging and optogenetic applications. Nat Commun 5:5258

54. Wu F et al (2015) Monolithically integrated μLEDs on silicon neural probes for high-resolution optogenetic studies in behaving animals. Neuron 88:1136–1148

55. Pappas TC et al (2007) Nanoscale engineering of a cellular interface with semiconductor nanoparticle films for photoelectric stimulation of neurons. Nano Lett 7:513–519

56. Berndt A, Lee SY, Ramakrishnan C, Deisseroth K (2014) Structure-guided transformation of channelrhodopsin into a light-activated chloride channel. Science 344:420–424

57. Han X et al (2009) Millisecond-timescale optical control of neural dynamics in the non-human primate brain. Neuron 62:191–198

58. Li X et al (2005) Fast noninvasive activation and inhibition of neural and network activity by vertebrate rhodopsin and green algae channelrhodopsin. Proc Natl Acad Sci U S A 102:17816–17821

59. Mateo C et al (2011) In vivo optogenetic stimulation of neocortical excitatory neurons drives brain-state-dependent inhibition. Curr Biol 21:1593–1602

60. Nagel G et al (2005) Light activation of channelrhodopsin-2 in excitable cells of Caenorhabditis Elegans triggers rapid behavioral responses. Curr Biol 15:2279–2284

61. Zhao S et al (2011) Cell type-specific channelrhodopsin-2 transgenic mice for optogenetic dissection of neural circuitry function. Nat Methods 8:745–752

62. Herman AM, Huang L, Murphey DK, Garcia I, Arenkiel BR (2014) Cell type-specific and time-dependent light exposure contribute to silencing in neurons expressing Channelrhodopsin-2. elife 3:e01481

63. Mahn M, Prigge M, Ron S, Levy R, Yizhar O (2016) Biophysical constraints of optogenetic inhibition at presynaptic terminals. Nat Neurosci 19:554–556

64. Christie IN et al (2013) fMRI response to blue light delivery in the naive brain: implications for combined optogenetic fMRI studies. NeuroImage 66:634–641

65. Long MA, Fee MS (2008) Using temperature to analyse temporal dynamics in the songbird motor pathway. Nature 456:189–194

66. Moser E, Mathiesen I, Andersen P (1993) Association between brain temperataure and dentate field potentials in exploring and swimming rats. Science 259:1324–1326

67. Stujenske JM, Spellman T, Gordon JA (2015) Modeling the spatiotemporal dynamics of light and heat propagation for in vivo optogenetics. Cell Rep 12:525–534

68. Sorokin JM et al (2017) Bidirectional control of generalized epilepsy networks via rapid real-time switching of firing mode. Neuron 93(1):194–210. https://doi.org/10.1016/j.neuron.2016.11.026

69. Galliano E et al (2013) Silencing the majority of cerebellar granule cells uncovers their essential role in motor learning and consolidation. Cell Rep 3:1239–1251. https://doi.org/10.1016/j.celrep.2013.03.023

70. Palmer LM et al (2010) Initiation of simple and complex spikes in cerebellar Purkinje cells. J Physiol 588:1709–1717. https://doi.org/10.1113/jphysiol.2010.188300

71. Vetter P, Roth A, Hausser M (2001) Propagation of action potentials in dendrites depends on dendritic morphology. J Neurophysiol 85:926–937

72. Krook-Magnuson E, Szabo GG, Armstrong C, Oijala M, Soltesz I (2014) Cerebellar directed optogenetic intervention inhibits spontaneous hippocampal seizures in a mouse model of temporal lobe epilepsy. eNeuro 1:e.2014

73. Lee KH et al (2015) Circuit mechanisms underlying motor memory formation in the cerebellum. Neuron 86:529–540. https://doi.org/10.1016/j.neuron.2015.03.010

74. Raman IM, Bean BP (1999) Ionic currents underlying spontaneous action potentials in isolated cerebellar Purkinje neurons. J Neurosci 19:1663–1674

75. Walter JT, Alvina K, Womack MD, Chevez C, Khodakhah K (2006) Decreases in the precision of Purkinje cell pacemaking cause cerebellar dysfunction and ataxia. Nat Neurosci 9:389–397. https://doi.org/10.1038/nn1648

76. Gao Z et al (2012) Cerebellar ataxia by enhanced Ca(V)2.1 currents is alleviated by Ca^{2+}-dependent K+−channel activators in Cacna1a(S218L) mutant mice. J Neurosci 32:15533–15546. https://doi.org/10.1523/JNEUROSCI.2454-12.2012

77. Jackman SL, Beneduce BM, Drew IR, Regehr WG (2014) Achieving high-frequency optical control of synaptic transmission. J Neurosci 34:7704–7714

78. Chow BY et al (2010) High-performance genetically targetable optical neural silencing by light-driven proton pumps. Nature 463:98–102

79. Govorunova EG, Sineshchekov OA, Janz R, Liu X, Spudich JL (2015) Natural light-gated anion channels: a family of microbial rhodopsins for advanced optogenetics. Science 349:647–650

80. Wietek J et al (2014) Conversion of channelrhodopsin into a light-gated chloride channel. Science 344:409–412

81. Chuong AS et al (2014) Noninvasive optical inhibition with a red-shifted microbial rhodopsin. Nat Neurosci 17:1123–1129

82. Zhang F et al (2007) Multimodal fast optical interrogation of neural circuitry. Nature 446:633–639

83. Alexander GM et al (2009) Remote control of neuronal activity in transgenic mice expressing evolved G protein-coupled receptors. Neuron 63:27–39. https://doi.org/10.1016/j.neuron.2009.06.014

84. Stachniak TJ, Ghosh A, Sternson SM (2014) Chemogenetic synaptic silencing of neural circuits localizes a hypothalamus → midbrain pathway for feeding behavior. Neuron 82:797–808. https://doi.org/10.1016/j.neuron.2014.04.008

85. Levitz J et al (2013) Optical control of metabotropic glutamate receptors. Nat Neurosci 16:507–516

86. Gutierrez DV et al (2011) Optogenetic control of motor coordination by Gi/o protein-coupled vertebrate rhodopsin in cerebellar Purkinje cells. J Biol Chem 286:25848–25858. https://doi.org/10.1074/jbc.M111.253674

87. Uusisaari M, De Schutter E (2011) The mysterious microcircuitry of the cerebellar nuclei. J Physiol 589:3441–3457. https://doi.org/10.1113/jphysiol.2010.201582

88. Uusisaari M, Knopfel T (2011) Functional classification of neurons in the mouse lateral cerebellar nuclei. Cerebellum 10:637–646. https://doi.org/10.1007/s12311-010-0240-3

89. Borgius L, Restrepo CE, Leao RN, Saleh N, Kiehn O (2010) A transgenic mouse line for molecular genetic analysis of excitatory glutamatergic neurons. Mol Cell Neurosci 45:245–257. https://doi.org/10.1016/j.mcn.2010.06.016

90. Ruigrok TJ, Teune TM (2014) Collateralization of cerebellar output to functionally distinct brainstem areas. A retrograde, non-fluorescent tracing study in the rat. Front Syst Neurosci 8:23

91. Houck BD, Person AL (2015) Cerebellar premotor output neurons collateralize to innervate the cerebellar cortex. J Comp Neurol 523:2254–2271

92. Husson Z, Rousseau CV, Broll I, Zeilhofer HU, Dieudonne S (2014) Differential GABAergic and glycinergic inputs of inhibitory interneurons and Purkinje cells to principal cells of the cerebellar nuclei. J Neurosci 34:9418–9431. https://doi.org/10.1523/JNEUROSCI.0401-14.2014

93. Lefler Y, Yarom Y, Uusisaari MY (2014) Cerebellar inhibitory input to the inferior olive decreases electrical coupling and blocks subthreshold oscillations. Neuron 81:1389–1400

94. Najac M, Raman IM (2015) Integration of Purkinje cell inhibition by cerebellar nucleo-olivary neurons. J Neurosci 35:544–549

95. Bagnall MW et al (2009) Glycinergic projection neurons of the cerebellum. J Neurosci 29:10104–10110. https://doi.org/10.1523/JNEUROSCI.2087-09.2009

96. Sugihara I (2011) Compartmentalization of the deep cerebellar nuclei based on afferent projections and aldolase C expression. Cerebellum 10:449–463. https://doi.org/10.1007/s12311-010-0226-1

97. Rowland NC, Jaeger D (2008) Responses to tactile stimulation in deep cerebellar nucleus neurons result from recurrent activation in multiple pathways. J Neurophysiol 99:704–717. https://doi.org/10.1152/jn.01100.2007

98. Bengtsson F, Ekerot CF, Jorntell H (2011) In vivo analysis of inhibitory synaptic inputs and rebounds in deep cerebellar nuclear neurons. PLoS One 6:e18822. https://doi.org/10.1371/journal.pone.0018822

99. Kitai ST, McCrea RA, Preston RJ, Bishop GA (1977) Electrophysiological and horseradish peroxidase studies of precerebellar afferents to the nucleus interpositus anterior. I. Climbing fiber system. Brain Res 122:197–214

100. Hoebeek FE, Witter L, Ruigrok TJ, De Zeeuw CI (2010) Differential olivo-cerebellar cortical control of rebound activity in the cerebellar nuclei. Proc Natl Acad Sci U S A 107:8410–8415. https://doi.org/10.1073/pnas.0907118107

101. Blenkinsop TA, Lang EJ (2011) Synaptic action of the olivocerebellar system on cerebellar nuclear spike activity. J Neurosci 31:14708–14720. https://doi.org/10.1523/JNEUROSCI.3323-11.2011

102. Lu H, Yang B, Jaeger D (2016) Cerebellar nuclei neurons show only small excitatory responses to optogenetic olivary stimulation in transgenic mice: in vivo and in vitro studies. Front Neural Circuits 10:21

103. Zheng N, Raman IM (2010) Synaptic inhibition, excitation, and plasticity in neurons of the cerebellar nuclei. Cerebellum 9:56–66. https://doi.org/10.1007/s12311-009-0140-6

104. Person AL, Raman IM (2011) Purkinje neuron synchrony elicits time-locked spiking in the cerebellar nuclei. Nature 481:502–505. https://doi.org/10.1038/nature10732

105. Gauck V, Jaeger D (2000) The control of rate and timing of spikes in the deep cerebellar nuclei by inhibition. J Neurosci 20:3006–3016

106. Alvina K, Walter JT, Kohn A, Ellis-Davies G, Khodakhah K (2008) Questioning the role of rebound firing in the cerebellum. Nat Neurosci 11:1256–1258. https://doi.org/10.1038/nn.2195

107. Dykstra S, Engbers JD, Bartoletti TM, Turner RW (2016) Determinants of rebound burst responses in rat cerebellar nuclear neurons to physiological stimuli. J Physiol 594:985–1003. https://doi.org/10.1113/JP271894

108. Best AR, Regehr WG (2009) Inhibitory regulation of electrically coupled neurons in the inferior olive is mediated by asynchronous release of GABA. Neuron 62:555–565. https://doi.org/10.1016/j.neuron.2009.04.018

109. Kros L et al (2015) Cerebellar output controls generalized spike-and-wave discharge occurrence. Ann Neurol 77:1027–1049

110. Chen CH, Fremont R, Arteaga-Bracho EE, Khodakhah K (2014) Short latency cerebellar modulation of the basal ganglia. Nat Neurosci 17:1767–1775

111. Tervo DG et al (2016) A designer AAV variant permits efficient retrograde access to projection neurons. Neuron 92:372–382. https://doi.org/10.1016/j.neuron.2016.09.021

112. Houck BD, Person AL (2014) Cerebellar loops: a review of the nucleocortical pathway. Cerebellum 13:378–385. https://doi.org/10.1007/s12311-013-0543-2

113. Gao Z et al (2016) Excitatory cerebellar nucleocortical circuit provides internal amplification during associative conditioning. Neuron 89:645–657

114. Lin JY (2011) A user's guide to channelrhodopsin variants: features, limitations and future developments. Exp Physiol 96:19–25. https://doi.org/10.1113/expphysiol.2009.051961

115. Chen AI et al (2011) TrkB (tropomyosin-related kinase B) controls the assembly and maintenance of GABAergic synapses in the cerebellar cortex. J Neurosci 31:2769–2780. https://doi.org/10.1523/JNEUROSCI.4991-10.2011

Chapter 7

A Combinatorial Approach to Circuit Mapping in the Mouse Olfactory Bulb

Gary Liu, Jessica Swanson, Brandon Pekarek, Sugi Panneerselvam, Kevin Ung, Burak Tepe, Longwen Huang, and Benjamin R. Arenkiel

Abstract

Neurons form neural circuits through functional synapses. Inputs and outputs across these circuits contribute to the brain's intricate mechanisms of information processing and behavioral responses. By combining acousto-optic deflector-based scanning microscopy with optogenetics, whole cell electrophysiological recordings, and transgenic viral delivery, researchers can now investigate and map relevant neural circuits with high spatial and cell type-specific precision. The goal of this review is to provide the reader with guidelines and methods for using a combinatorial approach toward circuit mapping in rodent brain tissue.

Key words Olfactory bulb, Neural circuit, Slice electrophysiology, Optogenetic, Synapse, Acousto optic deflector, Virus

1 Introduction: Synaptic Connections Within the Olfactory Bulb

Synapses represent the functional connections between neuronal projections and their targets. Many types of synapses exist and can be broadly categorized by anatomy, postsynaptic effect, and mode of communication [1, 2]. Chemical synapses utilize neurotransmitters secreted into a junctional space called the synaptic cleft and are ultimately recognized by their receptors. Depending on the neurotransmitter and the receptor pair, the postsynaptic membrane potential will either depolarize or hyperpolarize, consequently altering postsynaptic action potential frequency [3]. To form neural circuits, many synapses are systematically interconnected for discrete information processing. These intricate systems are also able to adapt, which forms the basis of synaptic plasticity [4].

The mammalian olfactory system is an excellent model to study circuit connectivity and plasticity due to its well-defined anatomy, ease of access, and genetic tractability. The olfactory bulbs (OBs),

Roy V. Sillitoe (ed.), *Extracellular Recording Approaches*, Neuromethods, vol. 134,
https://doi.org/10.1007/978-1-4939-7549-5_7, © Springer Science+Business Media, LLC 2018

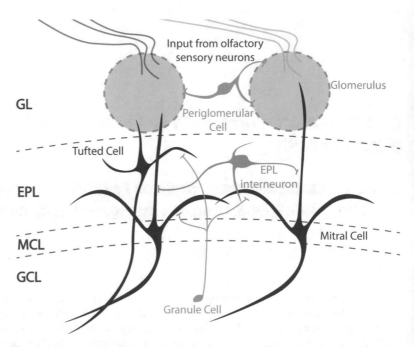

Fig. 1 Hierarchical and structural organization within the olfactory bulb. The olfactory bulb (OB) is anatomically segregated into distinct layers: glomerular layer (GL), external plexiform layer (EPL), mitral cell layer (MCL), and granule cell layer (GCL). Olfactory sensory neurons first project odor information onto mitral and tufted cells (M/Ts) within individual glomeruli of the OB. Reciprocal feedback inhibition from local interneurons (periglomerular cells, EPL interneurons, and granule cells) refine M/Ts odor-evoked responses. M/Ts then project odor information to other cortical areas

two bilaterally symmetric structures in the anterior CNS, receive odor information. First-order olfactory sensory neurons (OSNs) synapse onto the dendrites of mitral and tufted cells (M/Ts), the principle excitatory neurons of the OB, within spherical structural units of synaptic terminals called glomeruli [5] (Fig. 1). Individual M/Ts then project dendrodendritic synapses onto diverse interneuron populations, each identifiable by location, molecular identity, and connectivity pattern [6–10]. Single interneurons then project back onto many M/Ts, forming broad acting negative feedback loops to refine M/Ts odor-evoked responses. Through this hierarchal organization of olfactory neurons, a relay station for information processing is established: OSNs relay odor information to the dendrites of second-order M/Ts, which then receive local feedback inhibition from OB interneurons, including periglomerular (PG) cells, external plexiform layer (EPL) interneurons, and granule cells (GCs) [11] (Fig. 1). Olfactory sensory information is then projected by M/Ts to higher-order CNS regions.

Table 1
Genetic drivers for cell type-specific manipulations within olfactory bulb

Cell type	Cell-specific transgenic driver	References
Mitral/tufted cell	Thy1-ChR2-YFP (Jax 007612)	Arenkiel et al. [12],
	Tbx21 Cre (Jax 024507)	Mitsui et al. [13],
	Pcdh21 Cre (Jax 008616)	Faedo et al. [14],
Granule cell	Dlx5/6 Cre (immature)	Quast et al. [8],
	(Jax 008199), CRHR1 Cre (mature) (generated in house)	Garcia et al. [9]
Periglomerular cell	Tyrosine hydroxylase Cre	Kosaka and Kosaka [15]
	(Jax 008701)	Parrish-Aungst et al. [16]
EPL interneuron	CRH Cre (Jax 012704)	Huang et al. [6]
	PV Cre (Jax 008069)	Miyamichi et al. [17]

1.1 Methods for Genetic Manipulations in the Olfactory Bulb

Neurons within the OB are stratified into distinct cellular layers and largely derive their classification from where their cell bodies are found. This anatomical patterning, along with expression of unique genetic programs, specifies functionality. Many transgenic mice lines and molecular markers are available to genetically target specific cell types within the mouse OB. The tools described herein afford researchers a high degree of specificity which can be readily applied toward anatomical and functional circuit mapping within the OB. Table 1 illustrates the most common genetic drivers for each major cell type.

M/T cell bodies populate the OB within the mitral and EPL layers, respectively. They each extend a single primary dendrite into one of many thousands of glomeruli to receive odor input and then project that information to the olfactory cortex. Their relatively large cell bodies (20 μm) are more than double that of granule cells and EPL interneurons [5]. Most M/Ts selectively express the T-box family transcription factor Tbx21 and the cell adhesion protein Pcdh21 [18]. Furthermore, line 18 of the Thy1-ChR2-EYFP transgene generated from BAC clone insertion specifically labels M/Ts in the OB [19].

Interneurons are the most common cells within the OB and can be generally divided into three subtypes: granule cells, periglomerular cells, and EPL interneurons [6–9, 15, 16]. Granule cells are comprised of both embryonically generated and adult-born cells. Adult-born granule cells are derived from neural progenitors located near the lateral ventricle and are continuously integrated into the existing bulbar network [20]. These cells undergo distinct developmental stages within the OB and are characterized by

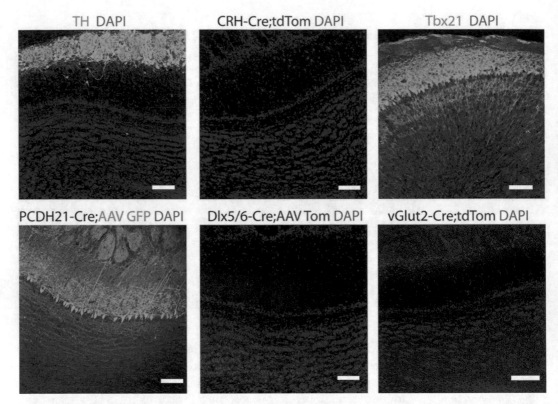

Fig. 2 Cell type-specific markers within the olfactory bulb. Periglomerular interneurons are labeled by tyrosine hydroxylase (TH), EPL interneurons by CRH, mitral cells by Tbx21 and PCDH21, adult-born granule cells by Dlx5/6 Cre, and all excitatory neurons by vGlut2. TH and Tbx21 are antibody stains; CRH and vGlut2 are Cre lines crossed to a conditional reporter mice. PCDH21 and Dlx5/6 Cre are injected by an AAV encoding Cre-dependent fluorescent reporter. Scale bar 100 μm

different genetic markers; distal-less homeobox 5/6 (Dlx5/6) labels immature cells, while corticotropin-releasing hormone receptor 1 (CRHR1) labels more mature granule cells (Fig. 2) [8, 9]. EPL interneurons within the OB express corticotropin-releasing hormone (CRH) and parvalbumin (PV) [6, 17]. The other prominent inhibitory cell type within the OB are the periglomerular cells (PG), which are located within the glomerular layer and express tyrosine hydroxylase [15, 16].

Because genes are dynamically expressed throughout development, traditional genetic crosses are sometimes not sufficient to target specific neuronal populations for circuit mapping. Stereotaxic injection of viral particles offers much improved temporal and spatial resolution [8]. It is also an easy, reproducible way to stably express genetic constructs in vivo without excessive trauma to the brain. Thus, combining transgenic animals with stereotaxic delivery of conditionally expressed adeno-associated viruses or lentiviruses is a common method for performing cell type-specific genetic manipulations.

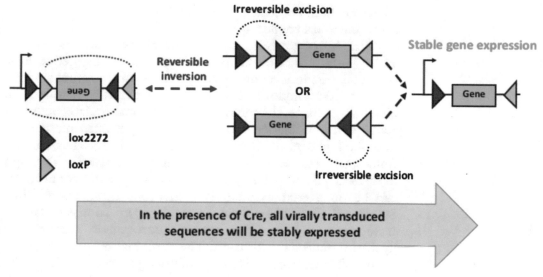

Fig. 3 The FLEX switch provides stable, permanent expression of a transgene of interest in cells expressing Cre recombinase

Cell type-specific reporter expression is obtained by using Cre recombinase-dependent constructs and transgenic mice expressing Cre recombinase under a cell type-specific promoter. A common Cre-dependent construct is the "flip-excision" or FLEX (alternatively named double inverted ORF, DIO) switch (Fig. 3) [21]. By introducing flip excision constructs into neurons that express Cre recombinase, the conditional reporter becomes irreversibly and stably expressed. For optogenetic circuit mapping, one can express ChR2 in a specific cell population by using the FLEX system and performing stereotaxic injection of adeno-associated viruses harboring a Cre-dependent ChR2 construct in Cre-expressing mice [22].

1.2 Methods to Study Neural Circuits in the Olfactory Bulb

Historically, synapses in the OB have been primarily studied using various microscopic techniques. One of the first staining techniques was the Golgi stain, pioneered by Camilo Golgi. Adaptations of this stain have led to increased reliability and clarity of synaptic structures [23]. Later, discovery of activity markers, such as immediate early genes (IEGs) in neurons, have provided increased cellular and temporal resolution. IEGs are induced naturally in neural circuits through learning and establishing long-term memory, but can be stimulated rapidly in response to specific extracellular stimuli [24, 25]. Further advances in fluorescence microscopy have allowed researchers to utilize double immunofluorescence with presynaptic protein markers, such as synaptophysin, to study synaptic connections as well [23]. The implementation of transgenic mice with fluorescent reporter proteins also enables long-term mapping of neural circuits within a viable animal model [24].

These methods are reliant on anatomical associations and do not show functional connections.

Alternatively, while many of the above methods are still widely used, slice electrophysiology and optogenetics have provided functional connectivity patterns within neural circuits. Optogenetics employs light-sensitive, genetically encoded ion channels such as channelrhodopsins to manipulate neuronal activity. In combination with the cell type-specific manipulations described above, it is now possible to interrogate direct, functional connections with induced activity. Through light-induced activation of targeted cells, synaptic connections can be uncovered by analysis of neurons downstream [26]. Single-cell recordings from slices of neuronal tissue enable researchers to investigate synaptic currents in response to upstream neuronal activation. The combination of optogenetics and slice electrophysiology is advantageous due to its ability to directly test functional connectivity with high temporal specificity and cellular resolution [27].

Traditionally, wide-field illumination has been a potent technique for stimulating a large field of neurons. This unbiased approach aided with different pharmacological blockers allow for researchers to reveal synaptic neurotransmitter and receptor identities [28]. Furthermore, recent advances in acousto-optic deflector (AOD) technology now allow narrow optical beams to be focused onto defined targets. By utilizing both a laser and AODs to control light positioning, researchers can focally stimulate cells within a 2D plane. This technique allows for the systematic stimulation along an X.Y grid to reveal the physical distances between interconnected cells [29]. As a comparison, traditional wide-field stimulation and AOD-based scanning microscopy are described in Table 2.

Table 2
Comparison between wide-field stimulation vs. AOD-based scanning microscopy.

	Advantage	Disadvantage
Traditional wide-field stimulation	High-intensity stimulation	Low resolution of stimulation (large area)
		Speed limited by shutter
AOD-based scanning microscopy	High-resolution spot stimulation	Uncommon/specialized/expensive equipment
	Bypass mechanical limitations of galvanometers	Script/software necessary to generate and track stimulation spots

2 Materials

2.1 Viral Injection Setup

All experimental animals for cell type-specific manipulations within the OB are detailed above. Adeno-associated virus 2 (AAV2) serotype DJ/8 [30] encoding flexed transgenes can be kept at 4°C for up to 6 months or indefinitely at −80°C. Avoid more than one freeze-thaw cycle. Injection setup includes Leica Angle Two computer-assisted stereotaxic system, a Zeiss Stemi 2000 Stereoscope, and a Nanoject II (Drummond Scientific Company) (Fig. 4). Although this protocol has been developed specifically for use with the Angle Two system, other stereotaxic setups can readily be used with slight modifications.

Injection pipettes are generated from 3.5″ Drummond glass micropipette pulled by glass pipette puller (Sutter Instruments). For OB injections, adjust pipette puller settings such that pipette tip is approximately 1 cm long. Once pulled, clip off ~1 mm of the micropipette tip. Back fill the pulled micropipette with sterile mineral oil, and place it into the Nanoject II auto-nanoliter injector. The micropipette must be emptied and filled with the desired volume of virus for injection using a Nanoject II (typically 200–600 nL depending on the injection site; the virus should be titered to at least 1×10^6 infectious particles/μL [31]). Other surgical supplies include nylon surgical sutures, sterile saline, surgical tools (dissection scissors, tweezers, hemostat, bulldog clamps), 30-gauge needles and syringes, pre- and postsurgical heating pads, vacuum system for biological fluids, antibiotic ointment (bupivacaine), and handheld drill (Vector Mega Torque with 0.5 mm dental drill bit). The surgery area and all tools to be used must first be disinfected using 70% ethanol and/or immersion in disinfectant solution (i.e., Germinator 500 or similar apparatus).

Fig. 4 Injection setup with anesthetized, head fixed mouse

2.2 Animal Preparations

Mice are weighed and sedated with an appropriate volume of a ketamine/xylazine mixture via intraperitoneal injection (100 and 10 mg/kg, respectively) using a 30-gauge needle. Once sedated, the animal's scalp is shaved via electronic razor and then disinfected with 70% isopropyl alcohol, followed by a Betasept wipe (4% chlorhexidine solution). The animal is then placed into the stereotaxic rig. To position the animal, fix one ear bar in place, followed by the second ear bar, ensuring that the ear bars are pushing into the skull firmly, without inflicting damage. It is imperative that the skull is stable and will not move when being drilled into later in the surgery. You can check this by tapping the skull with a tweezers. The skull should not move. Once the head has been firmly immobilized, place the animal's incisors into the nosepiece adaptor, which is connected to the ventilator, and tighten the noseband into place. Make sure the mouse is centered on the rig by gently pulling its tail to straighten the anterior-posterior axis. The animal is maintained under anesthesia during surgery using volatized isoflurate (VetEquip vaporizer, 1–3% isoflurane diluted with 100% oxygen depending on the physiological state of the animal as monitored by response to tail pinch). Ophthalmic ointment (Celluvisc) is applied to the eyes at the beginning of surgery to prevent dryness and postoperative eye pain.

3 Methods

3.1 Stereotaxic Viral Injection Into the Olfactory Bulb

Under microscopic guidance, make an incision in the scalp with surgical scissors, beginning at the posterior end of the head (between the ears) and cutting in the anterior direction toward the OB to expose the skull. Bulldog clamps can be used to hold the skin back, exposing the top of the skull completely. Following complete exposure, the top of the skull should be cleaned with sterile saline and a vacuum to remove excess fluids or tissue (it's helpful to keep the skull moist during surgery using sterile saline).

The two landmarks utilized for accurate stereotaxic targeting are (1) bregma, identified as the point where the coronal and sagittal sutures cross, and (2) lambda, which is the estimated point at which the sagittal and lambdoid sutures cross if they were a horizontal line (this also follows the interaural line, Fig. 5). Next, calibrate the Leica Angle Two™ computer-assisted stereotaxic system. For this, lower the prepared micropipette onto bregma, and select the "At Bregma" tab on the left-handed device software. After identifying bregma, raise the needle, and check the z coordinates by moving 1.5 mm to the left and right of bregma and lowering the micropipette to touch the skull carefully. The skull will need to be adjusted until the tilt is level,

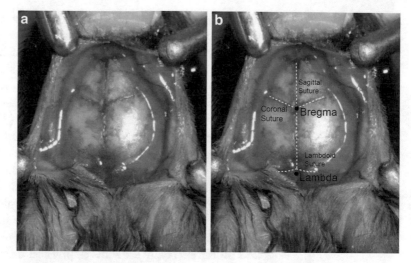

Fig. 5 The bregma point is identified as the intersection between the coronal and sagittal sutures. The lambda point is identified as the estimated point at which the lambdoid and sagittal sutures cross if they were a horizontal line, following the interaural line

with a margin of error of ±0.03 mm. After establishing the medial-lateral axis using bregma, lambda can be used to establish the anterior-posterior axis (note that bregma must always be identified before lambda). Once again, lower the micropipette to lambda and select the "At Lambda" tab on the left-handed device software. Adjust the skull until the dorsal/ventral coordinates of lambda and bregma are within ±0.10 mm of each other.

Once the skull has been correctly positioned with the assistance of the Leica Angle Two software, the injection site is located using coordinates from the Allen Brain Atlas (installed in the Leica Angle Two software). A small burr hole craniotomy (<1 mm) is then made over the injection site using a rotary tool with a 0.5 mm drill bit (using coordinates from the Allen Brain Atlas). The micropipette should be lowered into the injection site slowly (~3 mm/min) and, upon arrival at the target site, allow 5 min before beginning the injection to ensure settling of brain tissue. The injection paradigm should be established beforehand using the Nanoject II (e.g., if injecting 690 μL of virus, one should inject 69 μL of virus 10 times, with 30 s intervals between injections; to avoid excessive damage, injection rates should never exceed 200 μL/min). After the appropriate amount of virus has been injected, raise the needle 0.1 mm and wait 5 min before slowly raising the needle (~3 mm/min). Ensure that the tip of the micropipette does not become clogged and does not break during the surgery.

3.2 Postoperative Care

After completing the injection, the scalp must be sutured using surgical nylon monofilament in an interrupted pattern for skin closure. Topical antibiotic ointment (bupivacaine, 1–10 μL as needed to cover area) is applied as a local analgesic and to prevent infection, alongside a topical application of Neosporin immediately after surgery, then again within the next 24-h period. Mice are also treated with ketoprofen intraperitoneally (10 mg/kg) to decrease postoperative pain; this should be repeated 24 h postoperation as well. After all surgical procedures, the mice are moved to a cage that is partially positioned on a heat pad maintained at 37 °C to assist in thermal regulation until recovery. It typically takes 1–2 h for mice to recover from anesthesia. During this time, mice should be monitored every 15 min over the next 4 h or until animals recover consciousness. The postoperative mouse is then returned to a clean home cage and monitored daily for signs of physical distress. Any animals showing signs of prolonged discomfort, sickness, or unanticipated distress should be humanely euthanized. Typically, the time for sufficient expression of most AAVs, including channelrhodopsin-2, is 1–2 weeks postinjection.

Potential Problems and Trouble Shooting

- **Anesthesia**: Mice can display variability in how they respond to the ketamine/xylazine mixture. If mice are not fully anesthetized after the first dose of ketamine/xylazine, inject them with an additional 10% of the original dose intraperitoneally. During surgery, if the animal regains consciousness or becomes insufficiently anesthetized (according to a tail pinch), increase the dose of isoflurane (but monitor breathing closely).

- **Breathing abnormalities**: If the animal displays abnormal breathing (gasping, lack of breathing, etc.), this could indicate several different issues. One problem may be that their tongue is blocking airflow. To fix this, remove the animal from the rig and reposition their incisors on the nosepiece adaptor. Another possibility is that the nose bar is pressing their mouth shut, preventing them from breathing, and should be loosened. A third possibility (especially if they appear to be gasping) is that the dose of isoflurane is too high and needs to be lowered. If they still exhibit labored breathing after lowering the dose, shut off the isoflurane, providing them with 100% oxygen until breathing recovers.

- **Micropipette breaks or clogs**: A common problem is that the micropipette tip breaks when adjusting the mouse or touching the skull to identify bregma or lambda. If this occurs, you must replace the needle and reestablish bregma and lambda if the break is severe. If it's a small break, you can simply reestablish bregma and lambda. If the micropipette tip becomes clogged with blood, simply use scissors to snip the very tip off, and then reestablish Bregma and Lambda.

- **Viral expression**: If you experience problems with viral expression, it may be that the virus titer is too low. Alternatively, there may be issues with accurate targeting. To address this issue, try injecting a larger volume of virus, repackage the virus with a higher titer, or alter your coordinates to attempt to achieve better targeting.

3.3 Brain Extraction for Slice Electrophysiology

Use isoflurane to deeply anesthetize animal. To minimize tissue death, the brain must be extracted quickly and placed in oxygenated solution. Make an incision with scissors just below the rib cage to expose the diaphragm. Cut the diaphragm to allow access to the heart. Clip the right atrium while the left ventricle is perfused with ice-cold oxygenated (5% CO_2/95% O_2) artificial cerebrospinal fluid (ACSF, in mM: 125 NaCl, 25 glucose, 25 $NaHCO_3$, 2.5 KCl, 2 $CaCl_2$, 1.25 NaH_2PO_4, and 1 $MgCl_2$, pH 7.35, 305–315 mOsm) to allow blood drainage. Post perfusion, cut the skin at the base of the skull and move away the skin to expose the cranium by cutting the connective tissue around the ear canal and eye orbits. To remove the cranium, cut crosswise from the eye sockets through the nasal turbinate bones and lengthwise from each ear canal to the eye socket with bone scissors. With small dissection scissors, cut the occipital bone plate overlying the cerebellum from one ear canal to the other. With the same dissection scissors, split the skull with an anterior cut along the midline to the level of the olfactory bulbs. Remove the bone plates using forceps to expose and dissect out the brain.

3.4 Coronal Sectioning of Live Brain Tissue

Embed the brain in low melting point agarose and mount upright with cyanoacrylate before submerging in an ice-cold oxygenated sucrose-based solution (in mM, 87 NaCl, 2.5 KCl, 1.6 NaH_2PO_4, 25 $NaHCO_3$, 75 sucrose, 10 glucose, 1.3 ascorbic acid, 0.5 $CaCl_2$, 7 $MgCl_2$). Generate 300 μm coronal olfactory bulb slices on a vibratome (VT1200, Leica) with an oscillating blade advancing at 0.4 mm/s. Collect slices with a glass transfer pipette and moved to continuously oxygenated ACSF (37 °C) for 15 min for recovery. Acclimate slices to room temperature for at least 15 min prior to recordings.

3.5 Patching Onto Single Cells

Use borosilicate glass electrodes (Sutter Instruments) for whole-cell patch clamp recordings. Pull electrodes with tip resistances between 4 and 6 MΩ and fill with internal solution (for voltage clamp, in mM, 120 Cs-methanesolfunate, 6 CsCl, 2 $MgCl_2$, 0.05 $CaCl_2$, 20 HEPES, 0.02 EGTA, 10 phosphocreatine di(Na) salt, 4 Mg-ATP, 0.4 Na-GTP, 0.2% biocytin, pH 7.3 with CsOH, 290–300 mOsm; for current clamp, in mM, 120 potassium gluconate, 10 KCl, 10 phosphocreatine, 4 Mg-ATP, 0.4 Na-GTP, 10 HEPES, and 0.15% biocytin, pH 7.3 with KOH, 290–300 mOsm). Place coronal olfactory slice in a perfusion chamber (Warner Instruments) under a slice anchor (Warner Instruments) and continuously perfuse (1–2 mL/min) with oxygenated ACSF at RT. Obtain recordings on an Axon MultiClamp 700B amplifier and digitize at 10 kHz (Axon Digidata 1440A).

3.6 Light Stimulation and Cellular Recording

Visualize sections using a combination of infrared-Nomarski DIC optics and fluorescent microscopy (BX50WI, Olympus) via a 20×/1 NA water immersion objective (XLUMPLFLN-W, Olympus). Use a BLM-Series 473 nm blue laser system (Spectra Services) to generate laser light for stimulation. Laser intensity is attenuated with a neutral density filter (Thorlabs). Employ two orthogonally mounded acousto-optic deflectors (AODs, LS55-V Isomet) to produce arbitrary random-access 2D scan patterns. Use two deflector drivers (IntraAction Corp.) to generate acoustic waves. Use custom software in Matlab (MathWorks) to control output voltages through a D/A converter (National Instruments) connected to the deflector drivers. Implement a scan angle magnification telescope to increase scan range to fit the objective's field of view. Couple the AOD scanning beam into the upright microscope for output through the objective described above. For each session prior to circuit mapping, the laser output should be aligned to the 2D scanning grid. Alignment can be checked by creating a 5×5 grid for stimulation sites with 1 ms dwell duration. For circuit mapping, stimulation sites are programmed at grid points spaced by 30 μm in olfactory bulb slices with 10 ms duration and 100 ms intervals. Laser intensity should be optimized to meet the requirements for photoactivation in different scenarios such as genome-encoded vs. virus-encoded ChR2 expression.

4   Conclusions

The rodent OB affords an ideal brain model for studying synaptic connectivity. Not only is it genetically tractable, the OB is anatomically positioned such that engineered viruses can be easily implemented to further utilize available transgenic lines for cell type-specific manipulations. While this article has focused mainly on the use of AAV, other viral methods are also available with differing applications. For example, the canine adenovirus type 2 (CAV2) rapidly infects axon terminals. Efficient retrograde axonal transport allows genetic targeting of neuronal populations remote from the injection site [32, 33]. Alternatively, engineered rabies virus is a unique and powerful tool to genetically manipulate presynaptic partners of OB neurons [34].

With the combination of AOD-assisted scanning microscopy, focal areas of activation can be achieved. By stimulating different regions in a systematic way, connectivity patterns in relationship to physical distances between neurons can then be interrogated [7].

While AOD mapping strategies have afforded us the ability to two-dimensionally map connectivity patterns, the combination of AOD with two-photon light microscopy can further our understanding by generating a 3D connectivity map [35, 36].

There are currently many challenges in understanding brain tissue and its connectivity patterns, primarily due to a lack of knowledge regarding functionally distinct neuronal subtypes, patterns of connectivity, and neurotransmitter function. However, recent advances in cell type-specific manipulations for fluorescent reporter expression, optogenetic activation, and circuit mapping are now facilitating rapid advances in neuroscience research [37]. Furthermore, single-cell RNA sequencing has begun to show us the diversity of cell types within the CNS [38]. Researchers are now in the unique position to uncover novel neuronal populations, their synaptic partners, and their ultimate role in guiding brain function.

Acknowledgments

This research was supported by funding from the McNair Medical Institute and NINDS grant R01NS078294.

References

1. Katz LC, Shatz CJ (1996) Synaptic activity and the construction of cortical circuits. Science 274(5290):1133

2. Ko H, Hofer SB, Pichler B, Buchanan KA, Sjöström PJ, Mrsic-Flogel TD (2011) Functional specificity of local synaptic connections in neocortical networks. Nature 473(7345):87–91

3. Lüscher C, Jan LY, Stoffel M, Malenka RC, Nicoll RA (1997) G protein-coupled inwardly rectifying K+ channels (GIRKs) mediate post-synaptic but not presynaptic transmitter actions in hippocampal neurons. Neuron 19(3):687–695

4. Crabtree GW, Gogos JA (2014) Synaptic plasticity, neural circuits, and the emerging role of altered short-term information processing in schizophrenia. Front Synaptic Neurosci 6:28

5. Nagayama S, Homma R, Imamura F (2014) Neuronal organization of olfactory bulb circuits. Front Neural Circuits 8:98

6. Huang L, Garcia I, Jen HI, Arenkiel BR (2013) Reciprocal connectivity between mitral cells and external plexiform layer interneurons in the mouse olfactory bulb. Front Neural Circuits 7:32

7. Huang L, Ung K, Garcia I, Quast KB, Cordiner K, Saggau P, Arenkiel BR (2016) Task learning promotes plasticity of interneuron connectivity maps in the olfactory bulb. J Neurosci 36(34):8856–8871

8. Quast KB, Ung K, Froudarakis E, Huang L, Herman I, Addison AP et al (2017) Developmental broadening of inhibitory sensory maps. Nat Neurosci 20(2):189–199

9. Garcia I, Quast KB, Huang L, Herman AM, Selever J, Deussing JM et al (2014) Local CRH signaling promotes synaptogenesis and circuit integration of adult-born neurons. Dev Cell 30(6):645–659

10. Arenkiel BR et al (2011) Activity-induced remodeling of olfactory bulb microcircuits revealed by monosynaptic tracing. PLoS One 6(12):e29423

11. Imai T (2014) Construction of functional neuronal circuitry in the olfactory bulb. In: Seminars in cell & developmental biology, vol 35. Academic Press, London, pp 180–188

12. Arenkiel BR, Peca J, Davison IG, Feliciano C, Deisseroth K, Augustine GJ et al (2007) In vivo light-induced activation of neural circuitry in transgenic mice expressing channelrhodopsin-2. Neuron 54(2):205–218

13. Mitsui S, Igarashi KM, Mori K, Yoshihara Y (2011) Genetic visualization of the secondary

142   Gary Liu et al.

olfactory pathway in Tbx21 transgenic mice. Neural Syst Circuits 1(1):5

14. Faedo A, Ficara F, Ghiani M, Aiuti A, Rubenstein JL, Bulfone A (2002) Developmental expression of the T-box transcription factor T-bet/Tbx21 during mouse embryogenesis. Mech Dev 116(1):157–160

15. Kosaka T, Kosaka K (2008) Tyrosine hydroxylase-positive GABAergic juxtaglomerular neurons are the main source of the interglomerular connections in the mouse main olfactory bulb. Neurosci Res 60(3):349–354

16. Parrish-Aungst S, Shipley MT, Erdelyi F, Szabo G, Puche AC (2007) Quantitative analysis of neuronal diversity in the mouse olfactory bulb. J Comp Neurol 501(6):825–836

17. Miyamichi K, Shlomai-Fuchs Y, Shu M, Weissbourd BC, Luo L, Mizrahi A (2013) Dissecting local circuits: parvalbumin interneurons underlie broad feedback control of olfactory bulb output. Neuron 80(5):1232–1245

18. Nagai Y, Sano H, Yokoi M (2005) Transgenic expression of cre recombinase in mitral/tufted cells of the olfactorybulb, genesis,43.1

19. Wang H, Peca J, Matsuzaki M, Matsuzaki K, Noguchi J, Qiu L et al (2007) High-speed mapping of synaptic connectivity using photostimulation in Channelrhodopsin-2 transgenic mice. Proc Natl Acad Sci 104(19): 8143–8148

20. Alvarez-Buylla A, García-Verdugo JM (2002) Neurogenesis in adult subventricular zone. J Neurosci 22(3):629–634

21. Herman AM, Ortiz-Guzman J, Kochukov M, Herman I, Quast KB, Patel JM et al (2016) A cholinergic basal forebrain feeding circuit modulates appetite suppression. Nature 538(7624):253–256

22. Atasoy D, Aponte Y, Su HH, Sternson SM (2008) A FLEX switch targets Channelrhodopsin-2 to multiple cell types for imaging and long-range circuit mapping. J Neurosci 28(28):7025–7030

23. Perez-Costas E, Melendez-Ferro M, Roberts RC (2007) Microscopy techniques and the study of synapses. Modern Res Educat Topics Microsc 1:164–170

24. Guzowski JF, Timlin JA, Roysam B, McNaughton BL, Worley PF, Barnes CA (2005) Mapping behaviorally relevant neural circuits with immediate-early gene expression. Curr Opin Neurobiol 15(5):599–606

25. Sheng M, Greenberg ME (1990) The regulation and function of c-fos and other immediate early genes in the nervous system. Neuron 4(4):477–485

26. Nagel G, Szellas T, Huhn W, Kateriya S, Adeishvili N, Berthold P et al (2003) Channelrhodopsin-2, a directly light-gated cation-selective membrane channel. Proc Natl Acad Sci 100(24):13940–13945

27. Baker CA, Elyada YM, Parra A, Bolton MM (2016) Cellular resolution circuit mapping with temporal-focused excitation of soma-targeted channelrhodopsin. elife 5:e14193

28. Kravitz AV, Kreitzer AC (2011) Optogenetic manipulation of neural circuitry in vivo. Curr Opin Neurobiol 21(3):433–439

29. Wang K, Liu Y, Li Y, Guo Y, Song P, Zhang X et al (2011) Precise spatiotemporal control of optogenetic activation using an acousto-optic device. PLoS One 6(12):e28468

30. Holehonnur R, Luong JA, Chaturvedi D, Ho A, Lella SK, Hosek MP, Ploski JE (2014) Adeno-associated viral serotypes produce differing titers and differentially transduce neurons within the rat basal and lateral amygdala. BMC Neurosci 15(1):28

31. Osten P, Grinevich V, Cetin A (2007) Viral vectors: a wide range of choices and high levels of service. In: Conditional mutagenesis: an approach to disease models. Springer, Berlin, Heidelberg, pp 177–202

32. Soudais C, Laplace-Builhe C, Kissa K, Kremer EJ (2001) Preferential transduction of neurons by canine adenovirus vectors and their efficient retrograde transport in vivo. FASEB J 15(12):2283–2285

33. Bru T, Salinas S, Kremer EJ (2010) An update on canine adenovirus type 2 and its vectors. Virus 2(9):2134–2153

34. Arenkiel BR, Ehlers MD (2009) Molecular genetics and imaging technologies for circuit-based neuroanatomy. Nature 461(7266): 900–907

35. Denk W, Delaney KR, Gelperin A, Kleinfeld D, Strowbridge BW, Tank DW, Yuste R (1994) Anatomical and functional imaging of neurons using 2-photon laser scanning microscopy. J Neurosci Methods 54(2):151–162

36. Matsuzaki M, Ellis-Davies GC, Kasai H (2008) Three-dimensional mapping of unitary synaptic connections by two-photon macro photolysis of caged glutamate. J Neurophysiol 99(3):1535–1544

37. Cardin JA, Carlén M, Meletis K, Knoblich U, Zhang F, Deisseroth K et al (2010) Targeted optogenetic stimulation and recording of neurons in vivo using cell-type-specific expression of Channelrhodopsin-2. Nat Protoc 5(2):247–254

38. Saraiva LR, Ibarra-Soria X, Khan M, Omura M, Scialdone A, Mombaerts P et al (2015) Hierarchical deconstruction of mouse olfactory sensory neurons: from whole mucosa to single-cell RNA-seq. Sci Rep 5:18178

Chapter 8

In Vivo Recordings at the *Caenorhabditis elegans* Neuromuscular Junction

Shangbang Gao and Zhitao Hu

Abstract

The communication between neurons occurs via synapses, specialized connections with other cells or tissues (e.g., muscles). The function of the synapse is mainly determined by synaptic vesicle exocytosis from the presynaptic nerve terminal, which contains hundreds of vesicles that are filled with neurotransmitters. Electrophysiology provides a direct way to measure synaptic vesicle release and investigate synaptic properties. *C. elegans* neuromuscular junction (NMJ) has been established as a model system to study synaptic function because of the application of electrophysiological recording. Here we describe a recently developed patch-clamp recording technique in neuromuscular junction of *C. elegans* including details of the construction of typical electrophysiology rig, the proper dissection technique, and the recording of different types of synaptic currents.

Key words *C. elegans*, Neuromuscular junction, Synapse, Synaptic vesicle, Patch clamp, Electrophysiology

1 Introduction

1.1 Using *C. elegans* as a Model Organism to Study Synaptic Transmission

The patch clamp technique has been widely used to study cellular and membrane activity such as the opening and closing of ion channels, Ca^{2+} release from internal stores, and vesicle fusion in synapses. It is particularly useful in the nervous system to record synaptic transmission, a major way in which electrical signals travel from one neuron to another. Although recordings from large numbers of mutants or knockout mice have provided significant evidence regarding the synaptic vesicle fusion process [1–4], the underlying mechanism remains largely unknown. In the last 20 years, *C. elegans* has gained considerable popularity in the study of synaptic transmission because of the successful application of electrophysiological techniques in the neuromuscular junction (NMJ). Compared with mammals, worm exhibits great advantages when studying neurotransmission. For example, *C. elegans* mutants lack some genes which are lethal in mammals remain viable, allowing us to study the functional importance of these genes in vivo, as well as the effects of gene regulation on behav-

Roy V. Sillitoe (ed.), *Extracellular Recording Approaches*, Neuromethods, vol. 134,
https://doi.org/10.1007/978-1-4939-7549-5_8, © Springer Science+Business Media, LLC 2018

ior. Moreover, it is easier to make genetic double or triple mutants in order to investigate the role of cellular signaling pathways in regulating synaptic transmission. More importantly, many synaptic transmission defects in worm mutants have also been observed in mammals, indicating conserved mechanisms by which synaptic proteins regulate vesicle fusion [5–9]. Recently, individual neuron patching has been successfully established in *C. elegans* [10–13], making it possible to investigate the electrical properties of particular neurons and their relationship with behavior.

C. elegans body muscles form synapses with cholinergic and GABAergic motor neurons, thereby receiving both excitatory and inhibitory synaptic inputs. The presynaptic terminals release different neurotransmitters (e.g., acetylcholine and GABA) which then bind their receptors (e.g., AChR and $GABA_AR$) on the muscle membrane, in order to control muscle contraction and relaxation [14–18]. Electrophysiological recordings can directly measure the presynaptic vesicle fusion as well as the function of the postsynaptic receptors, providing comprehensive evaluation of synaptic function. As a fast and powerful tool, the aldicarb assay has been extensively used in *C. elegans* to screen mutants with potential synaptic transmission deficits in either cholinergic or GABAergic neurons [19–21]. Many genes have been identified as aldicarb resistant or hypersensitive in previous studies. However their role in synaptic transmission still needs to be confirmed by electrophysiological recordings. In this chapter, we outline the latest electrophysiology techniques at the *C. elegans* NMJ including our own modifications in terms of cell preparation and additional recording forms based on previous protocols [17, 22, 23].

2   Materials

Name	Company	Catalog #	
2.1 Sylgard	Dow Corning	3097366-1004	
2.2 Collagenase	Sigma	C-5138	1 mg/mL stock, store in −80 °C freezer
2.3 Histoacryl blue	McFarlane	15054BU	Store in −20 °C freezer. Good for the first step glue application
2.4 Topical tissue adhesive	Gluture	503763	Store in −20 °C freezer. Good for the second step glue application
2.5 Micropipette	World Precision Instruments	1B150F-4	Length 100 mm, 1.5 mm OD/0.84 mm ID Used for sample preparation and patch clamping
2.6 Coverslip	eBay		22 mm, circular, coated with Sylgard in the middle (about 1 cm area)

Name	Company	Catalog #	
2.7 High-vacuum silicone grease	Dow Corning	Z273554-1	Used to stick the coverslip to the inner well of the chamber
2.8 Chamber	Made by workshop		See Fig. 3
2.9 MicroFil	World Precision Instruments	MF28G67-5	Can also be handmade using the 200 μL tip

3 Equipment

3.1 Air Table and Faraday Cage

An air table and a faraday cage are both required for a typical electrophysiological rig to achieve stable recording and reduce noise. Several models of air table are available, and we use the TM63541 (TMC Micro-g) which provides excellent vibration control. The Faraday cage can be purchased commercially, but we recommend building it with a metal frame and caging. It can be designed to be larger than the air table, thereby generating more internal space, with the whole cage placed separately on the ground.

3.2 Microscope

We use a BX51WI (Olympus) upright microscope with 4× and 60× water immersion objectives. Other recommended options include Eclipse FN1 microscope (Nikon).

3.3 Moving Stage

We use the stage from Mike's Machine Company designed for the electrophysiology system. It has a large platform to house a number of manipulators. The stage is mounted separately from the microscope, and the small inner stage can move independently from the platform, allowing the user to quickly scan the entire field of the chamber.

3.4 Amplifier

HEKA and Axon Instruments are the two major companies that produce amplifiers, and both are widely used. Different patchers might have their own preference, but we use the HEKA EPC-10 double USB with PatchMaster as the data acquisition software. Action potentials of *C. elegans* body wall muscles shown in this chapter are recorded in the whole-cell configuration using a MultiClamp 700A amplifier and a Digidata 1440A, using Clampex 10 software and processed with Clampfit 10 (Axon Instruments, Molecular Devices). Data are digitized at 10–20 kHz and filtered at 2.6 kHz.

3.5 Manipulator

We use the MPC-325 (Sutter Instrument) which includes both left and right manipulators. The latter usually controls the patching pipette, whereas the former is used for the stimulus and puff pipettes.

3.6 LED Light Source

The LED source is introduced to provide the light stimulus to activate channelrhodopsin-2 (ChR2). We use the M470L3-C1 (Blue (470 nm) Collimated LED for Olympus BX & IX) from Thorlabs. A LED lamp driven by the Axon amplifier is also recommended (KSL-70, RAPP OptoElectronic, 470 nm, 8 mW/mm^2).

3.7 Stimulator

We use the A365 stimulus isolator which can be driven by the EPC-10 amplifier and the A362 battery charger. Both can be purchased from World Precision Instruments.

3.8 Pressure Puff Device

Drugs (e.g., acetylcholine, levamisole, glutamate, GABA/muscimol, sucrose) are applied by pressure puffs via an amplifier-driven Picospritzer III (Parker Instruments). The pressure is always the same in our system, but the puff duration varies with different drugs.

3.9 Dissection Scope

A high-quality stereo microscope is required for the dissection. Both Leica and Olympus have excellent models, and we use the SZ51microscope (Olympus) with an extra 1.5× auxiliary lens to reach a total of 60× magnification. Leica stereo microscopes like M50 and M80 are also recommended.

3.10 Micropipette Puller and Microforge

Several models of micropipette puller are available, and we use the PC-10 dual-stage vertical puller from Narishige or the P-1000 Flaming/Brown type model from Sutter Instrument Company. The PC-10 uses very solid kanthal wire as the heater which is not easily damaged when inserting the pipette. Compared with other models, the PC-10 is both small and affordable. The P-1000 has an extensive library of programs found in the popular Sutter Cookbook and is available to the user via the touchscreen display. The new pulled glass pipette has a rough and uneven tip which will damage the cell membrane. To avoid this, a microforge is required to polish the end of the pipette and create a smooth surface. We use the MF-830 model from Narishige.

4 Dissection

The major dissection steps are similar to those previously described by Janet Richmond [23]. The worms are maintained on an OP50 plate and transferred to a plate fed with HB101 bacteria in the L4 stage for recordings the next day. The main purpose of introducing the HB101 bacteria is to allow the sick animals to grow better and make them easier to dissect. After washing in the bath solution, the worm is transferred onto a circular glass coverslip (22 mm in diameter) coated with Sylgard 184 in the middle. It is maintained in a very small drop of bath solution (20 μL) on the Sylgard. The coverslip is then put on a cell culture bottle containing iced water, and the worm is quickly immobilized by the low temperature.

To glue down the worm, we first remove the extra solution with a mouth-controlled pipette, leaving very small volume under the worm. The same pipette is then used to adjust the worm body to an arc shape (the dorsal side is up and the ventral side is down). A tip-broken pipette with a small amount of glue at the end is used to touch the dorsal side of the worm gently and quickly. The glue then automatically spreads along the worm body. Usually we need to touch the worm several times at different spots on the dorsal side to solidify the connection between the worm and the glue. The reason we leave a small amount of solution under the worm is to prevent the glue crossing to the ventral side. A 15–20 μL bath solution is then quickly added back onto the worm.

We use another pipette with a very sharp end (called a cutting pipette) to cut the cuticle from a point a little behind the vulva to the bulb along the dorsal side. This step is the most important because the quality of the dissection directly determines the ability to obtain good data. Damage to the cell usually comes from shaking of the cutting pipette. We hold the pipette (about 1 cm from the sharp tip) between the thumb and index finger and place the hand with the pipette onto the iced bottle tightly before the cuticle incision. This allows us to hold the pipette in position without shaking, thereby avoiding unnecessary damage in the dissection process. Usually, this step requires a few weeks or months of practice, although it takes even longer to master the dissection of smaller worms.

After dissection, an extraction pipette (about 5–10 μm of the open size) controlled by the mouth via thin tubing is used to remove all the internal organs, including the intestine, gonad, and eggs. The cutting edge of the cuticle in the ventral side is glued down with a new pipette to expose the ventral nerve cord and the muscles. Positive pressure is applied in this pipette to prevent the solution entering the tip.

The coverslip is then taken off the ice bottle and placed into the inner well of the chamber with pre-applied silicone grease. Using the cap of a 15 mL centrifuge tube to apply gentle pressure to the coverslip and ensure that the seal between the coverslip and the chamber is good. A 20 μL collagenase (1 mg/mL dissolved in extracellular solution) is then added to the small drop of solution on the worm to digest the cells for 30 s (the final concentration of the collagenase applied to the cells is therefore about 0.5 mg/mL). We then transfer the chamber to the inner stage of the moving stage and adjust this to the correct position from where both the stimulus and patch pipettes can reach the cells easily. Washing several times with the bath solution via the perfusion system removes the collagenase completely, at which point the preparation is ready for recording.

5   Recordings

The recording condition varies with the nature of the recordings. Below we describe different types of recordings at the NMJ.

5.1   mEPSC and mIPSC (Also Called Tonic Release, Spontaneous Release, or Minis)

To record miniature excitatory and inhibitory postsynaptic currents (mEPSCs and mIPSCs, respectively), we use different recipes for the extracellular and internal solutions. Here we provide our recipes.

5.1.1   Hu Lab

The extracellular solution contains 127 mM NaCl, 5 mM KCl, 26 mM NaHCO$_3$, 1.25 mM NaH$_2$PO$_4$, 20 mM glucose, 1 mM CaCl$_2$, and 4 mM MgCl$_2$, bubbled with 5% CO$_2$ and 95% O$_2$ at 20 °C. Osmolarity is ~330 mOsm. The freshly made extracellular solution has a pH value between 4 and 5 which reaches 7.2 after bubbling with CO$_2$/O$_2$ for 20 min. The internal solution consists of 105 mM CH$_3$O$_3$SCs, 10 mM CsCl, 15 mM CsF, 4 mM MgCl$_2$, 5 mM EGTA, 0.25 mM CaCl$_2$, 10 mM HEPES, and 4 mM Na$_2$ATP, adjusted to pH 7.2 using CsOH and osmolarity ~315 mOsm. The Cs ion is introduced into the internal solution to make the gigaohm seal more solid. Under these conditions, we observe only excitatory acetylcholine EPSCs when holding the membrane potential at −60 mV (the reversal potential of GABA receptor channels) (Fig. 1a) or IPSCs when holding the membrane potential at 0 mV (the reversal potential of acetylcholine receptor channels) (Fig. 1b). The patch pipette (4–5 MΩ resistance) is pressed gently against the muscle cell (at the middle point), and suction is applied to form a high-resistance seal (gigaohm seal). Changing the holding potential to −60 mV and applying brief mouth suction rupture the patch, thereby forming the whole-cell patch. The mEPSCs can then be observed. A switch to 0 mV, mIPSCs should be recorded.

5.1.2   Gao Lab

We use another protocol for isolating the pure mEPSCs and mIPSCs through changing the recipes of the extracellular and internal solutions in combination with specific blocker [24]. Specifically, the pipette solution contains 115 mM K-gluconate, 25 mM KCl, 0.1 mM CaCl$_2$, 5 mM MgCl$_2$, 1 mM BAPTA, 10 mM HEPES, 5 mM Na$_2$ATP, 0.5 mM Na$_2$GTP, 0.5 mM cAMP, 0.5 mM cGMP, pH 7.2 with KOH, and ~320 mOsm. The bath solution consists of 150 mM NaCl, 5 mM KCl, 5 mM CaCl$_2$, 1 mM MgCl$_2$, 10 mM glucose, 5 mM sucrose, 15 mM HEPES, pH 7.3 with NaOH, and ~330 mOsm. Under these conditions, the reversal potentials are ~−30 mV and ~+20 mV for GABA and acetylcholine receptors, respectively. The pure mEPSC is recorded when holding the membrane potential at −30 mV. To isolate pure mIPSC events,

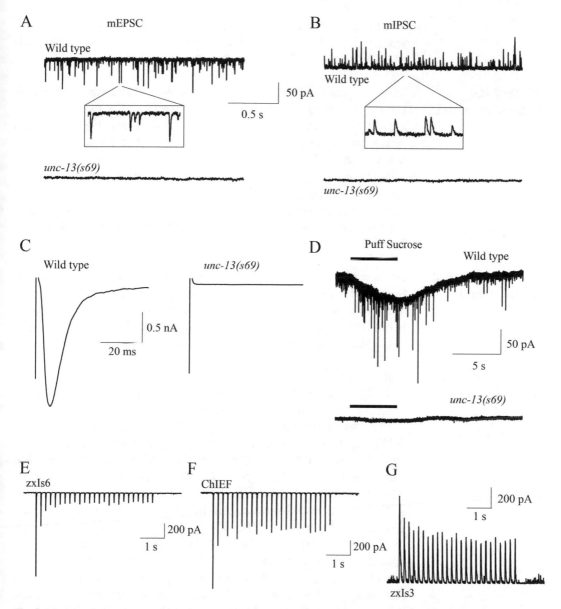

Fig. 1 Representative traces of the synaptic currents reflecting presynaptic vesicle exocytosis. (**a**, **b**) Miniature excitatory postsynaptic current (mEPSC, recorded at −60 mV) and inhibitory postsynaptic current (mIPSC, recorded at 0 mV). Insets show the zoomed view of the minis. Both the mEPSC and the mIPSC are eliminated in *unc-13 (s69)* mutants. (**c**) Electrical stimulus-evoked EPSCs in the wild type and *unc-13 (s69)* mutants. (**d**) Hypertonic sucrose-evoked EPSC in the wild type and *unc-13 (s69)* mutants. (**e–g**) Synaptic currents triggered by light stimulus trains (5 Hz, 470 nm) in strains expressing ChR2/ChIEF in cholinergic motor neurons or ChR2 in GABAergic motor neurons

recordings are performed with a holding potential of −10 mV, with 0.5 mM D-tubocurarine (d-TBC) included in the bath solution to block all acetylcholine receptors.

5.2 Electrical Stimulus-Evoked EPSC

The recording condition for electrical stimulus-evoked EPSCs is the same as those used in the mini recording. We place a second pipette (4–5 MΩ resistance, filled with bath solution) near the ventral nerve cord, then press it gently onto the cord, and apply a 0.4 ms, 85 µA square pulse to depolarize the neurons. It is important to place the stimulus pipette in the correct position (usually one muscle distance from the recording pipette) as the evoked current will be small if it is too far away or the gigaohm seal will be damaged if it is too close. A typical evoked EPSC is around 1.5–2.5 nA in amplitude (Fig. 1c).

5.3 Sucrose-Evoked Synaptic Current

Hypertonic sucrose is applied to estimate the size of the readily releasable vesicle pool (Fig. 1d). A puff pipette filled with 800 mM sucrose (dissolved in bath solution) is placed on top of the ventral nerve cord close to the end of the patched muscle. The same recipes as outlined above are used for the solutions.

5.4 Light-Activated Synaptic Current

ChR2 and its variants have been introduced into the NMJ to investigate synaptic activity and plasticity. By giving a short light stimulus (3–10 ms pulse, maximal light intensity of the LED), ChR2 can cause depolarization of either the cholinergic or the GABAergic motor neurons, thereby triggering the vesicle fusion (Fig. 1e–g). Several integrated lines are available, and we are using the following: zxIs6 [P*unc-17*::ChR2(H134R)::YFP + *lin-15*(+)], EG5793 [P*unc-17*::ChIEF::mCherry::*unc-54*UTR; *unc-119*(+)], and zxIs3 [p*unc-47*::chr2(h134r)::yfp; *lin-15*(+)] and hpIs166 [P*glr-1*::ChR2(H134R)::YFP; *lin-15*(+)]. All these strains can be ordered from the *Caenorhabditis* Genetics Center (http://cbs.umn.edu/cgc/home). The L4 stage worms are transferred to the plates with all trans-retina and kept in a dark box prior to recording the following day. To get a comparable evoked current with the electrical stimulus, we suggest the use of a high Ca^{2+} bath solution. For example, we use 2 mM Ca^{2+} for the zxIs6 and EG5793 lines and 5 mM Ca^{2+} for zxIs3 strain.

5.5 Drug-Activated Current

To measure the function of receptors on the muscle membrane, drugs (e.g., acetylcholine, GABA/muscimol, levamisole, dissolved in bath solution) are applied by puffing them directly onto the body of the muscle. The pipette filled with drug is placed at the end of the patched muscle cell. Puffing condition varies with different drugs (12 Psi, 50 ms for ACh/GABA/levamisole, and 1 s for muscimol) (Fig. 2a–c).

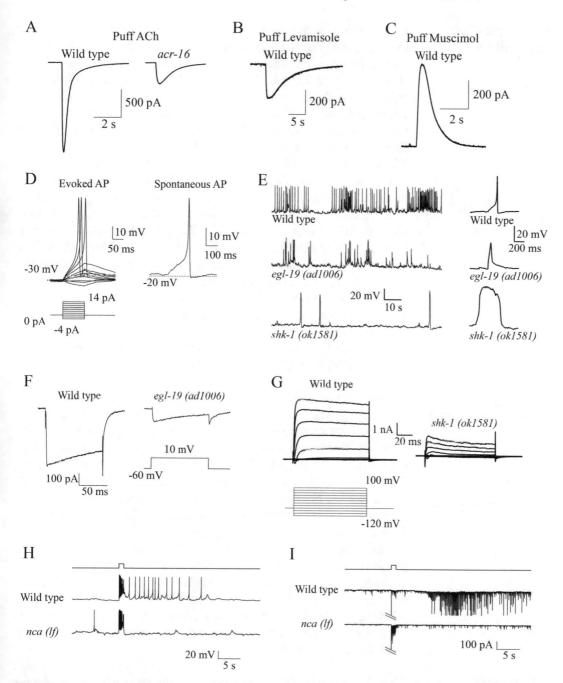

Fig. 2 Representative recordings of postsynaptic currents generated by muscles. (**a**) ACh-activated currents in the wild type and mutants lacking the nicotine receptor (*acr-16*). (**b**) Levamisole-activated current in the wild type. (**c**) Current induced by muscimol, a GABA agonist. (**d**) Action potentials (APs) fired by body wall muscles. Both evoked APs (triggered by current steps from −4 to 14 pA in 2pA increments, *left*) and spontaneous APs (*right*) are shown. (**e**) Trains of spontaneous APs in wild type and mutants lacking L-type Ca²⁺ channels (*egl-19*) or Kv1 K⁺ channels (*shk-1*). Kinetics changes of single APs are shown on the right. (**f**) L-type Ca²⁺ currents were recorded in the wild type and *egl-19* mutants, evoked by current steps from −60 to 10 mV. (**g**) Kv1 K⁺ currents were recorded in the wild type and *shk-1* mutants, evoked by current steps from −120 to 100 mV, at 20 mV increments. (**h, i**) Blue light (470 nm, 1 s)-induced persistent action potential firing (holding at 0pA) and persistent rhythmic PSCs bursting (holding at −60 mV), which are both terminated in the *nca(lf)* mutants. The transgenic strain is *hpIs166*, which expresses ChR2 in all premotor interneurons and some other neurons

5.6 Muscle-Fired Action Potential

C. elegans muscles fire classical all-or-none action potentials [14, 15]. We record spontaneous action potential firing (generated by tonic excitatory motor neurons and/or endogenous muscle Ca^{2+} oscillations), and electrical/optogenetic/pharmacological stimulations evoked action potential firing (Fig. 2d, e). To record body wall muscle action potentials, the pipette solution contained 115 mM K-gluconate, 25 mM KCl, 0.1 mM $CaCl_2$, 5 mM $MgCl_2$, 1 mM BAPTA, 10 mM Hepes, 5 mM Na_2ATP, 0.5 mM Na_2GTP, 0.5 mM cAMP, 0.5 mM cGMP, and pH7.2 with KOH, ~320 mOsm. cAMP and cGMP are included to maintain the activity and the longevity of the preparation. The extracellular solution consists of 150 mM NaCl, 5 mM KCl, 5 mM $CaCl_2$, 1 mM $MgCl_2$, 10 mM glucose, 5 mM sucrose, 15 mM HEPES, and pH7.3 with NaOH, ~330 mOsm. The recording conditions and operation are the same as those of the minis except switching to the current-clamp configuration. We hold the membrane potential at 0 pA to record spontaneous action potentials and optogenetically/pharmacologically evoked action potentials. Otherwise, we use depolarizing step currents (5–50 pA, dependent on the requirement of different experiments) to evoke muscle action potential firing.

5.7 L-Type Ca^{2+} Current and K^+ Current

To evaluate the function of voltage-dependent L-type Ca^{2+} channels (encoded by *egl-19*) on the body wall muscles, the Ca^{2+} current can be measured by depolarizing the membrane potential from −60 to +10 mV (Fig. 2f). The extracellular solution contains 140 mM TEA-Cl, 5 mM $CaCl_2$, 1 mM $MgCl_2$, 3 mM 4-aminopyridine, 10 mM glucose, 5 mM sucrose, and 15 mM HEPES (pH 7.4 adjusted by CsOH, ~330 mOsm). The pipette solution consists of 140 mM CsCl, 10 mM TEA-Cl, 5 mM $MgCl_2$, 5 mM EGTA, and 10 mM HEPES (pH 7.2 adjusted by CsOH, ~320 mOsm). The channel activity can be blocked by application of Nemadipine-A in the bath solution. To evaluate the function of voltage-gated K^+ channels on the body wall muscles, the K^+ current is evoked by voltage steps from −120 to +100 mV, at 20 mV increment (Fig. 2g). The solutions used for recording K^+ currents are the same as those used to record action potentials.

5.8 Synaptic Plasticity at the NMJ

The *C. elegans* NMJ is also a good model to study the molecular mechanism of synaptic plasticity. A variety of synaptic plasticity patterns have been found in previous studies, including aldicarb-induced spontaneous synaptic transmission enhancement [21], optogenetic stimulation-generated persistent rhythmic PSC bursts (holding at −60 mV), and persistent action potential firing (holding at 0 pA) (Fig. 2h, i) [25], as well as short-term potentiation (STP) (data not shown) and depression (STD).

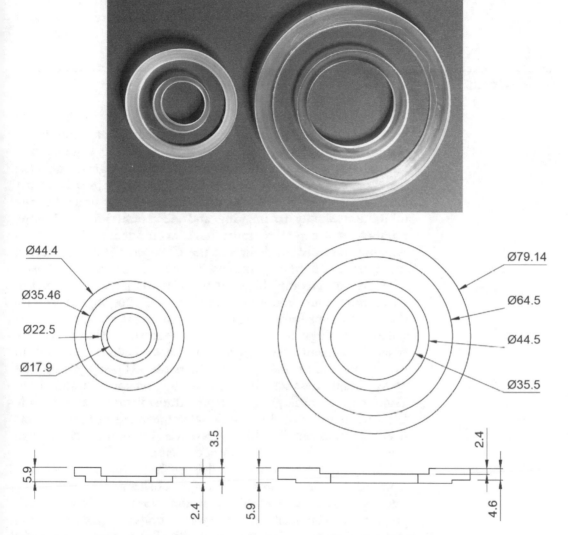

Fig. 3 The recording chamber and the chamber adaptor used for the electrophysiology rig. The adaptor is designed to fit the inner stage in the moving stage

6   Data Analysis

To analyze the data, we use both Igor (WaveMetrics) and Clampfit (Molecular Devices) software. The minis are analyzed by an Igor macro called "amperometric spike analysis" originally written by Dr. Eugene Mosharow from Columbia University. The software "Mini Analysis" is also recommended. For all other recordings (e.g., stimulus-evoked EPSC, drug-activated current, sucrose-evoked response, and light-activated EPSC), we use the Igor built-in macros to quantify all the parameters, including amplitude, charge transfer, rise time, inactivation decay, and synaptic recovery. Action potential frequency and amplitude are analyzed by Clampfit (Molecular Devices).

7   Discussion

Over the past 50 years, electrophysiological studies have provided a vast wealth of information concerning the functional properties of synapses. What has been lacking is a genetically accessible model to combine electrophysiological, genetic, imaging, and behavioral approaches. The *C. elegans* NMJ has been widely used as a platform to study synaptic function, due to its accessibility to imaging and electrophysiology. A large number of synaptic mutants have been identified based on synaptic transmission defects at the *C. elegans* NMJ, displaying its importance for the study of basic and fundamental mechanisms of neurotransmission. Most of the synaptic genes identified in *C. elegans* are also conserved in mammal [7, 26–28]. Therefore, the mechanisms learned from *C. elegans* NMJ are applicable to synaptic biology in other model organisms. More recently, synaptic transmission defects have also been implicated in psychiatric and neurological disorders [29–31], suggesting that many brain diseases are caused by synaptic dysfunction. Therefore, recordings of synaptic transmission can not only allow us to understand the underlying mechanisms of the vesicle fusion process but should also provide significant insights into the pathogenesis of neurological diseases.

Compared with other models such as the fly NMJ and mammalian cell cultures, the major challenge in performing electrophysiological recordings in the worm involves the cell preparation. The small size of the worm limits the glue application and the dissection. Our experience highlights ways which make these processes relatively easy to control: (1) placing the coverslip on an iced bottle immobilizes the worm quickly; (2) removing the solution around the worm before the first application of glue prevents the solution entering the pipette, and breaking the tip of the glue pipette allows the glue to come out naturally and quickly without mouth pressure; and (3) doing the dissection in a very small drop of solution allows the cutting pipette to be held close to the tip without touching the solution, thereby reducing the shaking caused by your hand. It usually takes a couple of months to learn and fully master this technique. The patching process is relatively easy and quite similar to that of single cell recordings in mammals. Recording from the same muscle for each worm is recommended, although there is no direct evidence to show that the synaptic transmission differs between distinct body wall muscles.

Acknowledgments

We thank Rowan Tweedale for reading the manuscript and providing feedback. This article was funded by Australian Research Council Discovery Project grant (DP 160100849) and National Health and Medical Research Council Project grant (APP1122351) to ZH and a grant from The National Nature Science Foundation of China (31671052) and the Junior Thousand Talents Program of China to SG.

References

1. Deng L, Kaeser PS, Xu W, Sudhof TC (2011) RIM proteins activate vesicle priming by reversing autoinhibitory homodimerization of Munc13. Neuron 69:317–331

2. Han Y, Kaeser PS, Sudhof TC, Schneggenburger R (2011) RIM determincs Ca²⁺ channel density and vesicle docking at the presynaptic active zone. Neuron 69:304–316

3. Jahn R, Sudhof TC (1999) Membrane fusion and exocytosis. Annu Rev Biochem 68:863–911

4. Koh TW, Bellen HJ (2003) Synaptotagmin I, a Ca²⁺ sensor for neurotransmitter release. Trends Neurosci 26:413–422

5. Gracheva EO, Burdina AO, Touroutine D, Berthelot-Grosjean M, Parekh H, Richmond JE (2007) Tomosyn negatively regulates CAPS-dependent peptide release at *Caenorhabditis elegans* synapses. J Neurosci 27:10176–10184

6. Madison JM, Nurrish S, Kaplan JM (2005) UNC-13 interaction with syntaxin is required for synaptic transmission. Curr Biol 15:2236–2242

7. Martin JA, Hu Z, Fenz KM, Fernandez J, Dittman JS (2011) Complexin has opposite effects on two modes of synaptic vesicle fusion. Curr Biol 21:97–105

8. Richmond JE, Davis WS, Jorgensen EM (1999) UNC-13 is required for synaptic vesicle fusion in *C. elegans*. Nat Neurosci 2:959–964

9. Hu Z, Tong XJ, Kaplan JM (2013) UNC-13L, UNC-13S, and Tomosyn form a protein code for fast and slow neurotransmitter release in *Caenorhabditis elegans*. Elife 2:e00967

10. Kang L, Gao J, Schafer WR, Xie Z, XZ X (2010) *C. elegans* TRP family protein TRP-4 is a pore-forming subunit of a native mechanotransduction channel. Neuron 67:381–391

11. O'Hagan R, Chalfie M, Goodman MB (2005) The MEC-4 DEG/ENaC channel of *Caenorhabditis elegans* touch receptor neurons transduces mechanical signals. Nat Neurosci 8:43–50

12. Ward A, Liu J, Feng Z, XZ X (2008) Light-sensitive neurons and channels mediate phototaxis in *C. elegans*. Nat Neurosci 11:916–922

13. Xie L, Gao S, Alcaire SM, Aoyagi K, Wang Y, Griffin JK, Stagljar I, Nagamatsu S, Zhen M (2013) NLF-1 delivers a sodium leak channel to regulate neuronal excitability and modulate rhythmic locomotion. Neuron 77:1069–1082

14. Gao S, Zhen M (2011) Action potentials drive body wall muscle contractions in *Caenorhabditis elegans*. Proc Natl Acad Sci U S A 108:2557–2562

15. Liu P, Ge Q, Chen B, Salkoff L, Kotlikoff MI, Wang ZW (2011) Genetic dissection of ion currents underlying all-or-none action potentials in *C. elegans* body-wall muscle cells. J Physiol 589:101–117

16. Nagel G, Brauner M, Liewald JF, Adeishvili N, Bamberg E, Gottschalk A (2005) Light activation of channelrhodopsin-2 in excitable cells of *Caenorhabditis elegans* triggers rapid behavioral responses. Curr Biol 15:2279–2284

17. Richmond JE, Jorgensen EM (1999) One GABA and two acetylcholine receptors function at the *C. elegans* neuromuscular junction. Nat Neurosci 2:791–797

18. White JG, Southgate E, Thomson JN, Brenner S (1986) The structure of the nervous system of the nematode Caenorhabditis elegans. Philos Trans R Soc Lond B Biol Sci 314:1–340

19. Mahoney TR, Luo S, Nonet ML (2006) Analysis of synaptic transmission in *Caenorhabditis elegans* using an aldicarb-sensitivity assay. Nat Protoc 1:1772–1777

20. Vashlishan AB, Madison JM, Dybbs M, Bai J, Sieburth D, Ch'ng Q, Tavazoie M, Kaplan JM (2008) An RNAi screen identifies genes that regulate GABA synapses. Neuron 58:346–361

21. Hu Z, Pym EC, Babu K, Vashlishan Murray AB, Kaplan JM (2011) A neuropeptide-medi-

ated stretch response links muscle contraction to changes in neurotransmitter release. Neuron 71:92–102

22. Goodman MB, Lindsay TH, Lockery SR, Richmond JE (2012) Electrophysiological methods for *Caenorhabditis elegans* neurobiology. Methods Cell Biol 107:409–436

23. Richmond J (2009) Dissecting and recording from the *C. elegans* neuromuscular junction. J Vis Exp. https://doi.org/10.3791/1165

24. Maro GS, Gao S, Olechwier AM, Hung WL, Liu M, Ozkan E, Zhen M, Shen K (2015) MADD-4/Punctin and Neurexin Organize *C. elegans* GABAergic Postsynapses through Neuroligin. Neuron 86:1420–1432

25. Gao S, Xie L, Kawano T, Po MD, Guan S, Zhen M, Pirri JK, Alkema MJ (2015) The NCA sodium leak channel is required for persistent motor circuit activity that sustains locomotion. Nat Commun 6:6323

26. Bai J, Hu Z, Dittman JS, Pym EC, Kaplan JM (2010) Endophilin functions as a membrane-bending molecule and is delivered to endocytic zones by exocytosis. Cell 143:430–441

27. Gracheva EO, Burdina AO, Holgado AM, Berthelot-Grosjean M, Ackley BD, Hadwiger G, Nonet ML, Weimer RM, Richmond JE (2006) Tomosyn inhibits synaptic vesicle priming in *Caenorhabditis elegans*. PLoS Biol 4:e261

28. Weimer RM, Gracheva EO, Meyrignac O, Miller KG, Richmond JE, Bessereau JL (2006) UNC-13 and UNC-10/RIM localize synaptic vesicles to specific membrane domains. J Neurosci 26:8040–8047

29. Hu Z, Hom S, Kudze T, Tong XJ, Choi S, Aramuni G, Zhang W, Kaplan JM (2012) Neurexin and neuroligin mediate retrograde synaptic inhibition in *C. elegans*. Science 337:980–984

30. Tong XJ, Hu Z, Liu Y, Anderson D, Kaplan JM (2015) A network of autism linked genes stabilizes two pools of synaptic GABAA receptors. Elife 4:e09468

31. Luscher C, Isaac JT (2009) The synapse: center stage for many brain diseases. J Physiol 587:727–729

Chapter 9

Electroantennograms (EAGs) and Electroretinograms (ERGs) in the Genetic Dissection of Synaptic Function in *Drosophila melanogaster*

Balaji Krishnan and Yogesh P. Wairkar

Abstract

Electroantennograms (EAGs) and electroretinograms (ERGs) are extracuticular recording techniques that require stimulations from two distinct sensory stimuli and have been instrumental in mapping the role of seven transmembrane receptors and downstream signaling in the transduction of the sensory stimulus to synaptic signals. This review will provide an overview of the different approaches used with these two techniques and the underlying success seen in the implementation of vertebrate studies that led to translational outcomes. We will review systematically the classic approach and its modifications and emphasize how electrophysiological studies of visual and olfactory transduction provided insights that have been instrumental in discovering signaling mechanisms that are disrupted in diverse neurological states including neurodegeneration and circadian rhythms.

Key words Vision, Olfaction, Electroantennogram, Electroretinogram, Drosophila, Sensory reception, Extracuticular, Neurogenetics, Neurodegeneration

1 Introduction

The humble *Drosophila melanogaster*, the ordinary pomace fly, vinegar fly, or fruit fly, was chosen almost a century ago by T.H. Morgan to study genes, the cellular structures, which are determinants of heredity. Since then, vast amount of knowledge gathered about its genome has made *Drosophila melanogaster* one of the best invertebrate models to study the nervous system.

The study of neurogenetics in *Drosophila melanogaster* was pioneered by Seymour Benzer [1]. Subsequent studies by many fly geneticists have led to successful genetic and mechanistic dissection of many complex behaviors [2–13].

Complex behaviors are a result of extensive communication between many neurons in neural circuits. A thorough functional assessment of participating neurons by measuring their electrical properties is often crucial to obtain a complete understanding of

Roy V. Sillitoe (ed.), *Extracellular Recording Approaches*, Neuromethods, vol. 134,
https://doi.org/10.1007/978-1-4939-7549-5_9, © Springer Science+Business Media, LLC 2018

any behavior. For the current chapter, we will be focusing on the sensation of visual and olfactory cues from the environment by *Drosophila* and the electrophysiological techniques that are currently used in understanding sensory reception in these two systems.

Olfaction and vision are two important modalities used by insects, in general, and *Drosophila*, in specific, to navigate its environment and ensure its success in survival and reproduction.

Extremely creative ways of electrophysiological analysis of excitable cells in the central and peripheral systems have been established in both the adult and larval stages in *Drosophila* [14–26].

In this chapter, we will first describe electrophysiological methods to measure olfactory and visual responses using electroantennograms (EAGs) and electroretinograms (ERGs), respectively.

1.1 Olfactory Electrophysiology

For insects, olfaction or chemosensation (that can occur by the chemosensory hairs located all over its body) is an important sensory modality in locating food sources, identifying appropriate sites for oviposition, selecting mates, and avoiding predators. *Drosophila melanogaster* is a powerful model organism for investigating mechanisms of odor coding because it has an olfactory system that is anatomically similar, but simpler than that of vertebrates, which offers the following advantages: (1) it can be manipulated genetically, (2) the complete annotated sequence of its euchromatic genome is available, (3) the development and anatomy of its olfactory system are well characterized, and (4) simple behavioral and electrophysiological paradigms have been developed to test olfactory perception [27].

2 Organization of the Olfactory System

Drosophila has three principal sensory organs, one in the larval stages and two at adult stages [28]. In 1907, W.M. Barrows showed that fruit flies detect odors through their antennae [29]. The role of the antenna as olfactory sense organ was confirmed using the mutants, *antennaless* [30] and *antennapedia* [31]. The main olfactory sense organs in *Drosophila* are the pair of funiculus (third antennal segment, which is the terminal part attached to the second antennal segment, Fig. 1). Structurally, the second antennal segment contains the Johnstons' organ that is mechanosensory, while the first antennal segment is a stalk that connects the second antennal segment to the head capsule. The funiculus detects long-range volatile chemicals. The adult antenna contains on its third segment some 500 sensory hairs, which divide into three morphological classes: sensilla basiconica, trichoidea, and coeloconica [28]. Each sensillum contains two to four neurons. Odorants are believed to pass through pores in hairs to the sensillar lymph.

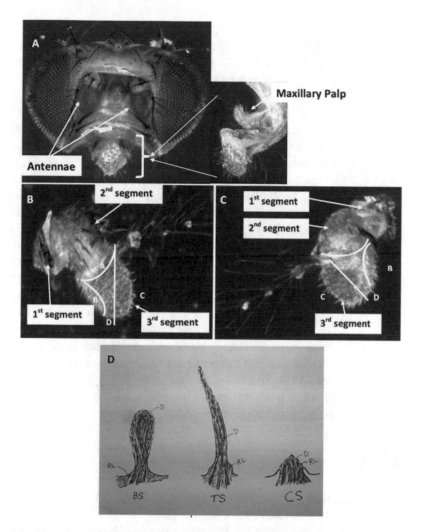

Fig. 1 *Drosophila melanogaster* **antennae.** (a) A micrograph illustrating relative positions of the fly sensory organs (on the head) that detect airborne chemicals. Organs labeled antennae (also called third antennal segment or funiculus) are known to detect airborne long-range volatile odorants, while maxillary palps assist in short-term odor detection. Both these organs are implicated in the reception of odorant molecules and form the most peripheral mechanism involved with the olfactory system. (b) Anterior view of right funiculus of *Drosophila melanogaster*. The illustration represents the antennal organization divided by imaginary lines to yield four regions A, B, C, and D denoted as basal, medial, anterior, and lateral regions or edges for ease of morphological definition [28]. The distribution of sensilla on this view of the antenna shows that there are regions of subtypes such as basiconic sensilla on the medial edge, trichoid sensilla on the anterior edge, and coeloconic sensilla interspersed. On the second segment, Johnston's organ is placed within the cuticle. The arista is the longest hair that is present on the third antennal segment but has its origin placed on the posterior side in this illustration. (c) Posterior view of the right funiculus of *Drosophila melanogaster*. Trichomes are hairy projections of the cuticle that are interspersed in the areas between the sensilla filling the funicular region [142]. (d) Structure of the three different antennal olfactory sensilla in *Drosophila melanogaster*. *BS* basiconic sensilla, *CS* coeloconic sensilla, and *TS* trichoid sensilla. Multiporous as indicated here, these sensilla are implicated in the perception of different odor types such as food odor by BS and pheromones by TS. *RL* outer lymph space, *D* dendrites. The more complex the structure, the more odor types it detects [143]. Stubby BS with its high level of perforation is thought to have increased surface area for odor molecules to penetrate, while TS extend away from the cuticle and increase specificity and direction of the pheromone reception. CS function has been mapped relatively recently even though it is one of the ancient forms of sensilla discovered in many insect species [144, 145]. A detailed list of the odorant specificity and sensitivity is provided elsewhere [145]

Olfactory sensilla are interspersed with nonolfactory sensilla. A majority of the funicular sensilla occupy surface of the segment, but many others line the wall of a pit (the sacculus), that is, located proximally on its posterior side.

Maxillary palps (Fig. 1a), the other pair capable of detecting volatile odorants, albeit at close range, are located on the proboscis, the so-called mouth or the feeding organ of the fruit fly. The function of maxillary palps in olfactory function was determined physiologically and behaviorally by some elegant studies [32, 33]. It contains about 60 basiconic sensilla, whose ultrastructure is consistent with olfactory function, as well as trichoid sensilla, similar in morphology to moth pheromone-sensing olfactory sensilla [28].

Zonal organizations of heterogeneously distributed olfactory sensilla provide evidence for certain overlaps in the kind of odor detected. Thus, it is possible to measure local physiological responses by concentrating on specific regions of the funiculus [34]. Neurons from the funiculus project directly to the antennal lobes of the brain, which are composed of about 35 glomeruli [35, 36]. Neurons from different classes of sensilla project to different glomeruli, and the organization of these projections has been described in some detail [28]. From antennal lobes, there are projections to higher centers of the brain, including the mushroom bodies, which have been implicated in olfactory driven learning and memory [35–37].

3   Sensilla Architecture

Basiconic sensilla are single-walled multiparous chemosensory bristles [28] (see Fig. d). These are assigned to a large or a small subtype; only the former of the two is observed in maxillary palps. Large basiconic sensilla house four neurons, while small basiconic sensilla house two neurons and are clearly characterized as olfactory sensilla [34]. Evidence from transmission electron microscopy (TEM) indicates that the pores on basiconic sensilla likely connect the external environment with the internal dendritic lymph [28]. Basiconic sensilla are associated with reception of volatile odorants generally associated with food (acetates, alcohols) and predators (aldehydes).

Sensilla trichoidea have longitudinal cuticular ridges [38]. These sensilla also contain near their base a transverse invagination, which extends partway around the sensillum. The upper portion contains small indentations and pores and support possibility of olfactory role. A subset of these sensilla responds in a dose-dependent manner to cis-vaccenyl acetate, an aggregation pheromone in *Drosophila melanogaster*, as measured by single-unit recordings that measure action potential [27], whereas the other subset did not. There were at least two neurons identified in this subtype that responded to trans-2-hexenal and 4-methylcyclohexanol. The subtypes could also be spatially segregated with the former subtype, also known as Type 1, on the dorsomedial range; while Type 2 forms the ventrolateral range of the anterior face on the third antennal segment. In this context, basiconic sensilla are distributed to the medial region (or edge) that lies opposite to the anterior edge of the antenna. As a result, trichoid sensilla have been more often associated with very highly species-specific odorant reception and perception. What is unknown is whether the receptor signaling in these sensilla that shows much higher specificity in odor reception is different in downstream signaling compared to the signaling in basiconic or coeloconic sensilla. Coeloconic sensilla are located on the areas between basiconic and trichoid sensilla and respond to a subset of odors including the two described for sensilla trichoidea [27].

4   Signaling in the Sensilla

Olfactory receptor neurons (ORNs) are present within the sensilla and extend dendrites, which are bathed in sensillar lymph. Present at high concentration in this fluid are members of a large family of odorant-binding proteins (OBPs) which have been the subject of extensive investigation in many insects but remain poorly understood [39–48]. OBPs were implicated in assisting odorant molecules to stimulate olfactory neurons [49]. The current hypothesis suggests a broader role for OBPs in transport and mitigation of odorants to specific sensilla [48].

OBPs may solubilize hydrophobic odorants in aqueous sensillum lymph, present these odorous ligands to the receptor, or assist in terminating odor response by removing ligands from the receptor [42–45, 47–50]. The first functional evidence that OBPs indeed participate in olfactory responses came from analysis of *Drosophila* mutants that lack LUSH OBP; its mutant phenotype is an abnormal attraction to high concentrations of ethanol and encodes the LUSH protein [51]. *Lush*−/− mutant flies fail to be repelled by high concentrations of ethanol. However, functional studies of vertebrate and nematode ORs expressed in heterologous systems indicate that receptors can be activated by ligands in the absence of OBPs. Since the discovery of the first OBP in 1998, 52 of these small secreted proteins have been identified in *Drosophila*, and there is still a lot unknown about how these proteins assist in binding of the ligands to the olfactory receptors and how, sometimes, despite transporting odorants, these regulate smell reception [48].

The other class of proteins in olfactory reception are the ORNs of which 60 have been characterized to encode 62 seven transmembrane domains through alternative splicing [37]. Only a single member of this family is expressed in the basiconic and trichoid sensilla [52, 53] and is important for specifying properties of the ORN such as the odor response spectrum, excitatory or inhibitory mode of response, termination kinetics, and the level of spontaneous activity (see Single-Unit Responses section for details on spontaneous activity). In *Drosophila*, there is also the expression of a co-receptor, Orco, that heterodimerizes with the OR and is essential for neuronal complex to the dendritic membrane [54]. Using narrow and broad tuning that depends on the concentration of the odorants, different ORs can mediate both excitatory and inhibitory responses to different odors in the same ORN [55], thus allowing for a combinatorial coding. The extent of the combinatorial coding is subject to the activation of transcription factors that can result in an epigenetic regulation of OR expression patterns in the sensilla. But it is also observed that the spatial distribution of the expression involves clusters of sensilla with the same ORs on a subset of the ORNs, intermingled with ORNs that express other ORs, thus, with an overlapping mosaic pattern of the ORs, completely dependent on pupal development [56, 57]. The ORs express a seven-transmembrane domain protein [58–62] with a membrane topology resembling that of G-protein-coupled receptors (GPCRs) that function as ligand-gated ion channels and have very little homology with vertebrate 500–1000 GPCR ORs [63, 64]. Moreover, *Drosophila* ORs can transduce information independent of G-protein activation, despite all studies showing the activation of G-proteins, which has kept the field of *Drosophila* olfaction very active for its intriguing complexity that is not seen in evolutionarily primitive genetic models, like *Caenorhabditis elegans*. In 25 years since the discovery of the multigene family of vertebrate odorant receptors, there are still many novel observations at the level of reception of odor by the receptors, such as the olfactory marker protein that assists in olfactory receptor neuron physiology [65] and single base pair changes affecting responsiveness of the olfactory receptors [66]. Vertebrate olfactory receptor studies have been assisted by the fact that chemoreception in the worm, *Caenorhabditis elegans*, shares many aspects of olfactory transduction [67–69]. Interestingly, functional studies of vertebrate and nematode ORs expressed in heterologous systems indicate that receptors can be activated by ligands in the absence of OBPs [70], adding yet another layer of difference between vertebrate and insect olfaction.

Despite these stark differences at level of olfactory reception, mechanisms of olfactory perception and especially strategies employed for odorant discrimination at multiple levels from the sensory organ to the central nervous system are similar [67, 69, 71]. While studying insect olfactory reception may appear seemingly irrelevant since it is markedly different from vertebrate olfaction, there are a few crucial reasons that make these studies invaluable: (a) insect olfaction/chemosensation is attributed to be the critical sensory modality toward the success in the variety of species (50% of all studies species on the planet) thereby increasing the evolutionary chances of survival [72]; (b) circadian oscillator mechanisms in peripheral chemosensation share a similarity to the vertebrate central oscillator, making the fruit fly sensory modality an excellent model to dissect the genetic pathway [8, 73–79]; and (c) insights into the etiology of several human neurological disorders [80, 81].

4.1 Electroantennograms The recording of odor-induced electroantennograms (EAGs) was first performed by Schneider and colleagues [82] in the antennae of the male silkworm moth, *Bombyx mori*. Since then, EAG has been measured in many insects using electrodes inserted into the tip and the base of the antenna. Extracellular recording from the funiculus has been extensively used to describe the electrical response of the *Drosophila* antennae to airborne chemicals [32, 77, 83–90]. In addition to defining signaling components of olfactory transduction, the technique has been effectively utilized to address signaling mechanisms of olfactory reception in circadian rhythms [73–76, 78, 79] and neurodegenerative diseases [80, 81]. Qualitative differences in response to various odorants and quantitative differences in amplitude of EAG response to different concentrations have been reported [56, 70, 91, 92]. Electrophysiological measurements at the antennal level give functional information about how odorant molecules are captured, and these external chemical signals transduced to a corresponding electrical signal in olfactory neurons. Electrophysiological methods such as whole antenna EAG and nerve and single sensilla recordings have been extensively used in insects, particularly for studying response of moths to pheromones [93]; see the next section for details. EAGs are believed to measure summed receptor potentials of neurons near recording electrode [33]. These have been very useful in determining whether mutants identified by defects in olfactory behavior have peripheral, as opposed to central, defects. Such assay revealed that mutants defective in visual system physiology (*norp A*-phospholipase C gene) also have olfactory defects [38]. Mutants for the principal *Drosophila* sodium channel gene *paralytic* (para) (also *olf D/smell-blind*) have EAG defects for a wide range of odorants; the electrophysiological phenotype placed it as an important component of peripheral olfactory transduction [94]. Mutants in *acj6* (abnormal chemosensory jump 6) were isolated in an olfactory behavioral screen; the gene encodes a POU-homeodomain protein that may regulate *Drosophila* olfactory receptor (DOR) gene expression (thus directing proper spatial expression) [56]. Extensive electrophysiological studies in this mutant show altered responses that corroborate the above result [27].

Odorant responses have been recorded extracellularly using EAG [83, 95] and single-unit recording techniques [27]. Single units measure action potentials in neurons from individual sensilla. Different regions of the antenna appear to be more responsive to some odorants than others [33, 84], and single neurons appear to be sensitive to multiple odorants [34]. Patch clamp recording techniques were applied to study mechanisms of odor transduction and development of odor sensitivity in *Drosophila melanogaster* [96]; these show the existence of diversity in transduction mechanism where mixture of odors either increased or decreased conductances in individual patches in a region-specific manner. Electrophysiological studies have therefore been invaluable in the elucidation of olfactory mechanisms manifested in antennae.

5 Procedure

Using an optimal dissecting microscope such as the Wild Heerbrugg S5 with 2× objective and 20× eyepiece, there is a large working distance (33 cm) that allows placement of four coarse micromanipulators (Narishige M 152) and a flexible fiber-optic source. The ground electrode is positioned at a steeper angle to impale the tip of an upright third antennal segment (see Fig. 2A(c), B), while the recording electrode is oriented at a more oblique angle since it is placed against the cuticle on the anteromedial side. The third manipulator is mounted with a glass micropipette pulled and then flamed to obtain a hook, which is smaller than the funiculus—this allows the third antennal segment to be lifted from its receptacle in the head. The fourth coarse manipulator is connected to a glass Pasteur pipette that acts as the airflow tube (Fig. 2A(c)) for the olfactometer. An olfactometer starts with a flowmeter (Cole-Parmer Instruments) that allows a flow of 250 mL/min generated by an air source (lab air supply or a quieter aquarium pump). A colorless, odorless tygon tubing connects the flowmeter to the inlet port of an electrically controlled six-way valve. Volatile odorants (ethyl acetate, isoamyl acetate, benzaldehyde, etc.) were diluted in light odorless mineral oil which undergoes serial dilution (1:10; vortexed for 20 s, glass Pasteur pipettes used) to give a range of 10^1–10^{10}. Tygon tubing was the best in preventing any adsorption of odor and facilitated convenient changing at regular intervals to prevent odor buildup (every month). All odorants are purchased from Aldrich since we noted consistent and distinct biological responses with the purity provided. Disposable plastic scintillation vials were used for making the preparations. The olfactometer consisted of a circular clear Plexiglas that has a central slot for housing the six-way valve. The inlet was at the bottom, and there were six small outlets that were directed to the top of the Plexiglas into a stainless steel minicylinder (inlet that carried the air

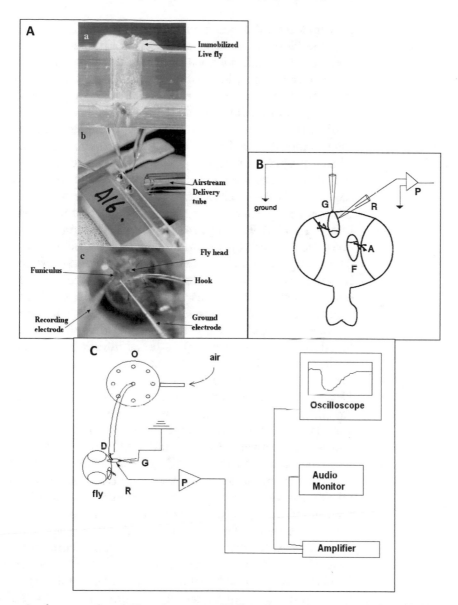

Fig. 2 Apparatus for measuring electroantennogram (EAG) responses (**A**). (*a*) Lateral view of immobilized fly in panel. (*b*) View of the experimental arrangement with electrodes in position in panel. (*c*) Magnified live fly preparation ready for measuring EAG responses in panel. (**B**) The preparation for recording EAG from the funiculus (F). The ground (G) impales the tip of the funiculus, while the recording electrode (R) is placed on the cuticle. The probe (P) feeds the signal to the computer. (**C**) Schematic of EAG—odor evoked changes in transepithelial potential. An indifferent/ground electrode (G) is impaled into the tip of funiculus in a live fly preparation (fly); recording electrode (R) is pressed against its medial edge. The recording electrode is connected to a probe (P) and in turn to a DC amplifier, which amplifies the response and displays it on the oscilloscope. The olfactometer (O) receives air input from a pump and passes it through a valve-based system with a possibility of delivering eight different odorant concentrations interrupting a stream of air. The tube (**D**) delivers the airstream and odor thereof onto the funiculus. The audio monitor assists in placement of the recording electrode by decreasing noise in the system when a stable contact is established

puff into the odor containing scintillation vial) that was wedged into the Plexiglas. Another stainless steel minicylinder (outlet that carried the air puff laden with odorant out of the scintillation) was also placed adjacent but not touching the first one. Both stainless steel minicylinders were surrounded by a lid that could be used to fasten the scintillation vial containing the diluted odorant. This lid with two holes was fastened to the Plexiglas using superglue. Everything was maintained airtight to prevent any odor leakages. The air puff containing the odorant was directed through a tygon tubing to one of the six inlet ports of a second six-way valve. The six-way valve then channeled the puff of air with the odorant dilution to the main outlet that was connected to a Pasteur pipet blowing on the fly head through a tygon tubing, thus ensuring a constant flow of air, interrupted by packets of odorant dilution. The arrangement facilitates removal of odor and prevents electrical responses that measure mechanical stimulation of the Johnston's organ. Electronics were used to the control the olfactometer as well as record the responses. These consisted of an amplifier such as Axopatch 200B (Molecular Devices) that was connected to a Windows Computer through a digital interface (Digidata 1322A, Molecular Devices). Stimulation was controlled through the protocol files generated in the Clampex 9.2 software (Molecular Devices), and the analog responses were digitized using the same system (Fig. 2C). The signals were low-pass filtered (3 kHz). The amplifier output was also routed to an audio monitor (Grass, AM9) and filtered between 300 Hz and 3 kHz to emit bursts of high-frequency sound that was used as indicator of circuit completion when recording under dark or low light conditions that compromised the visibility under the microscope. Circadian rhythm recordings were facilitated by audio monitoring. The recordings were done in the far-red range of the visible spectrum, where humans have visibility while red-pigmented compound eyes of *Drosophila melanogaster* do not [73, 74]. Micropipettes used for EAG recording consisted of borosilicate glass capillary tubes pulled to an internal tip diameter of 1–2 μm and are backfilled with 0.17 M sodium chloride (NaCl) solution. Chloridization of silver electrode wires for both ground and recording electrodes were immersed in concentrated bleach for about 15 min until the surface of the wires were oxidized to a dark-brown or dull-gray appearance that ensured a stable baseline in the EAG trace. Typical EAG signals evoked an amplitude change in the baseline that ranged in several mV (see Fig. 3a). Since it is important to maintain vibration isolation for preventing mechanical confounds, a system that is inexpensive such as a heavy marble balance table on partially inflated tire inner tubes or a comprehensive TMC Micro-g air table can be used with a grounded Faraday cage to reduce electrical noise from external sources. EAG recordings were done at a temperature of 25 °C with a relative humidity of 50–80%. Adult

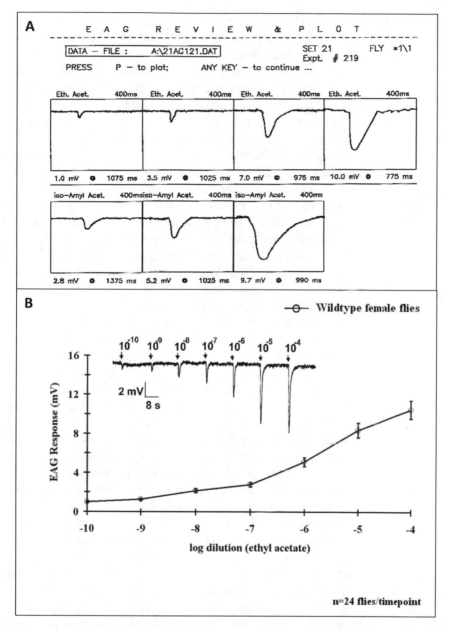

Fig. 3 (**a**) A representative screen output of an EAG program depicting the EAG waveforms. Important data about the data file, fly genotype, and experiment number are resent above the seven boxes that show the acquired data from sequential delivery of increasing concentration of odorant going from left to right and then from bottom left to right. Increasing concentrations of ethyl acetate (Eth. Acet.) applied for 400 ms results in a change in the baseline, where there is a small downward shift in the baseline that increases with a sharp drop in the on phase and a gradual recovery back to the baseline. The amplitude from the baseline to the tip of the drop is measured as the amplitude in mV, along with the total area represented in milliseconds for the recovery to baseline (shown below each box). The lower panel represents the delivery as it is done and observed on the oscilloscope before the processed output seen above. (**b**) Summary of dose dependence of peak EAG responses to ethyl acetate in 24 wild-type flies. Data collected from individual flies are pooled to provide a concentration response curve that demonstrates clearly that the amplitude of response is one way to assess the chemosensory transduction of reception of the odorant at the sensory organ. The traces above the graph show how the different dilution amplitudes from a single fly can be assessed

female *Drosophila* were used for EAG experiments, even though characterization was done with both male and female flies. Female flies provided a robust response because they were more resilient to the procedure used in immobilization (see below) prior to the recording. Flies were anesthetized in an empty prechilled vial in an ice bucket and transferred to a Whatman filter paper atop chilled aluminum block on ice. A pair of blunt forceps was used to hold the fly with both wings and place the flies in an "L"-shaped Plexiglas "fly holder" (see Fig. 2A) with 1-mm-diameter holes drilled every 10 mm along its length on the wide portion of the "L." With one fly per hole, a maximum of six flies were immobilized per fly holder allowing for $n = 6$ recordings at a minimum per odorant. With the body lying flat on the broader edge, the fly head is pushed into the hole, so that it protrudes out on the longest side. Low-melting-point wax (myristic acid, 58.5 °C) powder is taken on a small loop (smaller than the fly head) of fine silver wire and then heated to just reach and maintain a molten state using a rheostat. First, the wings are glued at the tips, then the ventral abdominal tip is glued to the opposite end of the hole. This sufficiently immobilizes the fly that the fly holder is now turned to have the fly head faces the objective. Using some more molten myristic acid, the head region just above the neck is glued to the fly holder, followed by the extended proboscis on the opposite side (see Fig. 2A). The heated silver wire never touches any part of the fly—this is where the males are more susceptible to dehydration – but is sufficiently close that the wax spreads by capillary action alone. The spacing of adjacent holes (3 mm apart) are sufficient to prevent odor stimulus from the experimental fly to not to affect the others. After ensuring that all flies for one fly holder get done within a few minutes (15–30 s per fly), the holders are kept in a Whatman filter paper moistened with water in a 14 cm petri dish for an hour recovery before commencing recordings. To facilitate recording the anterior region of the funiculus, the blunt hook electrode is lowered under the funiculus to lift it up and expose the underside rich in basiconic sensilla (see Fig. 1b, c), so that the long end and the tip of the antenna is the closest to the objective lens (upright, see Fig. 2A(c)). EAG recordings from the basiconic sensilla that are in majority at the anterior surface are now accessible to the recording electrode. The ground electrode is directed into the tip of the antenna, where the final advance into the tip is ensured by a gentle tap of the coarse movement knob; this prevents damage to the antennal segment but ensures contact with the lymph bathing the antennal segment. The recording electrode is then placed onto the dorsomedial surface of the third antennal segment such that the tip touches the extracuticular surface. As described earlier, this is where the audio monitor comes in handy because the change in the sound ensures that contact is achieved and therefore the advance of the recording electrode can be stopped.

Different odorant dilutions are typically applied once every 30 s to a minute to facilitate complete clearing – the process is greatly aided by having an activated charcoal filter or any other filter (active or passive, HEPA) assembly that can clear the odor—this unit can be placed as close to the outlet that removes the air from the room. There is a baseline change resulting in a downward change of extracuticular potential measured in mV (see Fig. 3a, b). The amplitude of this change is dependent on the concentration of the odorant used. The shape of the trace changes with different odorants, and ethyl acetate used in Fig. 3a, b, has the highest volatility of all the appetitive odors used and, thus, gives a larger potential that is useful for conducting studies where the sensory physiology can be affected by changes in the central nervous system of the fly [73, 74, 76–79]. The EAG responses did not saturate over the range of ethyl acetate concentrations used in these experiments. Some desensitization was observed during odorant application (over a period of several seconds; Fig. 4), but recovery from desensitization was invariably complete when odorant pulses were delivered at 30 s intervals.

5.1 Single-Unit Responses

While EAGs represent field extracuticular responses of the entire funiculus, single-unit responses are important for studying the individual response of each of the different sensilla. One of the first publications detailing this method came from the Carlson lab

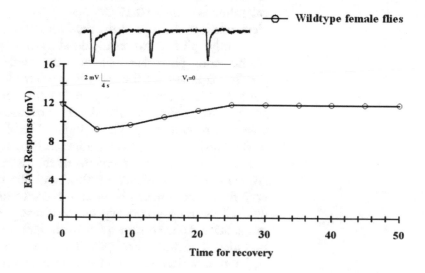

Fig. 4 Recordings showing typical desensitization and recovery from desensitization of responses to ethyl acetate (1:10^4 dilution in paraffin). Ordinate of graph represents interval between subsequent applications of 1:10^4 ethyl acetate. Note complete recovery from desensitization with 30 s intervals between odorant pulses

at Yale University [27]. For this procedure, it is important to isolate and suspend the fly antenna from the head of the intact fly such that there is 360° access to all the sensilla. This can be achieved by immobilizing the fly, by forcing it headfirst to a 200 μL microcentrifuge tip, and then by cutting the tip to expose the antennae and a second cutting behind the fly, which will be sealed to prevent the fly from escaping. The antenna is then teased out from the head cavity by using a thin wire and placed on a stack of coverslips, while the cut microcentrifuge is wedged to the glass slide containing the stack of coverslips. The exposed antenna is now secured by wedging a glass capillary on the second segment close to the head cavity.

In contrast to the optics in EAG, the single-unit recordings require a compound microscope that effectively reduces the working space from 33 mm to about 2–3 mm when the magnification increases from ~30× to ~2000×. Thus, the newer arrangement of limited working space around the funiculus calls for insertion of the ground/indifferent electrode into the eye and the absence of the hook electrode. The glass microelectrode used as the recording electrode needs to impale the base of the basiconic or trichoid sensilla and as a result needs to have a sturdier tip. A higher resistance than for EAG is needed to measure the action potentials (ranging from 1 to 4 types depending on the sensilla, [27]); a two-stage puller that yields a stubbier yet sharper electrode is optimal where the tip is <1 μm diameter. As before, the electrodes are filled with physiological saline solution (0.17 M NaCl). Insertion of the recording electrode into the base of the sensillum brings it in contact with the sensillar lymph that bathes the dendrite and facilitates measurement of the action potentials that form the single-unit recordings. The precise placement of the recording electrode calls for the need to have fine movements associated with the micromanipulators. Inexpensive unidirectional hydraulic movement accessories are available with several companies including Narishige that can be added to the existing coarse micromanipulators to achieve the level of precision in micrometer movements of the electrode. For single units, AC signals were amplified (1000×) and filtered (300 Hz–10KHz); each fly preparation can be used to record from at least three different ORNs. High-frequency burst of firing, if observed, with any preparation signaled signs of neuron damage and was immediately excluded. The number of spikes per odor pulse per 500 ms duration or 1 min duration was used for quantitative studies.

6   Studies on Olfactory Reception Using EAG and Single-Unit Responses

One of the earliest utilities of the EAG technique was the identification of odorant reception including olfactory receptor neuron characteristics, co-receptors, and odorant-binding proteins (reviewed in [37, 47, 48]). We have gained a comprehensive understanding of the similarities and differences between the vertebrate, nematode, and insect volatile chemosensory and olfactory reception and perception through the systematic studies of the sensory physiology using EAGs and single-unit recordings [36, 46, 69–71, 91].

In our group, we utilized the power of EAGs to characterize whether the larval chemosensory mutants that we identified in a screen showed adult deficits in olfactory reception (Figs. 5, 6, and 7). We could identify three classes of olfactory receptor mutants based on their response profile to different volatile odorants. The ability of EAG to quantitatively assess responses that are greater than the wild type (Fig. 5—*olf* 402) suggests that the nature of the mutation increases the responsivity of the ORN to a particular odorant. Further analysis of the structure and protein conformation will

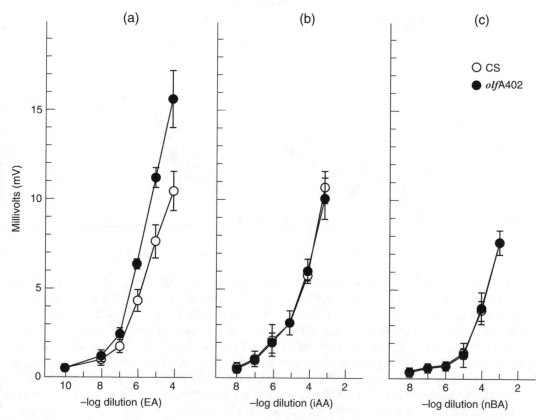

Fig. 5 EAG responses of *olf*A402. EAG amplitudes (*Y*-axis) for increasing dilutions (represented as logdilutions on the *X*-axis) demonstrate that the P-element insertion in this mutant (*black circles*) causes a specific increase (50%) in amplitudes toward ethyl acetate (**a**), but no change in isoamyl acetate (**b**) or n-butyl acetate (**c**) compared to wild-type female flies (*clear circles*). Six to eight flies were used for each data point, and the data is represented as mean and standard error of mean. Significance was calculated at $p < 0.05$. These data represent an olfactory reception mutant and demonstrates a unique example of how EAG is a versatile technique to identify increases as well as decreases (see next figure)

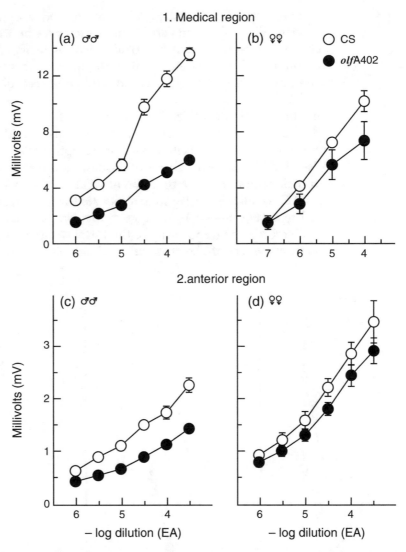

Fig. 6 EAG responses of *olf*A186. EAG amplitudes (*Y*-axis) for increasing dilutions (represented as log dilutions on the *X*-axis) demonstrate that the P-element insertion in this mutant (*black circles*) causes a decrease in response to ethyl acetate (EA) only in the male flies (**a**) not in the female flies (**c**) compared to wild type (*clear circles*). We assessed whether the EAG response seen in the medial region (predominated by basiconic sensilla) was also observed in the anterior region (with a greater number of trichoid sensilla implicated in sexual olfactory reception). We observed a similar decrement (**c**) in male flies, but not in the female flies (**d**), suggesting that the olfactory reception component affected by the P-element insertion may be present in neurons located in both these regions. Six to eight flies were used for each data point, and the data is represented as mean and standard error of mean. Significance was calculated at $p < 0.05$

provide insight into how the binding of the receptor with the odorant contributes toward the response. Another OR mutant (*olf*A534) shows deficits in less volatile odorants while leaving the response to the more volatile odorants intact (Fig. 7). We speculate that the reception of the OR may be altered or there could be

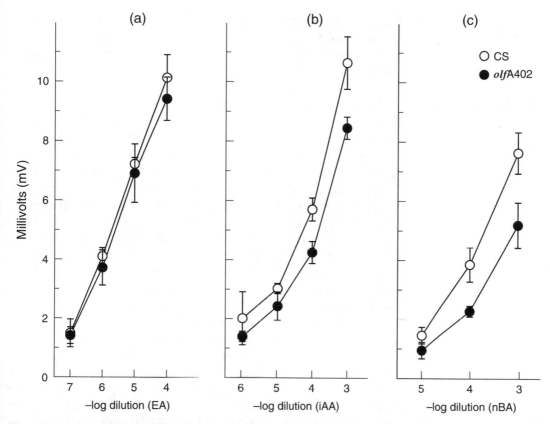

Fig. 7 EAG responses of *olf*A534. EAG amplitudes (*Y*-axis) for increasing dilutions (represented as log dilutions on the *X*-axis) demonstrate that the P-element insertion in this mutant (*black circles*) causes a decrease in response to the less volatile isoamyl acetate (iAA, **b**) and *n*-butyl acetate (*n*BA, **c**) while remaining unchanged for ethyl acetate (EA, **a**) compared to wild type (*clear circles*), suggesting that the olfactory reception component affected by the P-element insertion may be affecting OR/signaling that is complementary to *olf* A402 (Fig. 5). Six to eight flies were used for each data point, and the data is represented as mean and standard error of mean. Significance was calculated at $p < 0.05$

changes in downstream signaling in this event that could be altered, since it involves more than one odorant, thereby affecting the EAG response. More intriguingly, our EAG studies identified, for the first time, a sexually dimorphic response profile to food odorants (Fig. 6). Since we observed the change in male flies in the medial region enriched in basiconic sensilla as well as the anterior region, having more trichoid sensilla (involved in pheromone reception), single-unit responses (see next section) will be insightful in localizing the OR to the sensilla and the neuron. Such a finer understanding is necessary before pursuing further biochemical studies that will reveal insights into the protein conformation and/or signaling affected by the mutation.

But the functional studies have not been limited to just olfactory transduction; the studies have provided greater understanding of peripheral circadian rhythms and their influence on central circadian oscillators and behaviors [8, 73–76, 78, 91] and involvement of the sense of smell in the etiology of neurodegenerative states such as Parkinson's disease [80, 81] among many others that are just emerging.

6.1   Electroretinograms   Vision is very important for the orientation of flying insects and especially in *Drosophila*, which is a crepuscular species that are active during the twilight hours of the day. The visual system of *Drosophila melanogaster* is composed of two compound eyes (see Fig. 8a) made up of 750 highly ordered individual ommatidia [97]. Each ommatidium is a functional unit arranged in a hexagonal pattern, visible from the outside by the facets, which are formed like lenses. With eight photoreceptor cells (PRCs) arranged into an elongated barrel-like structure (100 μm in length), where the distal 10% contains the dioptric apparatus, the proximal end contains the pigment and cone cells. The cell bodies of the outer PRCs (R1–R6) span the entire length of the retina and project axons to the lamina, the first optic neuropil. The axons of the central R7 and R8 cell synapse in the medulla, the second neuropil of the optic lobe. R1–R6 express a rhodopsin sensitive to blue light, while R7 detects UV and R8 either green or blue light [98–101]. A photon of light passes through the cornea and hits the rhodopsin in the rhabdomeres to convert 11-cis-3-hydroyretinal to all-trans-3-hydroxyretinal. The converted form is the isomer that activates the rhodopsin to form metarhodopsin, which then triggers GDP-GTP exchange on membrane-bound G-proteins (G_q family) that in turn activate phospholipase C (PLC) to break down phosphatidylinositol 4,5 biphosphate into inositol triphosphate and diacylglycerol (DAG). DAG is instrumental in gating the cation-selective Trp and Trpl channels that trigger the Ca^{++} and Na^+ influx, depolarizing the photoreceptor, releasing histamine at the synapse, and causing downstream hyperpolarization [102–104].

Drosophila retinal physiology was pioneered by stalwarts such as Yoshiki Hotta [1], William Pak [105, 106], and others [107, 108] to genetically dissect visual reception and perception, and it has been immensely successful in both the understanding visual systems (reviewed in [104, 109–118]) and the application of the understanding toward understanding human neurodegenerative states [119–123]. Like the EAG technique, the electroretinogram (ERG) uses an extracellular electrode to record the compound field potential from photoreceptors and downstream neurons within the fly eye in response to flashes of light. Transient spikes at the onset and offset of a light flash correspond to postsynaptic potentials in visual system neurons, while a sustained potential during the light stimulus results from the depolarization of photoreceptor cells. Such small signal frequency analysis is widely adopted for describing and predicting the response dynamics of various elements in visual systems. The ERG response can be

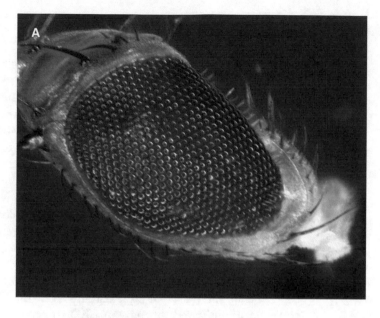

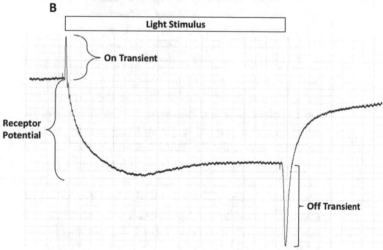

Fig. 8 (**a**) The compound eye of *Drosophila melanogaster*. (**b**) Typical electroretinogram (ERG) profile. The duration of the light stimulus is shown in the upper rectangular bar. The positive on-transient appears within 10 ms of the initiation of the light pulse. The negative receptor potential reaches peak amplitude between 0.5 and 1 s after initiation of the light pulse. A negative off-transient peak within 10 ms of the termination of the light pulse. Both the transients arose from the second-order lamina neurons; the right up-pointing arrow indicates a slow component arising from the pigment (glia) cells; the slowly rising corneal-negative peak amplitude also originates from the pigment cells

triggered by light ranging from the visible spectrum (400–600 nm) to ultraviolet (<400 nm)—all of which can be reproducibly evoked by fiber-optic source of light using the appropriate filters. ERG involves a small-amplitude sinusoidal modulation of the light stimulus which gives rise to a sinusoidal response of the same frequency but generally of different amplitude and phase.

7   Procedure

The fly holder preparation explained in the EAG section can be easily utilized here; however, care must be taken to ensure that the light stimulus applied to the experimental fly does not affect the others. As a result, it may be advisable to mount fewer flies farther away from one another. Alternatively, only one fly per fly holder would also help the cause. Two borosilicate glass electrodes (approximate tip size: 0.5 μm) are required for this preparation. Only the reference electrode, to be inserted into the fly's thorax, was broken with forceps to facilitate easy entry into the thorax. The recording electrode (3 M NaCl) and the reference electrode (0.9% w/v or 0.17 M NaCl) were used with chloridized wires as described earlier. Using a dissecting stereomicroscope, the reference electrode was inserted into the thorax of the fly, and the recording electrode either was placed against the cornea of the eye or was poked just through the cornea into the retina. Like the EAG experiments, the pulses of light coming from the fiber-optic source could be administered through an LED source (18,000 mcd white) that can be driven by the Clampex software using pulses of 5 V amplitude via the digitizer for precisely controlled duration with a very sharp on and off. The recording parameters were the same as described in EAG. The ERG of the *Drosophila* compound eye has a complex waveform (Fig. 8) and consists of the contributions from receptor cells and the second-order cells in the lamina. The corneal-positive on-transient and the negative off-transient arise in the lamina, and most of the sustained negative component reflects depolarization of receptor cells. Like the EAG potentials, the ERG signal has an opposite polarity to potentials within photoreceptors and laminar neurons. With the initiation of a light pulse stimulus, a corneal-positive "on" transient voltage spike reflects the hyperpolarization of laminar neurons. This is followed by a sustained corneal-negative potential, corresponding to depolarization of the photoreceptors and reaching a peak of 10–20 mV within approximately 100 ms of the onset of the light pulse. Termination of the light pulse elicits a corneal-negative "off" transient corresponding to the repolarization of the laminar neurons. The magnitude of the transients is a good indicator of the quality of the recording—the bigger the transients, the better the placement of the recording electrodes. Some of the changes that occur in these transients are seen in eye mutants—for example, the *white* mutant has a negative deflection following the on-transient—this occurs because the absence of the pigmentation prevents resolution between the ommatidia, thereby increasing the initial photoreceptor response [108, 124]. The classical *trp* (transient receptor potential) loss-of-function mutation shows no off-transients—as the name suggests, because of the mutation in the cation channel permeable to Na^+ and Ca^{++} [125]. *trp* mutants conduct less Ca^{++} current that alters the time that the remaining photoresponsive channels (the trpl channels) remain open, thus showing a robust on-transient followed by a reduced receptor potential, which quickly decays to baseline, without regard to the duration of the light pulse. These are some of the parameters that can be altered along with the intensity of light and the duration to find changes in the response as indicated here.

Thus, the functional role of the signaling components in visual transduction measuring the field potential ERG signals provides information on both sustained receptor potentials in primary sensory neurons (the photoreceptors) and the activity of downstream laminar neurons receiving synaptic input from the photoreceptors. The versatile multifaceted ERG, thus, opens the functional analysis to study specific mutations or differences in light stimulus to dissect each of the above components.

8   ERGs, Retinal Degeneration, and Neurodegeneration

ERGs have been instrumental in the study of retinal degeneration since the easiest measure would be seen in the changes of the transients and the receptor potential, namely, diminished responses with progression of the defect. It was first discovered in *Drosophila* that retinal degeneration occurs by mutations in rhodopsin and ever since the finding there are over 100 mutations in human rhodopsin that have been found to cause autosomal dominant retinitis pigmentosa (a group of diseases involving progressive degeneration of the retina that leads to severe visual impairment in the fifth to sixth decade of life) [126]. Since then, many forward genetic screens in *Drosophila* have identified a profusion of genes involved in retinal degeneration, and ERG genetic screening approach was utilized in finding over 200 defective lines that were studied by Pak and his group [111, 127]. Mapping and characterizing these genes revealed that mutations in any of the signal transduction components described above resulted in retinal degeneration [128–130].

An even more interesting aspect is the potential for translational application of these degeneration studies to understanding the mechanisms that impinge the progression of the neurodegenerative states [119, 131–133]. Most of the phototransduction pathway mutants are not lethal because the eye is not an essential organ in the fly and in addition, Bellen and his colleagues [123] circumvented any lethality in eye-associated genes by utilizing eye-specific mosaic clone [134, 135] and conducting ERG analysis of previous generated set of mutant flies containing neurodegeneration-implicated genes [136, 137] in young and old flies. Some examples of the human homologs that showed ERG deficits include Sicily gene which is implicated in Leigh syndrome [138], mitochondrial associated regulatory factor (Marf), mitofusin in Charcot-Marie-Tooth disease [139], frataxin in Friedreich's ataxia [140], and mitochondrial methionyl-tRNA synthetase in autosomal recessive spastic ataxia with leukoencephalopathy (ARSAL) [141].

In conclusion, vision and olfaction are sensory modalities that are conserved over evolution and provide effective ways of navigating the environment and ensure success in survival and reproduction. *Drosophila* offers the unique advantage of studying genetics, electrophysiology, and behavior that cannot be done using a neural-connectivity-wise simpler organism like *C. elegans* or a more complex vertebrate species like the preclinical rodents. As described in this review, multiple electrophysiological approaches can be utilized to ask questions about connectivity at the sensory levels and function at the cellular levels. In the past, these procedures have uncovered many aspects of signal transduction that have provided valuable insights into neurological functions such as circadian rhythms and disease states such as neurodegeneration. Despite decades of study, there are still a lot of questions that need to be answered, and these electrophysiological tools will prove invaluable in unraveling the complexity of *Drosophila* olfactory and visual systems.

Acknowledgments

We would like to acknowledge Rajalaxmi Natarajan, PhD, for her invaluable contributions in proofreading and editing the chapter for clarity and content.

References

1. Hotta Y, Benzer S (1970) Genetic dissection of the Drosophila nervous system by means of mosaics. Proc Natl Acad Sci U S A 67(3): 1156–1163

2. Allada R, Chung BY (2010) Circadian Organization of Behavior and Physiology in Drosophila. Annu Rev Physiol 72:605–624. https://doi.org/10.1146/annurev-physiol-021909-135815

3. Bellen HJ, Tong C, Tsuda H (2010) 100 years of Drosophila research and its impact on vertebrate neuroscience: a history lesson for the future. Nat Rev Neurosci 11(7):514–522

4. Burnet B, Connolly KJ (1981) Gene action and the analysis of behaviour. Br Med Bull 37(2):107–113

5. Ferrus A, Kankel DR (1981) Cell lineage relationships in Drosophila melanogaster: The relationships of cuticular to internal tissues. Dev Biol 85(2):485–504

6. Hall JC (1982) Genetics of the nervous system in Drosophila. Q Rev Biophys 15(2):223–479

7. Hall JC, Greenspan RJ (1979) Genetic analysis of Drosophila neurobiology. Annu Rev Genet 13:127–195. https://doi.org/10.1146/annurev.ge.13.120179.001015

8. Hardin PE (2011) Molecular genetic analysis of circadian timekeeping in Drosophila. Adv Genet 74:141–173. https://doi.org/10.1016/B978-0-12-387690-4.00005-2

9. Sayeed O, Benzer S (1996) Behavioral genetics of thermosensation and hygrosensation in Drosophila. Proc Natl Acad Sci 93(12):6079–6084

10. Sokolowski MB (2001) Drosophila: Genetics meets behaviour. Nat Rev Genet 2(11):879–890. http://www.nature.com/nrg/journal/v2/n11/suppinfo/nrg1101-879a_S1.html

11. Strauss R (2002) The central complex and the genetic dissection of locomotor behaviour. Curr Opin Neurobiol 12(6):633–638. https://doi.org/10.1016/S0959-4388(02)00385-9

12. Toma DP, White KP, Hirsch J, Greenspan RJ (2002) Identification of genes involved in Drosophila melanogaster geotaxis, a complex behavioral trait. Nat Genet 31(4):349–353

13. Tully T, Preat T, Boynton SC, Del Vecchio M (1994) Genetic dissection of consolidated memory in Drosophila. Cell 79(1):35–47. https://doi.org/10.1016/0092-8674(94)90398-0

14. Amrein H, Thorne N (2005) Gustatory perception and behavior in Drosophila melanogas-

ter. Curr Biol 15(17):R673–R684. https://doi.org/10.1016/j.cub.2005.08.021

15. Atkinson NS, Brenner R, Bohm RA, JY Y, Wilbur JL (1998) Behavioral and electrophysiological analysis of Ca-activated K-channel transgenes in Drosophila. Ann N Y Acad Sci 860:296–305

16. Broadie KS (1994) Synaptogenesis in Drosophila: coupling genetics and electrophysiology. J Physiol Paris 88(2):123–139

17. Caldwell JC, Eberl DF (2002) Towards a molecular understanding of Drosophila hearing. J Neurobiol 53(2):172–189. https://doi.org/10.1002/neu.10126

18. Cully DF, Wilkinson H, Vassilatis DK, Etter A, Arena JP (1996) Molecular biology and electrophysiology of glutamate-gated chloride channels of invertebrates. Parasitology 113(Suppl):S191–S200

19. Flourakis M, Allada R (2015) Patch-clamp electrophysiology in Drosophila circadian pacemaker neurons. Methods Enzymol 552:23–44. https://doi.org/10.1016/bs.mie.2014.10.005

20. Ganetzky B, CF W (1986) Neurogenetics of membrane excitability in Drosophila. Annu Rev Genet 20:13–44. https://doi.org/10.1146/annurev.ge.20.120186.000305

21. Kernan MJ (2007) Mechanotransduction and auditory transduction in Drosophila. Pflugers Arch: Eur J Physiol 454(5):703–720. https://doi.org/10.1007/s00424-007-0263-x

22. Kurtz R, Kalb J, Spalthoff C (2008) Examination of fly motion vision by functional fluorescence techniques. Front Biosci 13:3009–3021

23. Matthies HJ, Broadie K (2003) Techniques to dissect cellular and subcellular function in the Drosophila nervous system. Methods Cell Biol 71:195–265

24. Ocorr K, Vogler G, Bodmer R (2014) Methods to assess Drosophila heart development, function and aging. Methods 68(1):265–272. https://doi.org/10.1016/j.ymeth.2014.03.031

25. Seelig JD, Jayaraman V (2011) Studying sensorimotor processing with physiology in behaving Drosophila. Int Rev Neurobiol 99:169–189. https://doi.org/10.1016/b978-0-12-387003-2.00007-0

26. Wright NJ (2014) Evolution of the techniques used in studying associative olfactory learning and memory in adult Drosophila in vivo: a historical and technical perspective. Invertebrate Neurosci 14(1):1–11. https://doi.org/10.1007/s10158-013-0163-z

27. Clyne P, Grant A, O'Connell R, Carlson JR (1997) Odorant response of individual sensilla on the Drosophila antenna. Invert Neurosci 3(2):127–135. https://doi.org/10.1007/BF02480367

28. Stocker RF (1994) The organization of the chemosensory system in Drosophila melanogaster: a review. Cell Tissue Res 275(1):3–26

29. Barrows WM (1907) The reactions of the Pomace fly, Drosophila ampelophila loew, to odorous substances. J Exp Zool 4(4):515–537. https://doi.org/10.1002/jez.1400040403

30. Hillman R (1966) Histophenotype of the antennaless mutation in Drosophila melanogaster. J Hered 57(3):81–83

31. Postlethwait JH, Schneiderman HA (1969) A clonal analysis of determination in Antennapedia a homoeotic mutant of Drosophila melanogaster. Proc Natl Acad Sci U S A 64(1):176–183

32. Ayer RK Jr, Carlson J (1991) acj6: a gene affecting olfactory physiology and behavior in Drosophila. Proc Natl Acad Sci U S A 88(12):5467–5471

33. Ayer RK Jr, Carlson J (1992) Olfactory physiology in the Drosophila antenna and maxillary palp: acj6 distinguishes two classes of odorant pathways. J Neurobiol 23(8):965–982. https://doi.org/10.1002/neu.480230804

34. Siddiqi O (1987) Neurogenetics of olfaction in Drosophila melanogaster. Trends Genet 3:137–142. https://doi.org/10.1016/0168-9525(87)90204-6

35. Carlson JR (2001) Functional expression of a Drosophila odor receptor. Proc Natl Acad Sci U S A 98(16):8936–8937. https://doi.org/10.1073/pnas.171311198

36. Vosshall LB, Stocker RF (2007) Molecular architecture of smell and taste in Drosophila. Annu Rev Neurosci 30(1):505–533. https://doi.org/10.1146/annurev.neuro.30.051606.094306

37. Joseph RM, Carlson JR (2015) Drosophila chemoreceptors: a molecular interface between the chemical world and the brain. Trends Genet 31(12):683–695. https://doi.org/10.1016/j.tig.2015.09.005

38. Riesgo-Escovar JR, Piekos WB, Carlson JR (1997) The Drosophila antenna: ultrastructural and physiological studies in wild-type and lozenge mutants. J Comp Physiol A, Sensory, Neural, Behav Physiol 180(2):151–160

39. Schofield PR (1988) Carrier-bound odorant delivery to olfactory receptors. Trends Neurosci 11(11):471–472

40. Nef P (1993) Early events in olfaction: diversity and spatial patterns of odorant receptors. Receptors Channels 1(4):259–266

41. Pelosi P (1994) Odorant-binding proteins. Crit Rev Biochem Mol Biol 29(3):199–228. https://doi.org/10.3109/10409239409086801

42. Pelosi P, Maida R (1995) Odorant-binding proteins in insects. Comp Biochem Physiol B Biochem Mol Biol 111(3):503–514

43. Ziegelberger G (1996) The multiple role of the pheromone-binding protein in olfactory

transduction. Ciba Found Symp 200:267–275; discussion 275–280

44. Fan J, Francis F, Liu Y, Chen JL, Cheng DF (2011) An overview of odorant-binding protein functions in insect peripheral olfactory reception. Genet Mol Res 10(4):3056–3069. https://doi.org/10.4238/2011.December.8.2

45. Leal WS (2013) Odorant reception in insects: roles of receptors, binding proteins, and degrading enzymes. Annu Rev Entomol 58:373–391. https://doi.org/10.1146/annurev-ento-120811-153635

46. Martin F, Boto T, Gomez-Diaz C, Alcorta E (2013) Elements of olfactory reception in adult Drosophila melanogaster. Anat Rec 296(9):1477–1488. https://doi.org/10.1002/ar.22747

47. Brito NF, Moreira MF, Melo AC (2016) A look inside odorant-binding proteins in insect chemoreception. J Insect Physiol 95:51–65. https://doi.org/10.1016/j.jinsphys.2016.09.008

48. Larter NK, Sun JS, Carlson JR (2016) Organization and function of Drosophila odorant binding proteins. eLife 5:e20242. https://doi.org/10.7554/eLife.20242

49. Park SK, Shanbhag SR, Wang Q, Hasan G, Steinbrecht RA, Pikielny CW (2000) Expression patterns of two putative odorant-binding proteins in the olfactory organs of Drosophila melanogaster have different implications for their functions. Cell Tissue Res 300(1):181–192

50. Steinbrecht RA (1996) Structure and function of insect olfactory sensilla. Ciba Found Symp 200:158–174; discussion 174–175

51. Kim MS, Repp A, Smith DP (1998) LUSH odorant-binding protein mediates chemosensory responses to alcohols in Drosophila melanogaster. Genetics 150(2):711–721

52. Couto A, Alenius M, Dickson BJ (2005) Molecular, anatomical, and functional organization of the Drosophila olfactory system. Curr Biol 15(17):1535–1547. https://doi.org/10.1016/j.cub.2005.07.034

53. Hallem EA, Ho MG, Carlson JR (2004) The molecular basis of odor coding in the Drosophila antenna. Cell 117(7):965–979. https://doi.org/10.1016/j.cell.2004.05.012

54. Larsson MC, Domingos AI, Jones WD, Chiappe ME, Amrein H, Vosshall LB (2004) Or83b encodes a broadly expressed odorant receptor essential for Drosophila olfaction. Neuron 43(5):703–714. https://doi.org/10.1016/j.neuron.2004.08.019

55. Hallem EA, Carlson JR (2006) Coding of odors by a receptor repertoire. Cell 125(1):143–160. https://doi.org/10.1016/j.cell.2006.01.050

56. Vosshall LB, Amrein H, Morozov PS, Rzhetsky A, Axel R (1999) A spatial map of olfactory receptor expression in the Drosophila antenna. Cell 96(5):725–736

57. Song E, de Bivort B, Dan C, Kunes S (2012) Determinants of the Drosophila odorant receptor pattern. Dev Cell 22(2):363–376. https://doi.org/10.1016/j.devcel.2011.12.015

58. Robertson HM, Warr CG, Carlson JR (2003) Molecular evolution of the insect chemoreceptor gene superfamily in Drosophila melanogaster. Proc Natl Acad Sci U S A 100(Suppl 2):14537–14542. https://doi.org/10.1073/pnas.2335847100

59. Dweck HK, Ebrahim SA, Farhan A, Hansson BS, Stensmyr MC (2015) Olfactory proxy detection of dietary antioxidants in Drosophila. Curr Biol 25(4):455–466. https://doi.org/10.1016/j.cub.2014.11.062

60. Dweck HK, Ebrahim SA, Thoma M, Mohamed AA, Keesey IW, Trona F, Lavista-Llanos S, Svatos A, Sachse S, Knaden M, Hansson BS (2015) Pheromones mediating copulation and attraction in Drosophila. Proc Natl Acad Sci U S A 112(21):E2829–E2835. https://doi.org/10.1073/pnas.1504527112

61. Stensmyr MC, Dweck HK, Farhan A, Ibba I, Strutz A, Mukunda L, Linz J, Grabe V, Steck K, Lavista-Llanos S, Wicher D, Sachse S, Knaden M, Becher PG, Seki Y, Hansson BS (2012) A conserved dedicated olfactory circuit for detecting harmful microbes in Drosophila. Cell 151(6):1345–1357. https://doi.org/10.1016/j.cell.2012.09.046

62. Kurtovic A, Widmer A, Dickson BJ (2007) A single class of olfactory neurons mediates behavioural responses to a Drosophila sex pheromone. Nature 446(7135):542–546. https://doi.org/10.1038/nature05672

63. Buck L, Axel R (1991) A novel multigene family may encode odorant receptors: A molecular basis for odor recognition. Cell 65(1):175–187. https://doi.org/10.1016/0092-8674(91)90418-X

64. Bear DM, Lassance JM, Hoekstra HE, Datta SR (2016) The evolving neural and genetic architecture of vertebrate olfaction. Curr Biol 26(20):R1039–R1049. https://doi.org/10.1016/j.cub.2016.09.011

65. Dibattista M, Reisert J (2016) The odorant receptor-dependent role of olfactory marker protein in olfactory receptor neurons. J Neurosci 36(10):2995–3006. https://doi.org/10.1523/jneurosci.4209-15.2016

66. Yu Y, de March CA, Ni MJ, Adipietro KA, Golebiowski J, Matsunami H, Ma M (2015) Responsiveness of G protein-coupled odorant receptors is partially attributed to the activation mechanism. Proc Natl Acad Sci U S A 112(48):

14966–14971. https://doi.org/10.1073/pnas.1517510112

67. Leinwand SG, Chalasani SH (2011) Olfactory networks: from sensation to perception. Curr Opin Genet Dev 21(6):806–811. https://doi.org/10.1016/j.gde.2011.07.006

68. Prasad BC, Reed RR (1999) Chemosensation: molecular mechanisms in worms and mammals. Trends Genet 15(4):150–153

69. Nakagawa T, Vosshall LB (2009) Controversy and consensus: noncanonical signaling mechanisms in the insect olfactory system. Curr Opin Neurobiol 19(3):284–292. https://doi.org/10.1016/j.conb.2009.07.015

70. Vosshall LB (2001) The molecular logic of olfaction in Drosophila. Chem Senses 26(2):207–213

71. Kaupp UB (2010) Olfactory signalling in vertebrates and insects: differences and commonalities. Nat Rev Neurosci 11(3):188–200; http://www.nature.com/nrn/journal/v11/n3/suppinfo/nrn2789_S1.html

72. Mayhew PJ (2007) Why are there so many insect species? Perspectives from fossils and phylogenies. Biol Rev Camb Philos Soc 82(3):425–454. https://doi.org/10.1111/j.1469-185X.2007.00018.x

73. Krishnan B, Dryer SE, Hardin PE (1999) Circadian rhythms in olfactory responses of Drosophila melanogaster. Nature 400(6742):375–378. https://doi.org/10.1038/22566

74. Krishnan B, Levine JD, Lynch MK, Dowse HB, Funes P, Hall JC, Hardin PE, Dryer SE (2001) A new role for cryptochrome in a Drosophila circadian oscillator. Nature 411(6835):313–317. https://doi.org/10.1038/35077094

75. Hardin PE, Krishnan B, Houl JH, Zheng H, Ng FS, Dryer SE, Glossop NR (2003) Central and peripheral circadian oscillators in Drosophila. Novartis Found Symp 253:140–150; discussion 150-160

76. Tanoue S, Krishnan P, Krishnan B, Dryer SE, Hardin PE (2004) Circadian clocks in antennal neurons are necessary and sufficient for olfaction rhythms in Drosophila. Curr Biol 14(8):638–649. https://doi.org/10.1016/j.cub.2004.04.009

77. Ge H, Krishnan P, Liu L, Krishnan B, Davis RL, Hardin PE, Roman G (2006) A Drosophila nonvisual arrestin is required for the maintenance of olfactory sensitivity. Chem Senses 31(1):49–62. https://doi.org/10.1093/chemse/bjj005

78. Tanoue S, Krishnan P, Chatterjee A, Hardin PE (2008) G protein-coupled receptor kinase 2 is required for rhythmic olfactory responses in Drosophila. Curr Biol 18(11):787–794. https://doi.org/10.1016/j.cub.2008.04.062

79. Chatterjee A, Roman G, Hardin PE (2009) Go contributes to olfactory reception in Drosophila melanogaster. BMC Physiol 9:22. https://doi.org/10.1186/1472-6793-9-22

80. Poddighe S, Bhat KM, Setzu MD, Solla P, Angioy AM, Marotta R, Ruffilli R, Marrosu F, Liscia A (2013) Impaired sense of smell in a Drosophila Parkinson's model. PLoS One 8(8):e73156. https://doi.org/10.1371/journal.pone.0073156

81. De Rose F, Corda V, Solari P, Sacchetti P, Belcari A, Poddighe S, Kasture S, Solla P, Marrosu F, Liscia A (2016) Drosophila mutant model of Parkinson's disease revealed an unexpected olfactory performance: morphofunctional evidences. Parkinson's Dis 2016:3508073. https://doi.org/10.1155/2016/3508073

82. Schneider D, Kasang G, Kaissling KE (1968) Bestimmung der Riechschwelle vonBombyx mori mit Tritium-markiertem Bombykol. Naturwissenschaften 55(8):395–395. https://doi.org/10.1007/BF00593307

83. Alcorta E (1991) Characterization of the electroantennogram in Drosophila melanogaster and its use for identifying olfactory capture and transduction mutants. J Neurophysiol 65(3):702–714

84. Dubin AE, Heald NL, Cleveland B, Carlson JR, Harris GL (1995) Scutoid mutation of Drosophila melanogaster specifically decreases olfactory responses to short-chain acetate esters and ketones. J Neurobiol 28(2):214–233. https://doi.org/10.1002/neu.480280208

85. Martin F, Charro MJ, Alcorta E (2001) Mutations affecting the cAMP transduction pathway modify olfaction in Drosophila. J Comp Physiol A 187(5):359–370

86. Stortkuhl KF, Kettler R (2001) Functional analysis of an olfactory receptor in Drosophila melanogaster. Proc Natl Acad Sci U S A 98(16):9381–9385. https://doi.org/10.1073/pnas.151105698

87. Park KC, Ochieng SA, Zhu J, Baker TC (2002) Odor discrimination using insect electroantennogram responses from an insect antennal array. Chem Senses 27(4):343–352

88. Rollmann SM, Mackay TF, Anholt RR (2005) Pinocchio, a novel protein expressed in the antenna, contributes to olfactory behavior in Drosophila melanogaster. J Neurobiol 63(2):146–158. https://doi.org/10.1002/neu.20123

89. Drimyli E, Gaitanidis A, Maniati K, Turin L, Skoulakis EM (2016) Differential electrophysiological responses to odorant isotopologues in Drosophilid Antennae. eNeuro 3(3). https://doi.org/10.1523/eneuro.0152-15.2016

90. Martin F, Alcorta E (2016) Measuring activity in olfactory receptor neurons in Drosophila:

focus on spike amplitude. J Insect Physiol 95:23–41. https://doi.org/10.1016/j.jinsphys.2016.09.003

91. Vosshall LB (2000) Olfaction in Drosophila. Curr Opin Neurobiol 10(4):498–503. https://doi.org/10.1016/S0959-4388(00)00111-2

92. Vosshall LB (2008) Scent of a fly. Neuron 59(5):685–689. https://doi.org/10.1016/j.neuron.2008.08.014

93. Kaissling KE (1996) Peripheral mechanisms of pheromone reception in moths. Chem Senses 21(2):257–268

94. Ayyub C, Rodrigues V, Hasan G, Siddiqi O (2000) Genetic analysis of olfC demonstrates a role for the position-specific integrins in the olfactory system of Drosophila melanogaster. Mol Gen Genet 263(3):498–504

95. Boeckh J, Kaissling KE, Schneider D (1965) Insect olfactory receptors. Cold Spring Harb Symp Quant Biol 30:263–280

96. Dubin AE, Harris GL (1997) Voltage-activated and odor-modulated conductances in olfactory neurons of Drosophila melanogaster. J Neurobiol 32(1):123–137

97. Mishra M, Knust E (2013) Analysis of the Drosophila compound eye with light and electron microscopy. Methods Mol Biol 935:161–182. https://doi.org/10.1007/978-1-62703-080-9_11

98. Stark WS, Walker KD, Eidel JM (1985) Ultraviolet and blue light induced damage to the Drosophila retina: microspectrophotometry and electrophysiology. Curr Eye Res 4(10):1059–1075

99. McCann GD, Arnett DW (1972) Spectral and polarization sensitivity of the dipteran visual system. J Gen Physiol 59(5):534–558

100. Minke B, Wu C, Pak WL (1975) Induction of photoreceptor voltage noise in the dark in Drosophila mutant. Nature 258(5530):84–87

101. Harris WA, Stark WS, Walker JA (1976) Genetic dissection of the photoreceptor system in the compound eye of Drosophila melanogaster. J Physiol 256(2):415–439

102. Hardie RC, Raghu P (2001) Visual transduction in Drosophila. Nature 413(6852):186–193. https://doi.org/10.1038/35093002

103. Montell C (1999) Visual transduction in Drosophila. Annu Rev Cell Dev Biol 15:231–268. https://doi.org/10.1146/annurev.cellbio.15.1.231

104. Montell C (2012) Drosophila visual transduction. Trends Neurosci 35(6):356–363. https://doi.org/10.1016/j.tins.2012.03.004

105. Pak WL, Grossfield J, White NV (1969) Nonphototactic mutants in a study of vision of Drosophila. Nature 222(5191):351–354

106. Pak WL, Grossfield J, Arnold KS (1970) Mutants of the visual pathway of Drosophila melanogaster. Nature 227(5257):518–520

107. Heisenberg M (1971) Separation of receptor and lamina potentials in the electroretinogram of normal and mutant Drosophila. J Exp Biol 55(1):85–100

108. Stark WS, Wasserman GS (1972) Transient and receptor potentials in the electroretinogram of Drosophila. Vision Res 12(10):1771–1775

109. Matsumoto H (2012) Proteomics of Drosophila compound eyes: early studies, now, and the future--light-induced protein phosphorylation as an example. J Neurogenet 26(2):118–122. https://doi.org/10.3109/01677063.2012.691923

110. Dolph P, Nair A, Raghu P (2011) Electroretinogram recordings of Drosophila. Cold Spring Harbor Protoc 2011(1):pdb.prot5549. https://doi.org/10.1101/pdb.prot5549

111. Pak WL (2010) Why Drosophila to study phototransduction? J Neurogenet 24(2):55–66. https://doi.org/10.3109/01677061003797302

112. Behnia R, Desplan C (2015) Visual circuits in flies: beginning to see the whole picture. Curr Opin Neurobiol 34:125–132. https://doi.org/10.1016/j.conb.2015.03.010

113. Borst A (2009) Drosophila's view on insect vision. Curr Biol 19(1):R36–R47. https://doi.org/10.1016/j.cub.2008.11.001

114. Hardie RC, Juusola M (2015) Phototransduction in Drosophila. Curr Opin Neurobiol 34:37–45. https://doi.org/10.1016/j.conb.2015.01.008

115. Minke B (2012) The history of the prolonged depolarizing afterpotential (PDA) and its role in genetic dissection of Drosophila phototransduction. J Neurogenet 26(2):106–117. https://doi.org/10.3109/01677063.2012.666299

116. Paulk A, Millard SS, van Swinderen B (2013) Vision in Drosophila: seeing the world through a model's eyes. Annu Rev Entomol 58:313–332. https://doi.org/10.1146/annurev-ento-120811-153715

117. Schopf K, Huber A (2016) Membrane protein trafficking in Drosophila photoreceptor cells. Eur J Cell Biol 96(5):391–401. https://doi.org/10.1016/j.ejcb.2016.11.002

118. Yadav S, Cockcroft S, Raghu P (2016) The Drosophila photoreceptor as a model system for studying signalling at membrane contact sites. Biochem Soc Trans 44(2):447–451. https://doi.org/10.1042/bst20150256

119. Hindle S, Afsari F, Stark M, Middleton CA, Evans GJ, Sweeney ST, Elliott CJ (2013) Dopaminergic expression of the Parkinsonian

gene LRRK2-G2019S leads to non-autono-mous visual neurodegeneration, accelerated by increased neural demands for energy. Hum Mol Genet 22(11):2129–2140. https://doi.org/10.1093/hmg/ddt061

120. Hindle SJ, Elliott CJ (2013) Spread of neuronal degeneration in a dopaminergic, Lrrk-G2019S model of Parkinson disease. Autophagy 9(6):936–938. https://doi.org/10.4161/auto.24397

121. Ziegler AB, Brusselbach F, Hovemann BT (2013) Activity and coexpression of Drosophila black with ebony in fly optic lobes reveals putative cooperative tasks in vision that evade electroretinographic detection. J Comp Neurol 521(6):1207–1224. https://doi.org/10.1002/cne.23247

122. Jaiswal M, Haelterman NA, Sandoval H, Xiong B, Donti T, Kalsotra A, Yamamoto S, Cooper TA, Graham BH, Bellen HJ (2015) Impaired mitochondrial energy production causes light-induced photoreceptor degen-eration independent of oxidative stress. PLoS Biol 13(7):e1002197. https://doi.org/10.1371/journal.pbio.1002197

123. Chouhan AK, Guo C, Hsieh YC, Ye H, Senturk M, Zuo Z, Li Y, Chatterjee S, Botas J, Jackson GR, Bellen HJ, Shulman JM (2016) Uncoupling neuronal death and dysfunc-tion in Drosophila models of neurodegen-erative disease. Acta Neuropathol Commun 4(1):62. https://doi.org/10.1186/s40478-016-0333-4

124. CF W, Wong F (1977) Frequency characteris-tics in the visual system of Drosophila: genetic dissection of electroretinogram components. J Gen Physiol 69(6):705–724

125. Hardie RC, Minke B (1995) Phosphoinositide-mediated phototransduction in Drosophila photoreceptors: the role of Ca^{2+} and trp. Cell Calcium 18(4):256–274

126. Colley NJ (2012) Retinal degeneration in the fly. Adv Exp Med Biol 723:407–414. https://doi.org/10.1007/978-1-4614-0631-0_52

127. Pak WL (1995) Drosophila in vision research. The Friedenwald Lecture. Invest Ophthalmol Vis Sci 36(12):2340–2357

128. Knust E (2007) Photoreceptor morpho-genesis and retinal degeneration: lessons from Drosophila. Curr Opin Neurobiol 17(5):541–547. https://doi.org/10.1016/j.conb.2007.08.001

129. Shieh BH (2011) Molecular genetics of retinal degeneration: a Drosophila perspective. Fly 5(4):356–368. https://doi.org/10.4161/fly.5.4.17809

130. Xiong B, Bellen HJ (2013) Rhodopsin homeostasis and retinal degeneration: lessons from the fly. Trends Neurosci 36(11):652–660. https://doi.org/10.1016/j.tins.2013.08.003

131. Luan Z, Reddig K, Li HS (2014) Loss of Na(+)/K(+)-ATPase in Drosophila pho-toreceptors leads to blindness and age-dependent neurodegeneration. Exp Neurol 261:791–801. https://doi.org/10.1016/j.expneurol.2014.08.025

132. Palladino MJ, Hadley TJ, Ganetzky B (2002) Temperature-sensitive paralytic mutants are enriched for those causing neurodegeneration in Drosophila. Genetics 161(3):1197–1208

133. Phillips SE, Woodruff EA 3rd, Liang P, Patten M, Broadie K (2008) Neuronal loss of Drosophila NPC1a causes cholesterol aggre-gation and age-progressive neurodegenera-tion. J Neurosci 28(26):6569–6582. https://doi.org/10.1523/jneurosci.5529-07.2008

134. Stowers RS, Schwarz TL (1999) A genetic method for generating Drosophila eyes com-posed exclusively of mitotic clones of a single genotype. Genetics 152(4):1631–1639

135. Newsome TP, Asling B, Dickson BJ (2000) Analysis of Drosophila photorecep-tor axon guidance in eye-specific mosaics. Development 127(4):851–860

136. Haelterman NA, Jiang L, Li Y, Bayat V, Sandoval H, Ugur B, Tan KL, Zhang K, Bei D, Xiong B, Charng WL, Busby T, Jawaid A, David G, Jaiswal M, Venken KJ, Yamamoto S, Chen R, Bellen HJ (2014) Large-scale identification of chemically induced muta-tions in Drosophila melanogaster. Genome Res 24(10):1707–1718. https://doi.org/10.1101/gr.174615.114

137. Yamamoto S, Jaiswal M, Charng WL, Gambin T, Karaca E, Mirzaa G, Wiszniewski W, Sandoval H, Haelterman NA, Xiong B, Zhang K, Bayat V, David G, Li T, Chen K, Gala U, Harel T, Pehlivan D, Penney S, Vissers LE, de Ligt J, Jhangiani SN, Xie Y, Tsang SH, Parman Y, Sivaci M, Battaloglu E, Muzny D, Wan YW, Liu Z, Lin-Moore AT, Clark RD, Curry CJ, Link N, Schulze KL, Boerwinkle E, Dobyns WB, Allikmets R, Gibbs RA, Chen R, Lupski JR, Wangler MF, Bellen HJ (2014) A drosophila genetic resource of mutants to study mechanisms underlying human genetic diseases. Cell 159(1):200–214. https://doi.org/10.1016/j.cell.2014.09.002

138. Zhang K, Li Z, Jaiswal M, Bayat V, Xiong B, Sandoval H, Charng WL, David G, Haueter C, Yamamoto S, Graham BH, Bellen HJ (2013) The C8ORF38 homologue Sicily is a cytosolic chaperone for a mitochondrial com-plex I subunit. J Cell Biol 200(6):807–820. https://doi.org/10.1083/jcb.201208033

139. Sandoval H, Yao CK, Chen K, Jaiswal M, Donti T, Lin YQ, Bayat V, Xiong B, Zhang

K, David G, Charng WL, Yamamoto S, Duraine L, Graham BH, Bellen HJ (2014) Mitochondrial fusion but not fission regulates larval growth and synaptic development through steroid hormone production. Elife 3. https://doi.org/10.7554/eLife.03558

140. Chen K, Lin G, Haelterman NA, Ho TS, Li T, Li Z, Duraine L, Graham BH, Jaiswal M, Yamamoto S, Rasband MN, Bellen HJ (2016) Loss of Frataxin induces iron toxicity, sphingolipid synthesis, and Pdk1/Mef2 activation, leading to neurodegeneration. Elife:5. https://doi.org/10.7554/eLife.16043

141. Bayat V, Thiffault I, Jaiswal M, Tetreault M, Donti T, Sasarman F, Bernard G, Demers-Lamarche J, Dicaire MJ, Mathieu J, Vanasse M, Bouchard JP, Rioux MF, Lourenco CM, Li Z, Haueter C, Shoubridge EA, Graham BH, Brais B, Bellen HJ (2012) Mutations in the mitochondrial methionyl-tRNA synthetase cause a neurodegenerative phenotype in flies and a recessive ataxia (ARSAL) in humans. PLoS Biol 10(3):e1001288. https://doi.org/10.1371/journal.pbio.1001288

142. Demerec M (1965) Biology of Drosophila. Hafner Publishing Company

143. Carlson JR (1996) Olfaction in Drosophila: from odor to behavior. Trends Genet 12(5): 175–180

144. Yao CA, Ignell R, Carlson JR (2005) Chemosensory coding by neurons in the coeloconic sensilla of the Drosophila antenna. J Neurosci 25(37):8359–8367. https://doi.org/10.1523/jneurosci.2432-05.2005

145. Galizia CG, Sachse S (2010) Frontiers in neuroscience odor coding in insects. In: Menini A (ed) The neurobiology of olfaction. CRC Press/Taylor & Francis, Boca Raton, FL

Chapter 10

Combined Two-Photon Calcium Imaging and Single-Ommatidium Visual Stimulation to Study Fine-Scale Retinotopy in Insects

Ying Zhu and Fabrizio Gabbiani

Abstract

Individual neurons in several sensory systems receive synaptic inputs organized according to subcellular topographic maps, yet the fine structure of this topographic organization has not been studied in detail. The lobula giant movement detector (LGMD) neuron in the locust visual system is known to receive topographic (retinotopic) excitatory inputs on part of its dendritic tree. To study the fine structure of this retinotopic mapping of visual inputs onto the excitatory dendritic field of the LGMD, we designed a custom microscope allowing visual stimulation at the native sampling resolution of the locust compound eye (defined by a single ommatidium or facet) while simultaneously performing two-photon calcium imaging on excitatory dendrites. Our goal is to provide the reader with detailed guidelines on how to build a custom two-photon microscope and a single-facet stimulation setup and to show experimental results on the detailed retinotopic structure of the projection of excitatory inputs onto the LGMD based on these tools.

Key words Two-photon microscopy, Single-ommatidium visual stimulation, Lobula giant movement detector (LGMD), Calcium imaging, Retinotopic map, Locust

---

1 Introduction

1.1 Two-Photon Microscopy

Starting in the nineteenth century, staining techniques combined with optical imaging of brain tissue initiated the investigation of the morphology of individual neurons and opened the door to neuroscience research. Compared to morphology, studying the function and dynamic activity of neurons in vivo has been considerably more difficult. Extracellular recordings of neuronal spike patterns and intracellular recordings of membrane potential have been used as main tools to study the in vivo activity of the brain for many years. However, simultaneous intracellular recording of the membrane potential dynamics from all the thin and extensive dendrites of a neuron or from hundreds to thousands of neurons remains an outstanding challenge for neurophysiologists.

Roy V. Sillitoe (ed.), *Extracellular Recording Approaches*, Neuromethods, vol. 134,
https://doi.org/10.1007/978-1-4939-7549-5_10, © Springer Science+Business Media, LLC 2018

Two-photon-excited fluorescence laser-scanning microscopy (TPLSM), a nonlinear optical microscopy technique, was invented in 1990 to enable the study of neuronal function at the subcellular level in vivo [1]. In a quantum event, a fluorophore can absorb two photons simultaneously if each photon has half of the energy required to excite the fluorophore. A photon with lower energy has longer wavelength, and the resulting near-infrared excitation employed in TPLSM (700–1000 nm) causes less light scattering and photon damage in the tissue, allowing for deeper penetration depth when compared to the confocal microscope (up to 800 μm to 1 mm in the neocortex; [2]). Since the fluorescent molecule absorbs two photons simultaneously to reach its excited state, the probability of two-photon absorption is proportional to the square of the incident laser intensity. Therefore, only excitation light within the focal region contributes significantly to the fluorescent signal. This nonlinearity of absorption provides an intrinsic mechanism for optical sectioning by the incident beam. In confocal microscopy, the increase in excitation necessary to compensate for the signal loss caused by rejection of scattered and out-of-focus photons by the detector pinhole further exacerbates photobleaching and photodamage. In contrast, the three-dimensional optical sectioning effect of TPLSM decreases the photobleaching of tissue and avoids throughput loss by the pinhole of a confocal laser-scanning microscope. Continuous technological developments in recent years have been made to refine TPLSM technology, mainly to push limits of the imaging speed, resolution, and imaging depth. TPLSM combined with functional markers such as calcium indicators or voltage-sensitive dyes provides a powerful tool to study the function and dynamic activity of single neurons or neuron networks in vivo.

Although commercial two-photon microscopes are available, there are several advantages in building a custom instrument. One important advantage is the flexibility to choose all the components of the system including microscope, scanners, and laser, which allows for tailoring of the system to specific purposes. Price is another important consideration. Commercial systems are considerably more expensive than custom-built ones. Most importantly, the process of building a two-photon microscope provides deeper understanding of how the system works, how to maintain it, and how to modify it to interface with other experimental systems.

1.2 The Lobula Giant Movement Detector (LGMD) in Locusts

The LGMD neuron is a large visual interneuron located in the third neuropil of the optic lobe of the locusts [3]. The LGMD neuron is most sensitive to objects approaching on a collision course with the animal or their two-dimensional object simulations (i.e., looming stimuli; [4–6]) and plays an important role in visually

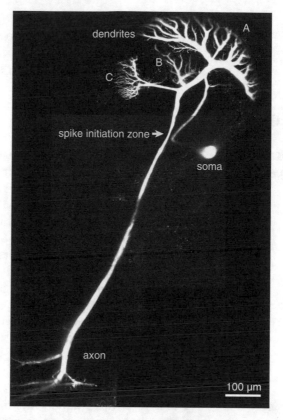

Fig. 1 Two-photon micrograph of an entire LGMD neuron stained with OGB1. The micrograph was obtained by stitching together four images of different parts of the LGMD neuron. Each of the constituent images was obtained in turn by maximal projection of a Z stack of images. The labels A, B, and C denote the three dendritic fields of the LGMD

evoked escape behavior [7, 8]. The LGMD output spikes are coupled 1:1 with those of its postsynaptic target, the descending contralateral movement detector (DCMD). The DCMD neuron has the largest axon in the locust nerve cord and projects to thoracic motor centers where it makes synapses with identified motor and interneurons implicated in the generation of jump and flight responses [9, 10].

The LGMD neuron has three dendritic fields (A–C), which receive visual input from the entire visual hemifield sampled by the ipsilateral compound eye (Fig. 1). The largest one, called field A, receives ~15,000 retinotopic excitatory inputs, whereas fields B and C receive ~500 nonretinotopic feedforward inhibitory inputs. The inputs to dendritic field A have been shown to activate calcium-permeable nicotinic acetylcholine receptors in Ref. [11]. This study employed wide-field charge-coupled device (CCD) camera imaging with a relatively low penetration depth and spatial resolution. As a consequence, fine-scale retinotopy was revealed indirectly, and the range of questions that could be addressed was limited compared to those that can be addressed with the experimental techniques described here.

2 Materials

2.1 Animal Preparations

Experiments were carried out on mature locusts (mostly female), 3–4 weeks past the final molt, taken from a crowded colony. Animals were mounted dorsal side up on a plastic custom holder and immobilized by wax and vacuum grease. The head was rotated 90° with the anterior side pointing downward. The entire head and neck were bathed in ice-cold locust saline [12], except for the right eye used for visual stimulation. The gut was removed. The head capsule was opened dorso-frontally between the two eyes. The muscles in the head capsule were removed. The cuticle and muscles on the neck were removed as well, leaving only the two nerve cords and four tracheas attached. The right optic lobe was desheathed mechanically with fine forceps. A metal holder coated with wax elevated and stabilized the brain and the right optic lobe.

2.2 Staining

The LGMD was uniquely identified through its characteristic spike pattern in 1:1 correspondence with that of its postsynaptic target, the DCMD, recorded extracellularly from the nerve cord with hook electrodes. At the beginning of an experiment, the LGMD neuron was impaled with a sharp intracellular electrode (230–300 MΩ) containing 3 µL of 2 M potassium chloride and 1 µL of the calcium indicator Oregon Green BAPTA-1 (OGB1, hexapotassium salt, 5 mM; Thermo Fisher Scientific, Waltham, MA). Iontophoresis of OGB1 was achieved with current pulses of −3 nA, alternating between 1 s on and 1 s off that lasted for 6 min. The pulses were delivered by an Axoclamp 2B amplifier (Molecular Devices, Sunnyvale, CA). Intracellular signals were amplified in bridge mode with an Axoclamp 2B and an instrumental amplifier (Brownlee Precision 440; NeuroPhase, Palo Alto, CA). Intracellular LGMD and extracellular DCMD signals were acquired using ScanImage and a data acquisition card with a sampling rate of 0.19 MHz (PCI 6110; National Instruments, Austin, TX).

2.3 Imaging, Single-Facet Stimulation, and Electrophysiology

After dissection and staining with OGB1, the animal was moved to the two-photon microscope setup. The excitation wavelength was 820 nm, and the emission filter had a pass band between 500 nm and 550 nm (ET525/50M-2P; Chroma Technology, Bellow Falls, VT). The single-facet stimulation setup was adjusted to focus on the locust's right eye. Each single-facet visual stimulus was an

"OFF" flash lasting 50 ms subtending a size of 22 μm at the retina and focused at the center of an ommatidium on the compound eye (ommatidium diameter: ~24 μm). The brightness of the display before and after the flash was 72.43 cd/m². The brightness of the screen at the level used for "OFF" stimuli was 1.63 cd/m². The brightness of the screen was calibrated by a photometer and linearized by loading a normalized gamma table to the MATLAB Psychtoolbox used to generate the stimuli (see below).

Before carrying out single-facet stimulation, we used a larger flash stimulus centered at the same location as the single-facet stimulus to approximately locate the activated dendritic branches on field A of the LGMD. In our experimental configuration, the dendritic branches of field A extend several tens of micrometers along the Z-axis of the microscope. Consequently, it is time-consuming to find the exact Z-plane of the activated branches under two-photon imaging due to its restricted depth of field (~2 μm). Instead, we employed the microscope in the single-photon excitation CCD mode to obtain a larger depth of field. After finding the activated branches, we switched to the two-photon mode while presenting the single-facet stimuli.

3 Methods

3.1 Building a Custom Two-Photon Microscope

A custom TPLSM is composed of a light source, beam expanders (consisting of two telescopes), a scanning system, dichroic filters, and a light detection module (Fig. 2). The whole system is laid out on an air table. Our construction detailed below mainly follows the principles exposed in Ref. [13] with a few modifications to adapt them for our own needs.

3.1.1 Light Source

For a continuous-wave laser, the probability of the near-simultaneous absorption (within ~0.5 fs) of two photons by a fluorescent molecule is extremely low. Therefore, a temporally resolved high flux of excitation photons is required, usually from a sub-picosecond pulsed laser. Pulsed lasers are not monochromatic. Shorter pulse widths corresponds to broader spectral widths according to the formula:

$$\Delta\lambda \sim \lambda_0^2 / \left(c \cdot \tau_{\text{pulse}}\right)$$

where λ_0 is the center wavelength, c is the speed of light, and τ_{pulse} is the pulse width. In TPLSM the center wavelength of the laser is typically in the near-infrared range of the electromagnetic spectrum, between 700 and 1000 nm. Commercially available titanium-sapphire (Ti:sapphire) systems are capable of generating pulses with a width of 140 fs and a corresponding spectral width of 10–20 nm.

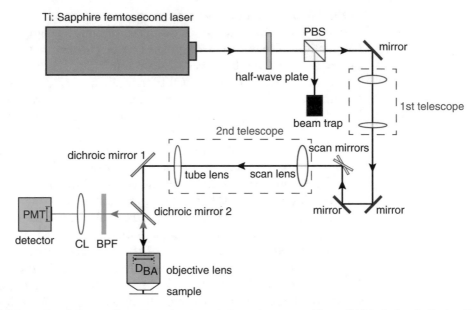

Fig. 2 Schematic of the configuration of a two-photon microscope. The solid black line indicates the laser beam path, with its propagation direction indicated by the black arrowheads. The solid green line indicates the beam path of the fluorescence signal, with its propagation direction indicated by the green arrowheads. Two telescopes are outlined by the blue dashed rectangles. The "scan mirrors" in yellow indicates two scanning mirrors with orthogonal rotational axes. PBS, polarized beam splitter; BPF, band-pass filter; CL, collection lens; PMT, photomultiplier tube; D_{BA}, diameter of the objective lens back aperture

This width is narrower than the width of the two-photon absorption band of many dyes. The average output power and the repetition rate of the laser can be monitored by a photodiode and a fast oscilloscope, respectively. A commercial Ti:sapphire system can provide an average laser power of up to ~3 W, in a wavelength range of 670–1070 nm with a repetition rate of 80 MHz that matches typical fluorescence lifetimes, thus balancing excitation efficiency and dye saturation [14]. A pulsed Ti:sapphire laser boosts the two-photon excitation rate 10^5-fold compared to a continuous-wave laser at the same average power.

Ti:sapphire lasers are linearly polarized. To control the laser power reaching the two-photon microscope, a half-wave plate mounted on a motorized rotation stage is placed at the laser outlet, which rotates the polarization angle of the laser. A polarized beam splitter is used to separate the s- and p-polarization components by reflecting the s-component and passing the p-component. The s-component is dumped to a beam trap, while the p-component is transmitted to the two-photon microscope. To decrease the photodamage caused by the excitation laser, the laser power passed through the objective lens is controlled to be as low as possible (normally below ~100 mW, with penetration of thicker tissue requiring higher power). A large fraction of the laser power is therefore unused and dumped to the beam trap.

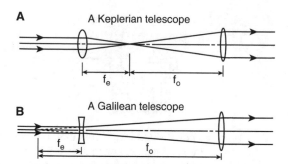

Fig. 3 Two basic types of telescopes. The separation between the two lenses is the algebraic sum of two focal lengths f_o and f_e. The magnification is MP = $|f_o|/|f_e|$. (a) A Keplerian telescope. In this case f_o and f_e are positive. (b) A Galilean telescope. Here f_e is negative and f_o is positive

3.1.2 Laser Beam Reshaping and Expansion

The optical resolution of a lens follows the Rayleigh criterion:

$$r \sim 0.61 \cdot \lambda / NA$$

where r is the minimum distance between resolvable points, λ is the wavelength of light, and NA is the numerical aperture of the lens. Because the radial intensity profile of the laser beam is Gaussian, making use of the full numerical aperture of the objective lens to obtain diffraction-limited focusing requires the laser beam diameter to slightly overfill the diameter of its back aperture (D_{BA}), although this decreases slightly the transmitted power. To take a concrete example, in our setup the beam diameter coming from the laser head is about 1.4 mm and is thus much smaller than D_{BA} (e.g., 18 mm for a 20× Olympus water-dipping objective lens). It is therefore necessary to expand the beam diameter. In addition, due to the limited size of the XY scanning mirrors (~6 mm), using two beam expanders is optimal with the first one expanding the beam from 1.4 to 6 mm before the mirrors and a second one expanding the beam from 6 to 20 mm after the mirrors. There are several ways to expand laser beams, such as the Keplerian (or astronomical) telescope and the Galilean telescope (Fig. 3). The magnification power of both telescopes is the ratio of the focal lengths of their two lenses: MP = $|f_o|/|f_e|$.

3.1.3 Scan System

The objective lens converts a collimated beam into a diffraction-limited focal spot on the focal plane of the objective lens. By using a scan system that changes the angle of incidence of the collimated beam at the back aperture, the focal spot can be moved throughout a desired region of the sample. A standard scan system consists of a scan head, which is typically composed of one or two scan mirrors, a scan lens, a tube lens, and an infinity-corrected objective lens (Fig. 4). The selection of the scan lens influences the imaging

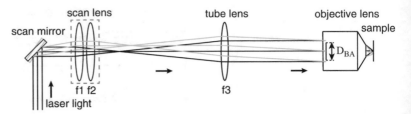

Fig. 4 Optical alignment of a scanning system. The scanning system consists of a scan mirror, a scan lens, a tube lens, and an infinity-corrected objective lens. Black and blue lines indicate the laser path corresponding to the scan mirror at the black and blue positions, respectively. The scan lens consists of two lenses with focal lengths $f1 = 150$ mm, $f2 = 100$ mm. The tube lens has a focal length $f3 = 250$ mm. The pivot point of the scan mirror is imaged onto the center of the back aperture of the objective lens. The scan lens together with tube lens also works as a telescope (beam expander)

quality. Off-axis beam deflection by the scan mirrors through a focusing lens generally results in aberrations, with the image spot being mapped onto a curved surface as opposed to a desirable flat plane. In general, to design a simple custom scan lens, a combination of two or three lenses with large diameters and long focal lengths works better than a single lens of the same effective focal length, because each lens in the combination has a larger curvature radius than that of a single lens. For example, in our system a combination of two 2″ diameter (∅) lenses with focal lengths of 150 and 200 mm (effective focal length = 60 mm) performs better than a single 1″ ∅ lens with a focal length of 60 mm.

Since the size of the scan mirrors is usually smaller than D_{BA}, an appropriate selection of the focal lengths of the scan and tube lenses is required to construct a second telescope system expanding the beam diameter to slightly overfill the back aperture of the objective lens. Two scan mirrors with rotational axes orthogonal to each other are used to generate a simple 2D scan pattern. When aligning these optical components, the conjugate plane of the back aperture of the objective lens needs to be placed halfway between the two scan mirrors. Among the various types of XY scanner designs currently available, such as resonant scanners, acousto-optical scanners and galvanometric scanners, the most common one is the galvanometric scanner, which scans the focused beam across the sample with an adjustable scan speed, allowing software "zoom in" and rotation of the scan axis. In our current system, we use galvanometric scanners from Cambridge Technology (Bedford, MA; model 6215HSM40B). These scanners are capable of rotations of up to ±20° at rates of up to 2 kHz at reduced deflection angles. The aperture size and the scan angle we use are 5 mm and ±5°, respectively. Under a 16 X Nikon objective lens, the deflection of the mirrors corresponds to a distance of 550 μm

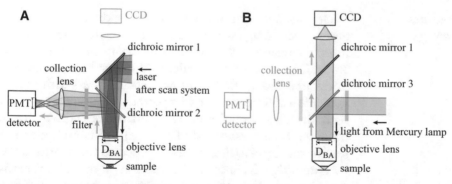

Fig. 5 Fluorescence detection optics. **(a)** Detection optics for two-photon-excited fluorescence. For three different scanning angles, the excitation laser beams are shown in different saturations of red and the emission beams in different saturations of green. The back aperture of the objective lens is focused onto the sensor plane of the PMT. **(b)** Detection optics for traditional wide-field excited fluorescence. Light from a mercury lamp is reflected by the dichroic mirror 3 and excites the sample. The emission beam passes through two dichroic mirrors and is focused onto a CCD

in the focal plane. The scan head is attached to a 2″ Ø post through a custom metal holder and a XY translational stage allowing precise adjustment of the position of the scan head.

3.1.4 Filters and Fluorescence Detection

After passing the scan system, the laser beam is reflected by a short pass dichroic mirror (see Note 1), $\lambda_{cut} = 700$ nm (mirror 1 in Fig. 5a). The fluorescence signal generated from the sample is isotropically emitted over the solid angle of an entire sphere enclosing it (4π). A portion of this fluorescence passes back through the objective lens and is deflected by a second dichroic mirror, which is used to separate excitation and fluorescence, $\lambda_{cut} = 560$ nm (long pass, mirror 2 in Fig. 5a). It then passes through a band-pass filter and a collection lens to the photomultiplier tube (PMT, H7422P-40; Hamamatsu, Bridgewater, NJ). To optimize the collection efficiency, the back aperture of the objective lens is imaged onto the PMT detection plane. The PMT is placed as close to the objective lens as possible to collect as many scattered photons as possible. The PMT is maintained at a constant temperature by monitoring a thermistor which regulates the current to a thermoelectric cooler housed in the PMT case. As the temperature of the PMT outer case rises due to the thermoelectric cooler, a heat sink with a fan attached to the outer case prevents a temperature rise in the PMT. The PMT connects to a preamplifier (current-to-voltage converter, C7319, Hamamatsu) whose output connects to the data acquisition board. It is easy to switch from the two-photon excitation mode to the traditional wide-field epifluorescence excitation mode by exchanging dichroic mirror 2 with dichroic mirror 3 (long pass, $\lambda_{cut} = 505$ nm), as illustrated in Fig. 5b. The whole-field fluorescence imaging system is excited by light from a mercury lamp, and the fluorescence is collected by a CCD camera (Rolera XR, QImaging, Surrey, BC, Canada).

3.1.5 Electronics and Software

Each galvanometric scanner mirror is driven by a MicroMax™ Series positioning system (Cambridge Technology) that receives as input an analog ±10 V signal and transforms it into a positioning angle. The scanner electronics and driver boards ensure that the horizontal and vertical scans are synchronized with each other to generate a coherent raster scan pattern. They are also synchronized with the data acquisition. The MicroMax™ driver boards were mounted on a breadboard (Thorlabs) and connected to the scanner mirrors by electrical cables crimped to the connectors supplied by Cambridge Technology. In addition, as the driver boards generate substantial heat during operation, a custom-made heat sink was connected to each one of them through a thermally conductive phase-change sheet (Alpha Novatech and Laird Technologies). The heat sinks allowed mounting of two fans on top of them that were connected to a power supply to enhance cooling (Cofan-USA and Astrodyne). The data acquisition board we use is a standard peripheral component interconnect (PCI) board (model PCI-6110, National Instruments, Austin, TX), which has two 16-bit analog outputs, eight digital I/O lines, and four simultaneously sampled analog inputs with acquisition rate of 5 MS/s. The data acquisition is carried out using the MATLAB-based software ScanImage r 3.8 (Vidrio Technologies, Ashburn, VA). ScanImage controls the PCI board that sends analog inputs to drive the two galvanometric scanning mirrors for two-photon imaging while simultaneously receiving analog inputs from the preamplifier connected to the PMT. ScanImage also controls the motorized micromanipulator MP285 (Sutter) through a serial port for adjusting the XY position of the animal holder and the rotation angle of the step motor that controls the Z-axis position of the objective lens. We ordered from Sutter and installed in the MP285 a microcontroller providing a firmware upgrade (KS3.4) to correctly interface with ScanImage. A custom serial cable was fabricated to connect the MP285 to the XY motors and the Z-step motor using a wiring diagram provided by Sutter. The stepper motor was fixed on the air table through a custom holder and connected to the manual fine-adjustment front Z focus of the microscope via the fast-feed knob provided by Olympus and a flexible shaft coupler. Microlimiting switches can be mounted next to the microscope body to turn off input to the step motor via the serial cable connection and thus limit travel in the Z direction so as not to damage the microscope in case of a mistake by the operator. In practice, experienced operators will not need them. The start of imaging acquisition can be activated by the onset of a stimulus via a trigger input to the PCI board.

3.1.6 Image Resolution

After the TPLSM was constructed, we calibrated the lateral and axial resolution of the image. The resolution of the microscope can be determined by imaging fluorescent yellow-green carboxyl-

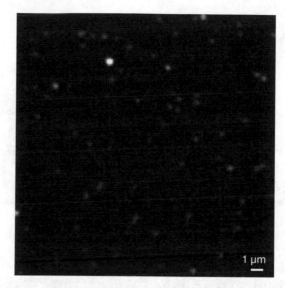

Fig. 6 Maximal Z-projected image of fluorescent microspheres used to calibrate the resolution of our two-photon laser-scanning microscope. The microspheres are 0.1 μm in diameter (Thermo Fisher Scientific, Waltham, MA)

ate modified microspheres (Thermo Fisher Scientific, Waltham, MA) whose size (~0.1 μm) is below the optical resolution limit. The beads are suspended in 2% agarose gel for imaging. To assess the imaging quality, a maximum Z-stack projection was formed (Fig. 6). By computing the full-width at half-maximal intensity of individual beads along the lateral and axial axes using a series of images taken at different depths, we determined the lateral and axial resolutions to be 0.5 and 2 μm, respectively, for a 0.8 NA, 16X magnification Nikon water-dipping objective lens.

3.1.7 Microscope Body   The body of the two-photon microscope is based on the upright Olympus BX51 WI microscope (Fig. 7a). The laser beam that passed through the scanning system and beam expanders enters the microscope from the right through the side input port of a module U-DP that can be purchased from Olympus separately and mounted between the filter turret and the binoculars pieces of the microscope. We drilled four holes in a 30 mm Thorlab cage plate and bolted it to the side input port of the U-DP module. This cage plate was connected to a 30–60 mm cage adapter. This allowed mounting a 60 mm cage system supported on one side by the U-DP module and on the other side by a 1.5″ Ø post positioned immediately before the scanning mirrors. The cage system facilitated the alignment and mounting of the scan lens and the tube lens. As explained in Sect. 3.1.4, two dichroic mirrors are placed inside the microscope. The beam reflects from the first dichroic mirror mounted in the U-DP module using an Olympus

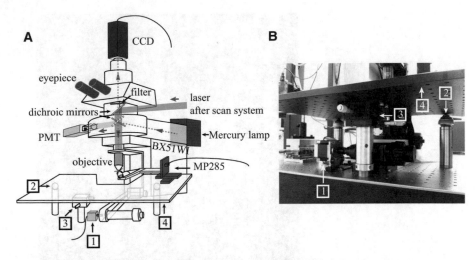

Fig. 7 Realization of the two-photon microscope body. **(a)** Schematics of the modified body of an upright Olympus BX51WI microscope. The two-photon excitation laser beam and emission beam are shown in red and green, respectively. Light coming from the mercury lamp is reflected by a different dichroic mirror (dashed), and its optical path is indicated by the gray dashed line. A band-pass filter (BG39, Schott Corp.) indicated in red on the emission path is used to prevent near-infrared laser light transmission to the eyepieces. A locust under the objective is fixed on an animal holder, which is attached to a motorized micromanipulator (MP285). (1)–(4) are labeling the same objects in (A) and (B). (1) step motor; (2) posts with freely rotating metal balls on top; (3) XY translational stage on a post attached to the table; (4) aluminum breadboard plate. **(b)** Image of part of the two-photon microscope body

filter cube and heads to the objective lens (Fig. 7a). We also mounted in the same filter cube a band-pass filter to prevent any accidental propagation of laser light to the eye pieces (Fig. 7a). The PMT is mounted on the left side of the microscope with a collecting lens placed in a cage system just in front of it. To allow light to exit the microscope on the left side, we drilled a 29 mm Ø hole aligned with the optical path of the microscope and four additional threaded holes for fixing the rods of a 30 mm cage system onto the microscope body (Fig. 7a). To reflect the fluorescent beam to the PMT, we custom fabricated a filter cube similar to the ones inside the microscope turret but rotated 90°. The second dichroic mirror was mounted in it. The mercury lamp used for conventional wide-field epifluorescence is located on the back of the microscope. In Fig. 7b, the dashed line shows the light path after switching to wide-field mode. The animal is placed on a custom holder, which is attached to a motorized micromanipulator that controls the X- and Y-axis position of the animal (MP285, Sutter Instruments, Novato, CA). In addition, an X, Y manual translational stage (two linear stages mounted orthogonally, 25 mm travel distances) mounted on a height-adjustable post that attaches to the air table is attached to an aluminum breadboard serving as an XY platform. The breadboard is also attached to two additional posts with freely rotating metal balls (ball transfer units) on top, to allow independent positioning of the breadboard and animal in its holder (Fig. 7).

3.2 Construction of a Single-Ommatidium Stimulation Setup

Locusts have compound eyes. Each locust compound eye consists of ~7500 ommatidia (facets) and receives visual input from an entire visual hemifield. Because the locust eye is of the apposition type [15], each facet represents the elementary sampling unit of the eye and determines its finest level of spatial resolution. To study the detailed structure of the retinotopic mapping onto the LGMD dendrites, we designed an optical setup that delivers spatially resolved visual stimuli to individual facets [16, 17].

Our single-facet stimulation setup lies perpendicular to the objective lens of the two-photon microscope (Fig. 8a). The space

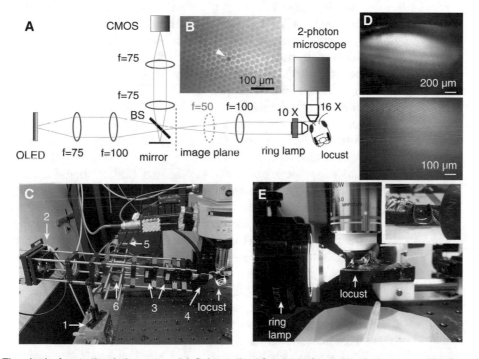

Fig. 8 The single-facet stimulation setup. **(a)** Schematics of a single-facet stimulation setup. The focal length of different lenses is denoted by f and their units are in mm. The $f = 100$ mm lens before the 10× objective can be replaced with a $f = 50$ mm lens (dashed lines) to decrease magnification. **(b)** A single-facet off visual stimulus (dark) at the center of the locust eye is marked by a white arrow. **(c)** Realization of single-facet stimulation setup. The numbers denote different components of the system: 1, a 3D translational stage; 2, an OLED display; 3, an insertable lens mount on one of two lens holders; 4, a ring light mounted on the 10× objective lens; 5, a CMOS camera; and 6, a mirror. **(d)** Images of the locust compound eye taken by the CMOS camera. Upper panel, lower magnification image using the $f = 50$ mm lens before the 10× objective in (A). Lower panel, higher magnification image using the $f = 100$ mm lens before the 10× objective in (A). **(e)** The positioning of the locust between two objectives. The locust after dissection and staining was fixed on a black plastic holder immersed in locust saline. The ring light on the 10× objective was turned on. Inset, a black water holder surrounding the locust eye was made of black tapes and sealed by black wax and grease. The eye may be seen inside the holder

between the animal's right eye and the horizontally positioned objective lens (10×, 0.3 NA, water immersion; CFI Plan Fluor 10XW, Nikon Instruments, Melville, NY) was submerged in water to neutralize the lenses of each ommatidium that have a similar refractive index [18]. The objective lens was focused in the plane of the photoreceptors behind the cornea, visualized with a ring light mounted on the objective (Schott, Southbridge, MA). Visual stimuli were delivered by a miniature organic light-emitting diode (OLED) display (0.5 in., 800 × 600 pixels, 60 Hz refresh rate; OLiGHTEK SVGA050SG). A pellicle beam splitter (BS) directs half of the ring-lamp light reflected from the eye and half of the light from the OLED display toward the CMOS camera sensor where they are both focused by an additional lens, allowing simultaneous visualization of the stimulus and the compound eye. In a sample image taken by the CMOS camera (Fig. 8b), the darker facet at the center is being stimulated with an OFF-stimulus, which subtends a size of 22 μm at the retina. Visual stimuli were generated with the Psychtoolbox and MATLAB (The MathWorks, Natick, MA). By changing the magnification of the horizontal microscope, we can adjust the field of view on the locust eye. One easy way to do so is to use insertable lenses with different focal lengths, which are mounted in lens holders at different distances from the image plane (Fig. 8c). During an experiment, we can first roughly locate the region to be stimulated by using the lower magnification view and next change to higher magnification when delivering single-facet visual stimuli (Fig. 8d). To stimulate facets at different locations on the eye, the use of a three-dimensional translational stage (Newport, Irvine, CA) is helpful (Fig. 8c). After moving the visual stimulus to a certain facet of interest, the ring light is turned off to perform two-photon imaging. Because the horizontal objective is a water immersion lens, it is necessary to build a water tight holder around the locust eye (Fig. 8e inset, see Note 2). The holder we built was made of a curved piece of black tape held in place by black wax and sealed with vacuum grease. This holder also blocked most of the light from the visual display that would have otherwise propagated through the vertical objective lens of the two-photon microscope and therefore reduced the background noise.

3.3 Detailed Structure of the Retinotopic Mapping

We first acquired a low magnification image of dendritic field A (Fig. 9a, top). In response to a single-facet stimulus, several dendritic branches within the yellow rectangle were activated (Fig. 9a, top). On average, we found that ~4 distinct dendritic branches were activated by single-facet stimulation. We then focused on specific dendritic segments that were activated by the stimulus (Fig. 9a, bottom left). The relative fluorescence change was computed as $dF/F = (F(t) - F_0)/F_0$, where F_0 is the baseline fluorescence

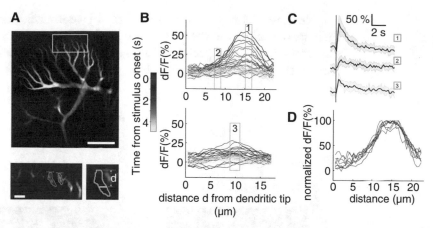

Fig. 9 Calcium responses in the LGMD's excitatory dendrite in response to single-facet stimuli. (**a**) Top, image of a LGMD neuron filled with OGB1. The yellow rectangle indicates the region that contains the dendritic branches generating calcium responses to single-facet stimuli (50 ms OFF flash), scale bar, 100 μm. Bottom left, expanded view of the dendritic branches in the yellow rectangle that were selected for further analysis in (**b–d**) (outlined in green and blue), scale bar, 20 μm. Bottom right, the average dF/F was computed in each successive 2-μm-wide strip (indicated by yellow line) starting from the distal tip of each dendritic branch. d denotes the distance from the center of each strip to the dendritic branch tip. (**b**) Spatial distribution of the average dF/F in each successive 2-μm-wide strip along the branches outlined in green and blue, respectively, in (**a**). Different colors represent different times after the onset of the stimulus. The time courses of dF/F within the green and cyan rectangles are illustrated in (**c**). (**c**) Time course of the average dF/F within the green rectangles (1, 2) and within the cyan rectangle (3) in (**b**). Black curve, mean of 10 trials; cyan and green shading, standard deviation. Red vertical line denotes the flash stimulus of 50 ms duration. (**d**) Normalized spatial distribution of the average dF/F in each successive 2-μm-wide strip along the branches outlined in green in (**a**). Each trace is normalized to the peak. The darkest red trace and the lightest red trace are at 0.4 s and 1.7 s after the onset of the stimuli, respectively. Adapted from Ref. [16], © 2016, The American Physiological Society

computed as the average of the first five frames before the stimulus and the last five frames of the recording. To obtain the spatial distribution of the relative fluorescence change and its evolution with time (Fig. 9b), we segmented each branch in 2-μm-wide strips and computed the average relative fluorescence change within each sequential strip (Fig. 9a, bottom right). On some branches, the peak amplitude of dF/F was much stronger than on others (Fig. 9b). On the same branch, different locations could have large differences in the amplitudes of dF/F (Fig. 9c). One possible cause for the differences in the amplitudes of relative fluorescence change at different locations is the diffusion of calcium from a single entry point along the branch. To address this issue, we examined the time evolution of the dF/F along a single branch. After normalizing each curve to its peak value, we detected almost no change over time in the shape of the spatial dF/F distribution (Fig. 9d). This effect is not expected from diffusion of calcium over the same distance [19].

Next, we delivered visual stimuli to adjacent facets and studied their spatial activation pattern on the LGMD dendrites. Three single-facet stimuli were presented along either the anterior-posterior axis or the dorsal-ventral axis (Fig. 10a, b). We found as noted

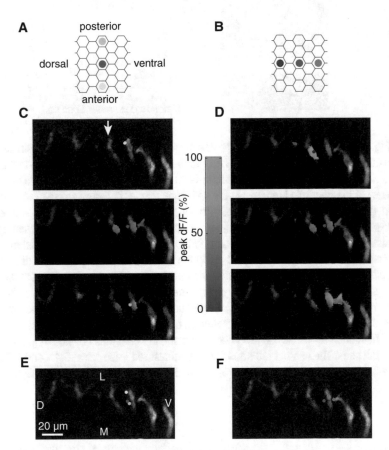

Fig. 10 Detailed structure of the retinotopic map on the LGMD excitatory dendrites. **(a)** Drawing illustrating the relative position of three single-facet stimuli along the anterior-posterior (A-P) eye axis (each stimulus is a 50-ms OFF flash). **(b)** Similar schematics of three single-facet stimuli along the dorsal-ventral (D-V) axis (50 ms OFF flashes). **(c)** Each panel illustrates the response to a single-facet stimulus along the A-P axis. From top to bottom: anterior, medial, and posterior stimuli, as illustrated in **(a)**. The intensity of green shading indicates for each pixel the peak dF/F amplitude at the corresponding dendritic location. In the top panel, note the slight activation of the branch immediately to the left of the dendritic branch with the strongest response (arrow). The small round spots, color-coded as the stimuli in A, represent the centers of mass (COMs) of the peak dF/F across all pixels. **(d)** Peak dF/F in response to three single-facet stimuli along the D-V axis as illustrated in **(b)**. Same plotting conventions as in **(c)**. **(e)** COM of peak dF/F responses for the three single-facet stimuli along the A-P axis, replotted from **(c)**. D, dorsal; V, ventral; M, medial; and L, lateral. **(f)** COM of peak dF/F responses for the three single-facet stimuli along the D-V axis, replotted from **(d)**. Adapted from Ref. [16], © 2016, The American Physiological Society

above that stimulating one facet activated more than one dendritic branch. In addition, stimulating adjacent facets resulted in overlapped activation patterns on the same dendritic branches. We computed the center of mass (COM) of the activated dendritic branches for each facet and plotted its location together with the activated dendritic regions (Fig. 10c, d). When plotting the COMs stimulated by three successive facets along the posterior-anterior axis, we found that they were mapped in the same order along the medial-lateral axis of the LGMD dendrite (Fig. 10e). Similarly, the dorsal-ventral axis on the eye mapped to the dorsal-ventral dendritic axis (Fig. 10f). Thus, the projection from adjacent facets to the LGMD excitatory dendrites preserved retinotopy.

As summarized above, we have shown that stimulating adjacent facets activated overlapping excitatory dendritic branches (Fig. 10c, d). Do adjacent facets share common synapses or do they make independent synapses over the overlapping dendritic region? To address the question, we took advantage of the fact that the excitatory inputs on dendritic field A activated by small visual stimuli (5°) habituate strongly in response to repeated stimulation, and this habituation is most likely to be located at the presynaptic terminal of the LGMD [20]. Therefore, if the same synaptic input is stimulated in rapid succession, the second stimulation is expected to elicit a weaker calcium response. If two facets share a common synaptic input, we would expect that stimulating the second facet resulting in weaker calcium signals after stimulating the first one. Conversely, if the two facets make independent synaptic inputs, then stimulation of the second facet should not dependent on the prior stimulation of the first facet.

In the following experiments, we thus focused on common dendritic branches activated by two adjacent facets. We stimulated each facet twice with an interval of 1 s between stimuli (Fig. 11a, b) and compared the responses to those elicited by alternating stimulation (Fig. 11c, d). The same experiment was repeated with 0.2 and 0.5 s interstimulus intervals. We compute a paired-pulse peak response ratio P2/P1 (Fig. 11e) to compare the response to the second pulse (P2) with the first pulse (P1, Fig. 11d). In comparison to the responses to sequential stimulation of the same facet, we found that paired-pulse ratios increased with the interstimulus interval, as expected from habituation of synaptic responses (Fig. 11e, Ea-a). We also found that at all three intervals of 0.2, 0.5 and 1 s, paired-pulse ratios of the two-facet stimulation protocol (Ea-b) were significantly higher (Fig. 11e, Ea-b). In addition, the paired-pulse ratios did not increase with increasing interstimulus interval. Taken together, these results suggest that different facets make independent synapses onto the excitatory dendrite of the LGMD.

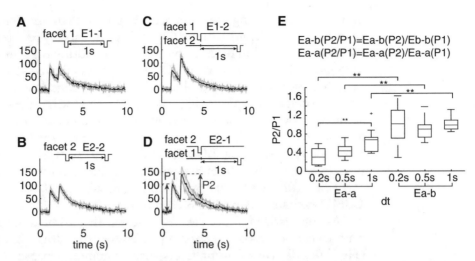

Fig. 11 Calcium signals are consistent with different facets making independent synapses onto the same dendritic branch. (**a** and **b**), responses to two single-facet stimuli separated by a time interval of 1 s at two adjacent facets 1 and 2, respectively. (**c** and **d**), sequential stimulation at facet 1 followed by facet 2 (C) or at facet 2 followed by facet 1 (D) with an interval of 1 s. In (D), the amplitudes of the first and second peaks are marked in red as P1 and P2. In (A)–(D), the time courses of average dF/F are taken from the portion of the dendritic branch activated by both facets 1 and 2. Black curves are means of five trials; gray shades are standard deviations of five trials. Insets are stimulation protocols. (**e**) The ratio of the second peak to the first peak either in the protocol using two different stimuli (Ea-b) or twice the same stimuli (Ea-a) with three different time intervals (dt) of 0.2, 0.5, and 1 s. **Significant difference at the $P < 0.005$ level (Mann-Whitney U-test; 10 Ea-a and 16 Ea-b stimulated facets and recording locations in $n = 4$ animals). Adapted from ref. [16], © 2016, The American Physiological Society

3.4 Checklist: Parts and Vendors for the TPLSM and Single-Facet Stimulation Setup

3.4.1 Mechanical

Anti-vibration air table: 4 ft. (width) × 6 ft. (length) × 8″ (thickness), 783 series, TMC (Technical Manifacturing Corp).

Microscope body: Olympus BX51 WI microscope.

Mirror mounts: 1″ Ø precision kinematic mirror mounts with three adjusters for ±4° pitch and yaw adjustment, Thorlabs, Newton, NJ.

Posts and clamping forks for posts: mirrors were attached to 1/2″ Ø posts. Cage systems and scan head were attached to 1.5″ Ø posts from Thorlabs.

The aluminum breadboard used as XY platform for mounting the animal holder and other equipment was attached to the air table via two ball transfer units mounted on 2″ Ø posts and a XY translational stage controlling a post of adjustable heights (BLP01, Thorlabs). Dimensions: 18″ × 24″ × 1/2″ (Thorlabs).

Ball transfer units: SMC 1/4, load capacity of 75 lbs., Ball Transfer Systems, LLC, Perryopolis, PA.

Assembly rods: 6 mm Ø ER assembly rods for 30 and 60 mm cage systems for mounting lenses of beam expanders, Thorlabs.

Translational stages: XY translational stage for mounting the breadboard/scan head are composed of two orthogonally mounted linear translational stages, 25 mm travel distance,

Thorlabs. The three-dimensional translational stage for the single-facet stimulation setup was model 562-XYZ, 13 mm travel distance, Newport, Irvine, CA.

3.4.2 Optical

Ti:sapphire laser: average power >2.5 W, 690–1020 nm, 80 MHz repetition rate, Chameleon Ultra, Coherent, Santa Clara, CA

Mirrors: silver coated, $\lambda/10$ flatness, Thorlabs

Lenses: achromatic doublets, antireflective coating for 650–1050 nm, 1″ or 2″ Ø, Thorlabs

Dichroic mirrors: long pass dichroic T560lpxrxt (λ_{cut} = 560 nm), short pass dichroic T700DCSPXR (λ_{cut} = 700 nm), long pass dichroic 41001 (λ_{cut} = 505 nm), Chroma Technology Corp, Bellows Falls, VT

Objective lenses: for single-facet stimulation (0.3 NA, CFI Plan Fluor, 10XW), for two-photon imaging (0.8 NA, CFI75, 16XW), Nikon Instruments, Melville, NY

Custom scan lens: two lenses with focal lengths of f = 100 mm and f = 150 mm (2″ Ø achromatic doublets, AR coating: 650–1050 nm) are mounted in a SM2 lens tube, Thorlabs

Wave plate: 1/2″ Ø mounted achromatic half-wave plate, 690–1200 nm, Thorlabs

Polarized beam splitter: broadband polarizing cube beam splitter, 25.4 mm, 620–1000 nm, 10FC16PB.5, Newport

Pellicle beam splitter: cube-mounted, CM1-BP145B1, Thorlabs

Scanners: 6215HSM40B, Cambridge Technololgy, Bedford, MA

OLED display: SVGA050SG, 0.5 in., 800 × 600 pixels, 60 Hz refresh rate, OLiGHTEK, Yunnan, China

3.4.3 Electronics

Motorized rotation stage (for mounting the wave plate): PRM1Z8, Thorlabs.

Step motor: two-phase, 1.8° (step angle), PK266-01A, Oriental Motor, Torrance, CA.

Motorized micromanipulator for controlling the XY motion of the animal holder and rotation of the step motor that controls the Z-axis motion of the objective lens: MP285, Sutter Instruments, Novato, CA, with KS3.4 firmware upgrade.

Data acquisition board: PCI-6110, National Instruments, Austin, TX.

Photomultiplier tube (PMT): H7422P-40, sensitive range 300–720 nm; temperature control and power supply for the PMT: C8137–02, Hamamatsu, Bridgewater, NJ.

Preamplifier for the PMT: current-to-voltage conversion, C7319, Hamamatsu. Input voltage to the preamplifier is provided by a DC power supply, E3630, Agilent, Santa Clara, CA.

3.4.4 Miscellaneous

Power supplies: Astrodyne SP-240-24 (one for each MicroMax™ driver), Astrodyne ESCC-1502 (for the four-driver cooling fans)

Heat sink: UB13070-25BM (Alpha Co., Japan)

Driver cooling fans: F-6025H12B-R, Co-FAN USA

Coupler for step motor: Helical Products Co, DSAC-100-11-8

Thermally conductive phase change film: T-pcm905C, 102×63.5 mm (Laird Technologies)

Crimp tools: Molex 63811–7000, 63811–8700

Microlimiting switches: 311SX3-T (Honeywell)

3.4.5 MP285 (Sutter) DB25 Wiring

1. Output to motors. Ground: Pins 3, 7, 11, 17, 25; Power XB: 4; Power XA: 16; Power YB: 8; Power YA: 20; Power ZB: 12; Power ZA: 24

2. Sensors to input: end of travel switch data, optical left (OPTL) and right (OPTR). X-OPTL: 14; X-OPTR: 2; Y-OPTL: 18; Y-OPTR: 6; Z-OPTL: 22; Z-OPTR: 10.

3. Power. +5VDC: 15, 19, 23; GND: 1, 5, 9.

4 Notes

1. The flatness of the dichroic mirrors used in the optical path of the laser is extremely important for the reflected wave front. If the mirror is not sufficiently flat, it may cause "defocusing" or other aberrations such as astigmatism to the reflected beam, although the transmitted wave front is often not affected. The manner in which a dichroic mirror is mounted will affect its flatness. Mounting using metal clips is not as good a choice for a thin dichroic mirror (<2 mm) as properly gluing it to a holder, since the heterogeneous torque stress caused by the clips can severely affect the flatness, which deteriorates the two-photon imaging quality. When we first built our two-photon setup, we failed to realize that it was the mounting of the dichroic mirror by metal clips that caused a weird shape of the point spread function. To debug the problem, we diverted the beam at different locations on the beam path by placing a mirror on the beam path, fed the reflected beam to a lens with a long focal length (f = 400 mm), and focused the beam onto the sensor of a CCD camera mounted on a motorized linear translational stage. The motorized stage allowed us to move the CCD camera slightly in and out of focus to check the shape of the focal spot. Eventually, we found an aberrated shape of focal spot only at locations right after one dichroic

mirror, which was mounted by means of metal clips. Changing the dichroic mirror with one glued to a holder solved the problem.

2. The shape of the water surface in the water cap between the locust eye and the horizontal objective lens affects the refraction of light. During an experiment, leakage or evaporation of water could change the stimulation location on the eye. Therefore, it is necessary to check whether the stimulus remains focused onto the same facet from time to time.

5 Conclusions

Two-photon microscopy provides high spatial resolution for in vivo experiments, in spite of tiny motion artifacts that often occur during live imaging. It also penetrates deeper and causes less photodamage than confocal microscopy. The design and construction of a custom two-photon microscope not only helps better understand the principles of two-photon microscopy but also enables further modifications and improvements to the system to adapt to more complex experimental requirements. Single-facet stimulation allows to accurately probe the visual system at the native spatial resolution of the locust compound eye. In conjunction with two-photon calcium imaging, single-facet stimulation proved to be a powerful tool to dissect the fine structure of the retinotopic mapping from the locust eye onto the excitatory dendrite of the LGMD, as well as the dynamics of calcium signals in response to time-varying visual stimuli. In particular, they revealed the overlapped but independent wiring scheme of adjacent facets onto LGMD dendrites. This detailed description is necessary to accurately model the mechanisms of synaptic integration within the LGMD and better understand the dendritic computations carried out by this neuron. These tools could also be applied to study other visual neurons in the locust and the visual system of other animals with compound eyes, such as the fly, honeybee, or dragonfly, for instance.

Acknowledgments

We would like to thank Dr. Peter Saggau for the technical assistance in building the two-photon microscope. Thanks are also given to Dr. Peter Saggau, Dr. Richard Dewell, and Dr. Hongxia Wang for the corrections and comments on this manuscript. This work was supported by grants from the National Science Foundation and the National Institutes of Health as well as a National Eye Institute/NIH Core Grant for Vision Research.

References

1. Denk W, Strickler JH, Webb WW (1990) Two-photon laser scanning fluorescence microscopy. Science 248:73–76

2. Theer P, Hasan MT, Denk W (2003) Two-photon imaging to a depth of 1000 μm in living brains by use of a Ti:Al$_2$O$_3$ regenerative amplifier. Opt Lett 28:1022–1024

3. O'Shea M, Williams JL (1974) The anatomy and output connection of a locust visual interneurone: the lobular giant movement detector (LGMD) neurone. J Comp Physiol 91:257–266

4. Rind FC, Simmons PJ (1992) Orthopteran DCMD neuron: a reevaluation of responses to moving objects. I. Selective responses to approaching objects. J Neurophysiol 68:1654–1666

5. Hatsopoulos N, Gabbiani F, Laurent G (1995) Elementary computation of object approach by a wide-field visual neuron. Science 270:1000–1003

6. Schlotterer GR (1977) Response of the locust descending movement detector neuron to rapidly approaching and withdrawing visual stimuli. Can J Zool 55:1372–1376

7. Fotowat H, Gabbiani F (2011) Collision detection as a model for sensory-motor integration. Annu Rev Neurosci 34:1–19

8. Santer RD, Rind FC, Stafford R, Simmons PJ (2006) Role of an identified looming-sensitive neuron in triggering a flying locust's escape. J Neurophysiol 95:3391–3400

9. Burrows M, Rowell CH (1973) Connections between descending visual interneurons and metathoracic motoneurons in the locust. J Comp Physiol 85:221–234

10. O'Shea M, Rowell CH, Williams J (1974) The anatomy of a locust visual interneurone; the descending contralateral movement detector. J Exp Biol 60:1–12

11. Peron SP, Jones PW, Gabbiani F (2009) Precise subcellular input retinotopy and its computational consequences in an identified visual interneuron. Neuron 63:830–842

12. Laurent G, Davidowitz H (1994) Encoding of olfactory information with oscillating neural assemblies. Science 265:1872–1875

13. Tsai PS, Nishimura N, Yoder EJ, Dolnick EM, White GA, Kleinfeld D (2002) Principles, design, and construction of a two-photon laser-scanning microscope for *in vitro* and *in vivo* brain imaging. In: Frostig RD (ed) *In vivo* optical imaging of brain function. CRC Press, Boca Raton, FL

14. Helmchen F, Denk W (2005) Deep tissue two-photon microscopy. Nat Methods 2:932–940

15. Shaw SR (1969) Optics of arthropod compound eye. Science 164:88–90

16. Zhu Y, Gabbiani F (2016) Fine and distributed subcellular retinotopy of excitatory inputs to the dendritic tree of a collision-detecting neuron. J Neurophysiol 115:3101–3112

17. Jones PW, Gabbiani F (2010) Synchronized neural input shapes stimulus selectivity in a collision-detecting neuron. Curr Biol 20:2052–2057

18. Franceschini N, Kirschfeld K (1971) Pseudopupil phenomena in the compound eye of Drosophila. Kybernetik 9:159–182

19. Helmchen F, Nägerl UV (2016) Biochemical compartmentalization in dendrites. In: Stuart G, Spruston N, Häuser M (eds) Dendrites, 3rd edn. Oxford Univ. Press, Oxford

20. O'Shea M, Rowell CH (1976) The neuronal basis of a sensory analyser, the acridid movement detector system. II. Response decrement, convergence, and the nature of the excitatory afferents to the fan-like dendrites of the LGMD. J Exp Biol 65:289–308

Chapter 11

Extracellular Loose-Patch Recording of Purkinje Cell Activity in Awake Zebrafish and Emergence of Functional Cerebellar Circuit

Jui-Yi Hsieh and Diane M. Papazian

Abstract

The cerebellum calibrates movements for accuracy and precision and is critical for motor learning. Purkinje cells, which are the sole output neurons of the cerebellar cortex, play an essential role in fine-tuning motor behavior. Determining how sensory stimuli, after processing by higher brain centers, modulate Purkinje cell activity and how changes in Purkinje cell firing improve the accuracy of movement is crucial for understanding cerebellar function. Due to its rapid development, small size, and transparency, the zebrafish is a promising system for functional brain mapping and for investigating the neural control of behavior in a vertebrate animal. To aid in understanding the role of the cerebellum in behavior, we have developed an approach for recording the electrical activity of cerebellar Purkinje cells in live, awake zebrafish using extracellular loose-patch electrodes. Using this method, we have found that zebrafish Purkinje cells fire simple and complex spikes similar to those seen in mammalian Purkinje cells. Spontaneous firing and a functional cerebellar circuit emerge early in zebrafish development in parallel with complex, visually guided behaviors. This preparation has significant advantages for investigating the effects of sensory stimuli on Purkinje cell firing, mapping the functional effects of afferent inputs to the cerebellum, and exploring how the cerebellum fine-tunes motor behavior.

Key words Purkinje cell, Cerebellar circuit, Zebrafish, Complex spike, Pacemaking activity, Brain mapping

1 Introduction

The cerebellum is responsible for insuring the accuracy and precision of movement and for learning new motor skills. The mammalian cerebellum has a trilaminar architecture composed of the granule cell, Purkinje cell, and molecular layers. Input to the cerebellum is carried by mossy fibers that synapse onto granule cells. Granule cells send their axons into the molecular layer where they form excitatory parallel fiber synapses onto the dendrites of GABAergic Purkinje cells, which are the sole output neurons of the cerebellar cortex. Purkinje cells also receive excitatory input

Roy V. Sillitoe (ed.), *Extracellular Recording Approaches*, Neuromethods, vol. 134,
https://doi.org/10.1007/978-1-4939-7549-5_11, © Springer Science+Business Media, LLC 2018

from inferior olivary neurons via climbing fibers that, according to the current consensus view, provide error correction signals that fine-tune movement and promote motor learning by triggering plastic changes in synaptic strength at parallel fiber-Purkinje cell synapses [1]. Output from Purkinje cells is received by neurons of the deep cerebellar nuclei.

In mammals, Purkinje cells are spontaneously active, firing simple, Na^+-based spikes in a regular pattern [2]. Pacemaking activity is controlled by voltage-gated ion channels with specialized gating properties including Nav1.6, which conducts resurgent Na^+ current in the interspike interval to trigger the next action potential, and Kv3.3, which opens in a depolarized voltage range with rapid activation and deactivation kinetics to maintain the supply of available Na^+ channels [3–6]. The frequency of spontaneous tonic firing is modulated by activity at parallel fiber synapses [7]. Climbing fiber input evokes complex spiking, a second action potential waveform in Purkinje cells characterized by an initial large amplitude spike followed by a prolonged depolarization during which several spikelets may occur [8]. Coincident activation of parallel and climbing fiber inputs results in long-term depression at parallel fiber synapses [1, 8]. This plasticity is thought to contribute to motor learning [1, 8].

Little is known about how changes in Purkinje cell output fine-tune motor behavior. Investigating the effects of afferent inputs on Purkinje cell firing and the cerebellar control of behavior requires animal models in which in vivo experiments are feasible. In mice, significant strides have been made in using optogenetics to control Purkinje cell activity in vivo while monitoring movement and associative motor learning [9]. However, such experiments are technically challenging. Furthermore, the mouse cerebellum contains an estimated 140,000–200,000 Purkinje cells depending on age and gender [10, 11]. The complexity of the cerebellum in higher vertebrates and the heterogeneity of afferent inputs make it difficult to investigate cerebellar fine-tuning of movement during behaviors driven by sensory stimuli. As a result, there is an important role for simpler vertebrate systems to investigate the cerebellar control of behavior and to complement studies in mammals [12].

The zebrafish (*Danio rerio*), a lower vertebrate, is an emerging model system with significant advantages for neuroscience and functional brain mapping [13–16]. Zebrafish eggs are fertilized and develop outside the body of the mother, facilitating the analysis of all developmental stages [17, 18]. Nervous system development begins at ~9 h postfertilization (hpf) with the formation of the neural plate [17]. Primary neurogenesis begins soon after [17]. Synaptic transmission at the neuromuscular junction is evident by 21 hpf [19]. A host of molecular and genetic tools exist to control gene expression, make mutations, and target expression to particular cell types [14]. Germline transgenic animals are readily produced [20]. In addition, F0 transgenesis, in which embryos are injected with plasmid DNA and then raised to the appropriate age for experiments, provides a rapid way to change gene expression in a mosaic pattern, including in single cells [20]. Zebrafish are optically clear early in life, and the window of clarity can be extended with albino strains or treatment with phenylthiourea [17, 18]. As a result, time-lapse imaging, optogenetics, and calcium imaging using genetically encoded indicators are easily performed in live animals [14, 15, 21].

One limitation of zebrafish for neuroscience research is that electrophysiological analysis of brain function has been relatively rare. Many studies, particularly for brain mapping, rely on calcium imaging to detect neuronal activity [15, 22]. Although calcium imaging has the advantage that global mapping of brain activity can be performed using advanced microscopic techniques [23, 24], some drawbacks remain. Without electrophysiological analysis, calcium signals cannot be directly correlated with the underlying electrical activity [25]. In some cases, increases in calcium may not signal excitation. Initial rises in calcium, whether due to influx across the plasma membrane or release from intracellular stores, may decrease excitability by opening calcium-activated potassium channels [26]. Furthermore, calcium indicators may lack the sensitivity to detect small, localized changes in calcium corresponding to single- or low-amplitude spikes. As a result, the absence of calcium signals cannot be interpreted as the lack of firing.

To investigate the role of the cerebellum in motor behavior and to facilitate functional mapping of afferent cerebellar inputs, we have developed a simple method for loose-patch, extracellular recording of electrical activity in Purkinje cells in living, awake zebrafish. The zebrafish cerebellum contains three lobes, one of which, the corpus cerebelli, has the same trilaminar architecture consisting of granule cell, Purkinje cell, and molecular layers in the same orientation as found in the mammalian cerebellum (Fig. 1) [27–29]. In zebrafish, Purkinje cells are born, migrate, and send out neurites on the 3rd day postfertilization (dpf) [27, 29, 30]. In larval zebrafish, the corpus cerebelli contains ~200–300 Purkinje cells [30]. This is approximately 500- to 1000-fold fewer than found in the mouse cerebellum [10, 11], greatly simplifying analysis of the role of Purkinje cells in motor behavior and motor learning. The zebrafish cerebellum is compact. Purkinje cells synapse onto eurydendroid cells located in the granule cell layer, which fulfill the roles of deep cerebellar nuclei neurons in higher vertebrates (Fig. 1) [27–29].

This chapter describes our approach for recording the electrical activity of Purkinje cells in living, awake zebrafish [31]. We have used this method to investigate the development of the functional cerebellar circuit in zebrafish.

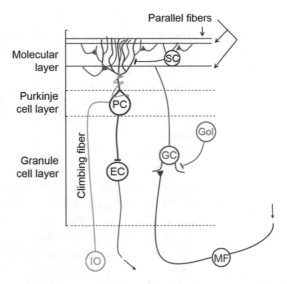

Fig. 1 Conserved trilaminar architecture of cerebellar circuit in zebrafish. A schematic diagram of the zebrafish corpus cerebelli lobe of the cerebellum shows cell types and connections identified anatomically [27]. PC, Purkinje cells; GC, granule cells; IO, inferior olivary neurons, EC, eurydendroid cells (the zebrafish equivalent of deep cerebellar nuclei neurons); SC, stellate cells; Gol, Golgi cells; MF, mossy fibers. At axon ends, triangles indicate excitatory synapses; bars indicate inhibitory synapses. Basket cells have not yet been identified in the zebrafish cerebellum [27–29]. Wiring diagram adapted from [27]

2   Materials and Methods

2.1 Electrophysiology Rig

An electrophysiology rig used for brain slices is suitable for recording from Purkinje cells in live zebrafish [31]. Our rig is equipped with a HEKA Elektronik EPC 10 patch clamp amplifier controlled by a desktop computer running Pulse software. Experiments are performed under an upright epifluorescence Olympus BX51WI microscope equipped with LUMPlanFL N 40X/0.80 and UMPlanFL N 10X/0.3 water immersion objectives. Alternatively, we have used a rig equipped with an upright laser scanning confocal microscope and an Axopatch 200B amplifier controlled by pClamp software. Experiments are performed on an anti-vibration table equipped with a Faraday cage. The patch electrode is advanced toward the preparation using a motorized micromanipulator (Sutter Instruments, #MP-225). Although our experiments are typically performed at ambient temperature, the temperature of the recording chamber (Siskiyou PC-R #15280000E) can be controlled using an in-line solution heater and automatic temperature controller (Warner Instruments, #SH-27B and #TC-324B).

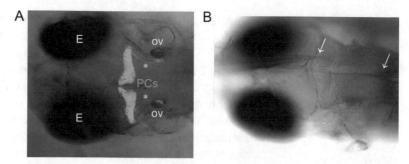

Fig. 2 Transgenic fish expressing membrane-bound Venus specifically in cerebellar Purkinje cells. (**a**) An image stack acquired using an Olympus Fluoview FV300 laser scanning confocal microscope shows a dorsal view of the head of a la118Tg:*Tg(aldoca:gap43-Venus)* transgenic zebrafish at 5 dpf [31]. The image stack has been superimposed on the corresponding bright field image. Anterior is to the left. PCs, Purkinje cells; E, eyes; ov, otic vesicles; *, cerebellovestibular axon tracks. (**b**) The bright field image shows a transgenic fish in recording configuration at 5 dpf [31]. A recording electrode (*white arrow*) forms a loose patch onto a cerebellar Purkinje cell. Also shown is a theta electrode (*yellow arrow*) that has been inserted into the inferior olive for direct electrical stimulation

2.2 Expression of Fluorescent Proteins in Zebrafish Purkinje Cells

When learning to record from Purkinje cells in live zebrafish, the cells are most easily identified if they express a fluorescent reporter protein. Two transgenic lines have been made that express membrane-bound Venus, a yellow fluorescent protein, specifically in Purkinje cells under the control of the *aldolase Ca* (*aldoca*) promoter [31, 32] (Fig. 2a). *Aldoca* encodes zebrin-II, a well-characterized Purkinje cell marker [33]. All Purkinje cells in zebrafish are thought to express zebrin-II [27]. The rk22Tg:*Tg(aldoca:gap43-Venus)* line, which was generated in the standard Oregon AB laboratory strain, is pigmented [32]; the la118Tg:*Tg(aldoca:gap43-Venus)* line, which was generated in the albino Tüpfel longfin nacre (TLN) strain, is unpigmented (Fig. 2a) [31]. The albino line is suitable for performing live imaging, calcium imaging, and optogenetics in the same animals used for electrophysiology. Alternatively, transient F0 transgenesis can be used to express a fluorescent protein in a subset of Purkinje cells [20]. In this approach, a Tol2 plasmid encoding a fluorescent protein under the control of the *aldoca* promoter is injected into single-celled embryos, which grow up to express the plasmid in a mosaic pattern [20, 32]. With experience, unlabeled Purkinje cells can be reliably recognized by their morphology and position near the dorsal surface of the cerebellum.

2.3 Breeding Zebrafish to Obtain Animals for Experiments

NIH policy requires that experiments using zebrafish that are 3 dpf or older (after hatching) be approved by the Institutional Animal Care and Use Committee (IACUC) (https://oacu.oir.nih.gov/sites/default/files/uploads/arac-guidelines/zebrafish.pdf).

A breeding colony is used to produce zebrafish of the appropriate age for experiments (3–14 dpf). Male and female breeders (3–12 months of age) are housed in separate tanks in the aquarium facility at 28.5 °C using a 14 h/10 h light/dark cycle [17, 18]. Breeders should not be crossed more frequently than once per week. In the early evening, one male and one female zebrafish each are transferred to the separated compartments of breeding chambers filled with aquarium water [17]. Breeding chambers consist of an inner tank with a perforated bottom, a removable divider to form two compartments, an outer tank, and a lid (Aquaneering, #ZHCT100). The next morning, within 30 min after the lights come on in the aquarium facility, the divider in each chamber is removed to initiate mating. Fertilized eggs fall through the perforated bottom of the inner tank. After mating, the adults are returned to their home tanks. Embryos are collected from the outer tank by pouring the water through a fine sieve and gently transferred to a 100 mm × 14 mm Petri dish containing E2 medium (in mM: 15 NaCl, 0.5 KCl, 0.5 $MgSO_4$, 0.15 KH_2PO_4, 0.04 Na_2HPO_4, 0.13 $CaCl_2$, 0.07 $NaHCO_3$) [17]. Under a stereomicroscope (Zeiss Stemi 2000C equipped with a transmitted light base and an adjustable mirror), a disposable transfer pipette is used to move the embryos to fresh Petri dishes at a density of ~50 animals per dish, leaving debris and unfertilized eggs behind. Approximately 5 h later, any embryos that look abnormal or have not reached 50% epiboly, a developmental stage in which a smooth layer of cells covers half the surface of the embryo [17], are removed and euthanized by immersion in 1% sodium hypochlorite. Embryos are raised in E2 medium containing 0.01% methylene blue to suppress fungal growth [17]. Embryos develop in a 28.5 °C incubator with a 14 h/10 h light/dark cycle matching that of the aquarium facility. Illumination is provided by wide-spectrum LED strips on a timer. Each day, abnormal and dead animals and discarded chorions, if any, should be removed, and fresh medium, pre-warmed to 28.5 °C, should be added. Starting at 4 dpf, zebrafish are bathed in pre-warmed aquarium water rather than E2 medium. Starting at 5 dpf, zebrafish are fed twice per day using GP 100–200 μm larval diet, available at http://www.brineshrimpdirect.com/. The aquarium water should be replaced after the second feeding.

2.4 Exposing the Cerebellum

Because electrophysiological recordings are made in live animals, the following procedure constitutes survival surgery [31]. A zebrafish (3–14 dpf) is anesthetized by immersion in 0.02% FDA-approved MS-222 (Tricaine-S, Western Chemical). Dissection to expose the cerebellum is performed under the Zeiss stereomicroscope on a glass cover slip held in place in the recording chamber by a dab of petroleum jelly. A small amount of Krazy Glue® is applied to the middle of the cover slip. The insensate fish is gently transferred to the glue using Dumont #5SF forceps (Fine Science Tools). The animal's head should be placed dorsal side up outside the spot of glue, whereas the tail should be released onto the glue.

The zebrafish is then paralyzed by gently filling the chamber with external recording solution (134 mM NaCl, 2.9 mM KCl, 2.1 mM CaCl$_2$, 1.2 mM MgCl$_2$, 10 mM HEPES, 10 mM glucose, pH 7.5) containing 10 μM (+)-tubocurarine hydrochloride [31]. To minimize the likelihood of getting glue on the animal's head, the solution is introduced at the rostral end of the fish so that it flows in the caudal direction. Note that this solution does not contain MS-222, which abolishes electrical activity in zebrafish Purkinje cells. After removal of MS-222, the animal remains insensate for 5–10 min. The cerebellum should be exposed immediately so that the zebrafish remains anesthetized during the dissection.

The skin and skull overlying the cerebellum are removed using two pairs of Dumont #5SF forceps. The head is anchored by grasping the skin immediately caudal to the brain with one pair of forceps. The second pair of forceps is then used to peel off the skin and skull from the brainstem to forebrain using one smooth movement. The exposed brain should appear translucent and should be free of pigment and blood vessels.

With practice, the entire procedure can be performed in less than 2 min. When perfecting the technique, it is important to keep the animal in solution containing 0.02% MS-222 during the dissection to ensure that anesthesia is maintained.

2.5 Loose-Patch Recording of Electrical Activity in Purkinje Cells in Living, Awake Zebrafish

After the dissection, the recording chamber is carefully transferred to the rig and fixed to the microscope stage. Recordings are made in external solution containing 10 μM (+)-tubocurarine hydrochloride. Zebrafish recover from anesthesia in 5–10 min after MS-222 is removed, so the animals are awake during the recording. Patch pipettes are made from borosilicate glass (World Precision #1B150F-4) using a Flaming-Brown microelectrode puller (Sutter Instruments, #P-97) and filled with filtered external solution lacking (+)-tubocurarine hydrochloride. Pipette resistance should be 7–10 MΩ. With the amplifier in voltage clamp mode, a pellet-type Ag/AgCl reference electrode is lowered into the bath, and the voltage is adjusted to 0 mV. Applying gentle positive pressure to the patch electrode, it is advanced at an angle of 30° relative to horizontal toward a Purkinje cell in the corpus cerebelli lobe using the motorized micromanipulator. Patches are most readily made when the electrode approaches the cerebellum from the rostral or caudal directions rather than from the side (Fig. 2b). Upon contact with a cell, determined visually, the positive pressure is released. A gentle negative pressure is then applied to form a loose seal (20 MΩ to 2 GΩ) on the extracellular surface of the cell. Seal formation enhances the signal-to-noise ratio and maximizes the probability of making single-unit recordings. Pipette capacitance is compensated

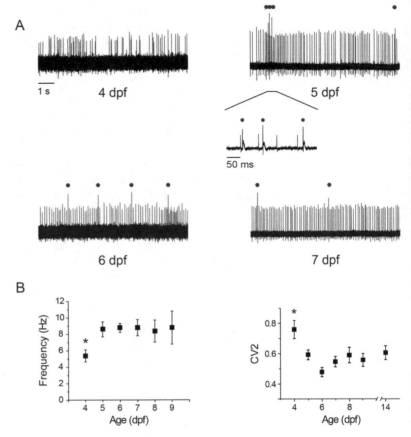

Fig. 3 Emergence of spontaneous firing in zebrafish Purkinje cells [31]. (**a**) Representative loose-patch recordings from zebrafish Purkinje cells at 4, 5, 6, and 7 dpf are shown. Red dots indicate complex spikelike events, shown on an expanded time scale at 5 dpf. (**b**) Average tonic firing frequency (*left*) and coefficient of variation of adjacent intervals (CV2, *right*) are shown as a function of developmental age. CV2 indicates the irregularity of firing, with lower values signifying more regular firing. Firing frequency and CV2 at 4 dpf differed significantly compared to subsequent days ($^*p < 0.05$)

if seal resistance is high but is unnecessary if seal resistance is low. The baseline should stabilize within 1–2 min of patch formation and remain stable during the recording. Data should be collected within ~1 h of the dissection. Recordings from individual cells can typically be maintained for at least 30 min. Data are recorded at 20–50 kHz and filtered at 3 kHz. Post hoc filtering of the data at 1 kHz using a low-pass Bessel filter is recommended to optimize automatic spike detection.

Zebrafish Purkinje cells are born at 3 dpf and begin firing action potentials spontaneously ~12 h later (Fig. 3a) [27, 31]. By 5 dpf, the frequency and regularity of spontaneous tonic firing have increased (Fig. 3a, b). Waveforms that resemble complex spikes, consisting of a large amplitude spike followed by a prolonged,

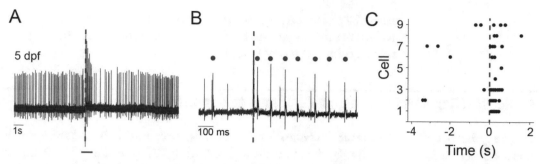

Fig. 4 Complex spikes evoked by direct electrical stimulation of inferior olive [31]. (a) While recording from a cerebellar Purkinje cell, a theta electrode filled with external solution (see Fig. 1b) was used to stimulate the inferior olive at the time indicated by the dashed red line. (b) An expanded view of the recording corresponding to the interval marked by the red bar in part A. Red dots indicate complex spikes. (c) Symbols in raster plot indicate complex spikes recorded before and after direct stimulation of the inferior olive at dashed red line (time 0). Each row corresponds to an individual Purkinje cell

lower-amplitude depolarization, are observed in the majority of cells by 5 dpf (Fig. 3a). By 6–7 dpf, a pattern of tonic firing and complex-like spiking has been established and remains stable until at least 14 dpf, the latest time point that we have examined [31].

To investigate whether the complex spikelike events result from the activation of climbing fiber synapses, we applied direct electrical stimulation to the inferior olive while simultaneously recording from Purkinje cells [31]. Inferior olive stimulation rapidly and transiently increases the frequency of complex spikelike events, establishing that they indeed reflect the postsynaptic response to transmission at climbing fiber synapses (Fig. 4).

In our experiments, the majority of Purkinje cells are electrically active by 4 dpf. Cells that fail to fire spontaneously may have been damaged during the dissection or insertion of the patch pipette. Recordings with low signal-to-noise ratios, unstable baselines, or spikes that vary in amplitude or overlap (which may indicate that the recording contains signals from more than one cell) are discarded.

As shown in Sect. 3, sensory input alters the electrical activity of zebrafish Purkinje cells as early as 4 dpf [31]. As a result, it is essential to minimize unwanted and uncontrolled visual, vibrational, and auditory stimuli during recording sessions. Whether Purkinje cell activity varies with the circadian clock has not yet been studied in detail. It is known, however, that locomotor activity in zebrafish shows circadian rhythmicity as early as 4 dpf [34]. Possible circadian alterations in Purkinje cell firing should therefore be taken into account when scheduling experiments.

2.6 Euthanasia

At the end of the recording, the zebrafish is euthanized on its cover slip by immersion in 10X MS-222 (0.2%). Unused embryos (prior to hatching) and larvae (post-hatching) are euthanized by immersion in 1% sodium hypochlorite or 0.2%

MS-222, respectively. Animals should be discarded according to institutional regulations. Live zebrafish should not be released into the environment.

3   Results: Rapid Wiring of Cerebellar Circuit in Zebrafish

3.1   Early Emergence of Functional Parallel Fiber Input

A notable advantage of the zebrafish preparation is that the animals are alive and awake during recording. As a result, the effects of sensory stimuli on Purkinje cell firing are readily characterized. We exploited this to investigate the development of functional connectivity in the cerebellar circuit [31]. Zebrafish exhibit a robust optokinetic response by ~3.3 dpf, which indicates that the visual system is active during cerebellar development and the emergence of Purkinje cell excitability [31, 35]. We therefore used visual stimulation to determine when parallel fiber inputs to Purkinje cells become active [31]. Zebrafish are adapted to a broad-spectrum LED (1W, 6500K, white LED, Thorlabs) for 2 min while recording Purkinje cell activity. The light is then extinguished. We found that sudden darkness alters Purkinje cell firing at 4 dpf (Fig. 5a). Thus, afferent pathways conveying visual information to the cerebellum are active during the emergence of spontaneous firing in Purkinje cells. Similar results are obtained on subsequent days (Fig. 5b). In response to sudden darkness, most Purkinje cells exhibit a dramatic increase in the frequency of tonic firing (Fig. 5c, d) [31]. In mammals, the frequency of tonic firing is modulated by parallel fiber input [7]. Therefore, in our experiments, visual input is most likely conveyed by mossy fiber pathways that synapse onto cerebellar granule cells, resulting in the activation of parallel fiber synapses.

Although the stimulus increases firing frequency in the majority of Purkinje cells, in some cells sudden darkness has no effect or results in a decrease in firing frequency (Fig. 5c, d) [31]. Most of our recordings have been made in the medial region of the cerebellum. It is possible that Purkinje cells located in more lateral regions of the cerebellum would respond differently to the stimulus.

The duration of higher frequency firing and the latency of the response to sudden darkness change between 4 and 5 dpf. At 4 dpf, most cells return nearly to the baseline firing rate within ~5 s after the stimulus (Fig. 6a) [31]. In contrast, on subsequent days, firing at an increased rate persists for a prolonged period (>10 s) (Fig. 6a) [31]. At 5 dpf and subsequently, the latency between sudden darkness and higher frequency firing is reduced (Fig. 6b). Maturation and refinement of synaptic input and the development of additional circuit components are likely to contribute to these changes in the response [31].

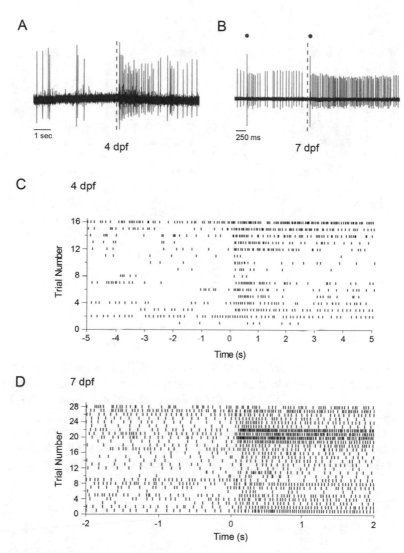

Fig. 5 Effect of visual input on tonic firing [31]. (**a**, **b**) Purkinje cell firing was recorded at (**a**) 4 dpf or (**b**) 7 dpf before and after sudden darkness, imposed at dashed red line. Red dots at 7 dpf correspond to complex spikes. (**c**, **d**) Raster plots of Purkinje cell firing at (**c**) 4 dpf or (**d**) 7 dpf before and after sudden darkness, imposed at time 0. Black and red symbols correspond to simple and complex spikes, respectively. Each row corresponds to an individual Purkinje cell. In **a** and **c**, data are presented on an expanded time base compared to **b** and **d** because firing frequency is lower at day 4 than at day 7 (see Fig. 3b)

3.2 Climbing Fiber Innervation and Winnowing

To investigate the development of functional climbing fiber synapses, we monitored the emergence of complex spiking with and without direct stimulation of the inferior olive (Fig. 7a, b) [31]. Note that complex spiking in the absence of olivary stimulation may reflect the activation of climbing fibers by

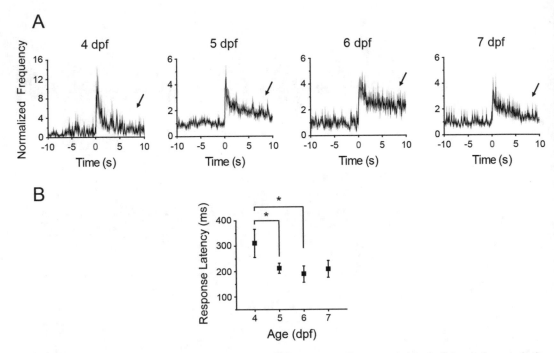

Fig. 6 Maturation and refinement of circuits [31]. (**a**) The average frequency of tonic firing before and after sudden darkness, imposed at time 0, was calculated for each 100 ms interval and normalized to the average firing frequency recorded for 10 s prior to the stimulus at 4, 5, 6, and 7 dpf. Black lines correspond to the average frequency; blue shading indicates the SEM. Data on different days were obtained from 6 to 14 cells in 5–11 animals. Arrows indicate prolonged elevation of firing frequency after the stimulus. (**b**) The latency between sudden darkness and the initial peak of elevated firing frequency is shown as a function of developmental age. *, the latency was significantly longer on day 4 than on days 5 and 6

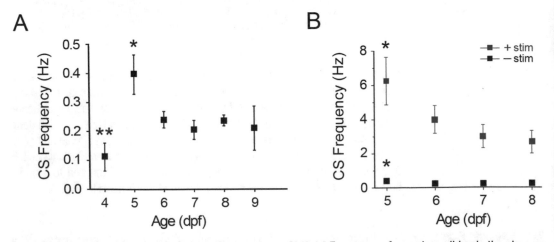

Fig. 7 Development of functional climbing fiber synapses [31]. (**a**) Frequency of complex spiking in the absence of inferior olive stimulation is shown at 4–9 dpf. Frequency of complex spiking was significantly higher at 5 dpf compared to subsequent days (*$p < 0.05$). Frequency of complex spiking was significantly lower at 4 dpf compared to 6 dpf (**$p < 0.05$). (**b**) Frequency of complex spiking after direct electrical stimulation of inferior olive is shown at 5–8 dpf (*red symbols*). Data from part A, obtained without inferior olive stimulation, is shown on the same axes for comparison (*black symbols*). With or without olivary stimulation, complex spike frequency on day 5 was significantly higher than on subsequent days. *$p < 0.05$

uncontrolled sensory stimuli. At 4 dpf, a minority of Purkinje cells (~35%) exhibit complex spiking [31]. In cells that fire complex spikes at 4 dpf, the frequency is low (Fig. 7a). On subsequent days, the majority of Purkinje cells exhibit complex spiking (~70% at 5 dpf, ~85% at 7 dpf), and the frequency of firing is increased (Fig. 7a). Olivary stimulation increases complex spiking at every time point (Fig. 7b).

Interestingly, the frequency of complex spiking, with or without direct stimulation of the inferior olive, is maximal at 5 dpf and then declines to a stable level by 6 dpf (Fig. 7) [31]. During cerebellar development in mammals, Purkinje cells are initially innervated by multiple climbing fiber inputs that undergo winnowing by an activity-dependent process until only the single strongest input remains [36, 37]. Analysis of complex spiking in zebrafish Purkinje cells suggests that this winnowing process is conserved. To gauge the strength of climbing fiber inputs, we investigated the latency between stimulation of the inferior olive and increased complex spiking. At 5 dpf, complex spike latencies are broadly distributed with the majority of responses occurring from 200 to 800 ms after the stimulus (Fig. 8) [31]. In contrast, on subsequent days, fewer long latency complex spikes are observed. By 8 dpf, the majority of spikes occur within 200 ms of inferior olive stimulation, and none occur with latencies greater than 400 ms [31]. Importantly, a broad distribution of latencies suggests that weak and strong inputs coexist. This reflects the fact

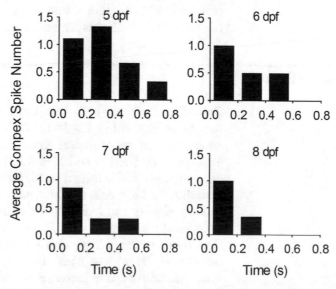

Fig. 8 Loss of long latency complex spikes after 5 dpf [31]. After direct electrical stimulation of the inferior olive, the average number of complex spikes per 200 ms interval was determined at 5–8 dpf. Only cells that exhibited complex spiking in the absence of electrical stimulation were included in this analysis

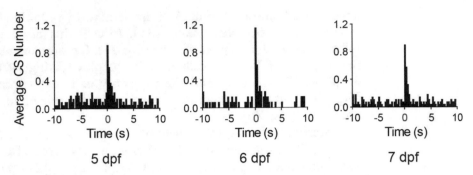

Fig. 9 Response of climbing fiber inputs to sudden darkness [31]. Graphs show the average number of complex spikes before and after sudden darkness, imposed at time 0 and at 5, 6, and 7 dpf. Not shown: at 4 dpf, sudden darkness did not alter complex spike frequency

that weaker inputs take longer to bring the postsynaptic cell to threshold. The loss of long latency but not short latency complex spikes strongly suggests that weaker inputs are eliminated between 5 and 7 dpf, while the strongest input is maintained.

3.3 Functional Innervation of Inferior Olive by Afferent Pathways Conveying Visual Information

To investigate when inferior olive neurons become responsive to sensory input, we monitored complex spike frequency in response to sudden darkness. Visual stimulation did not alter the frequency of complex spiking at 4 dpf, when relative few Purkinje cells receive functional climbing fiber input [31]. At 5 dpf, when the majority of Purkinje cells are innervated by climbing fibers, sudden darkness significantly increases complex spike frequency in most Purkinje cells (Fig. 9) [31]. Firing frequency remains elevated for ~1 s after the stimulus. Similar responses are seen on subsequent days.

4 Conclusions

Zebrafish Purkinje cells are born at 3 dpf and begin firing spontaneously within 12–24 h. We have shown previously that the emergence of spontaneous firing is closely correlated with the expression of Nav1.6 and Kv3.3 [31], ion channels that underlie pacemaking activity in mammalian Purkinje cells [3–6]. By 5 dpf, zebrafish Purkinje cells exhibit a pattern of tonic firing and complex spiking similar to that seen in mammals [31]. We find that a functional cerebellar circuit forms rapidly during zebrafish development. By 4 dpf, visual input, presumably carried to the cerebellum by mossy fiber pathways that activate parallel fiber synapses, dramatically increases the frequency of simple spiking in most medially located Purkinje cells. Climbing fibers of inferior olive neurons begin to form functional synapses onto Purkinje cells on day 4; by day 5, the majority of Purkinje cells exhibit complex spiking. Visual input modulates the frequency of complex spiking by 5 dpf. Our results suggest that zebrafish Purkinje cells, like those in mammals, initially receive multiple climbing fiber inputs of varying strengths which then undergo a winnowing process between 5 and 7 dpf to eliminate weaker inputs.

The rapid development of a functional cerebellum responsive to visual input is likely essential for zebrafish survival because the animals develop outside the body of the mother and must escape from predators and find food early in life. Feeding becomes imperative at 4–5 dpf when the yolk sac that nourishes the developing zebrafish is exhausted [17, 18]. Interestingly, formation of functional connectivity and visual responsiveness in the cerebellar circuit at 4–5 dpf correlates temporally with the emergence of prey capture, a complex, visually guided, feeding behavior that must be carefully calibrated to be successful [38]. Zebrafish are unable to capture prey at 3 dpf, but by 5 dpf, they can readily track and capture paramecia for food [38]. By 5 dpf, visual input modifies the activity of climbing fiber synapses, which may fine-tune movement to increase the accuracy of prey capture. The ability to record from Purkinje cells in live zebrafish will be useful in investigating the role of the cerebellum in feeding behavior. Purkinje cell output can be altered genetically or optogenetically, and the resulting changes in firing can be characterized electrophysiologically. Simultaneously, the effects of altered firing on the success and accuracy of prey capture can be quantified using a virtual behavior assay [39].

The ability to record Purkinje cell activity in live zebrafish is advantageous for functional mapping of afferent inputs to the cerebellum. Purkinje cells integrate multiple inputs and are uniquely responsible for generating the output of the cerebellar cortex. Channelrhodopsin-2, Archaerhodopsin-3 (Arch), or other optogenetic tools can be expressed in discrete brain regions using GAL4 lines or cell-type-specific promoters [14, 32]. Changes in Purkinje cell activity in response to activation or inhibition of potential input pathways can then be characterized electrophysiologically. It is important to note that currently available calcium indicators do not accurately report Purkinje cell activity in zebrafish. Whereas our results show that most Purkinje cells fire spontaneously and robustly within 12–24 h of their birth, calcium transients corresponding to this spontaneous activity have not been detected in imaging studies [23, 40, 41]. Intracellular recordings in larval zebrafish indicate that Purkinje cells fire low-amplitude spikes, which may result in small calcium transients that are below the sensitivity of the best available genetically encoded calcium indicators [25, 41, 42]. Instead, the signals detected in the zebrafish cerebellum using calcium indicators may correspond to large bursts of activity or coordinated network firing [23, 25, 40, 41].

New tools for neuroscience research in zebrafish continue to be developed, for example, vesicular stomatitis virus (VSV) for anterograde and retrograde transsynaptic tracing [43, 44]. Using the wide range of emerging experimental approaches, the stage is set for mapping afferent and efferent cerebellar pathways and investigating the role of the cerebellum in motor control and motor learning in zebrafish [14, 15, 25, 31, 39–41, 43, 44].

Acknowledgments

With the exception of Fig. 5a, c, the data in this paper were originally published by Hsieh et al. (2014) (Ref. 31). This work was supported by NIH grant R01NS058500 and a UCLA Stein-Oppenheimer Seed Grant to DMP. JYH was partially supported by the Jennifer S. Buchwald Graduate Fellowship in Physiology at UCLA. We are grateful to Drs. Thomas Otis, Joanna Jen, Fadi Issa, and Jijun Wan for advice and helpful discussions. We thank Dr. Masahiko Hibi for the gift of a pT2K *aldoca*:gap43-Venus plasmid.

References

1. Lang EJ, Apps R, Bengtsson F, Cerminara NL, De Zeeuw CI, Ebner TJ, Heck DH, Jaeger D, Jörntell H, Kawato M, Otis TS, Ozyildirim O, Popa LS, Reeves AMB, Schweighofer N, Sugihara I, Xiao J (2017) The roles of the olivocerebellar pathway in motor learning and motor control. A consensus paper. Cerebellum 16:230–252

2. Raman IM, Bean BP (1999) Ionic currents underlying spontaneous action potentials in isolated cerebellar Purkinje neurons. J Neurosci 19:1663–1674

3. Raman IM, Sprunger LK, Meisler MH, Bean BP (1997) Altered subthreshold sodium currents and disrupted firing patterns in Purkinje neurons of Scn8a mutant mice. Neuron 19:881–891

4. Khaliq ZM, Gouwens NW, Raman IM (2003) The contribution of resurgent sodium current to high-frequency firing in Purkinje neurons: an experimental and modeling study. J Neurosci 23:4899–4912

5. Martina M, Yao GL, Bean BP (2003) Properties and functional role of voltage-dependent potassium channels in dendrites of rat cerebellar Purkinje neurons. J Neurosci 23:5698–5707

6. Akemann W, Knöpfel T (2006) Interaction of Kv3 potassium channels and resurgent sodium current influences the rate of spontaneous firing of Purkinje neurons. J Neurosci 26:4602–4612

7. D'Angelo E, Mazzarello P, Prestori F, Mapelli J, Solinas S, Lombardo P, Cesana E, Gandolfi D, Congi L (2011) The cerebellar network: from structure to function and dynamics. Brain Res Rev 66:5–15

8. Schmolesky MT, Weber JT, De Zeeuw CI, Hansel C (2002) The making of a complex spike: ionic composition and plasticity. Ann N Y Acad Sci 978:359–390

9. Lee KH, Mathews PJ, Reeves AMB, Choe KY, Jami SA, Serrano RE, Otis TS (2015) Circuit mechanisms underlying motor memory formation in the cerebellum. Neuron 86:529–540

10. Woodruff-Pak DS (2006) Stereological estimation of Purkinje neuron number in C57BL/6 mice and its relation to associative learning. Neuroscience 141:233–243

11. Wittmann W, McLennan IS (2011) The male bias in the number of Purkinje cells and the size of the murine cerebellum may require Müllerian inhibiting substance/anti-Müllerian hormone. J Neuroendocrinol 23:831–838

12. Joshua M, Lisberger SG (2015) A tale of two species: neural integration in zebrafish and monkeys. Neuroscience 296:80–91

13. Rinkwitz S, Mourrain P, Becker TS (2011) Zebrafish: an integrative system for neurogenomics and neurosciences. Prog Neurobiol 93:231–243

14. Wyatt C, Bartoszek EM, Yaksi E (2015) Methods for studying the zebrafish brain: past, present and future. Eur J Neurosci 42:1746–1763

15. Feierstein CE, Portugues R, Orger MB (2015) See the whole picture: a comprehensive imaging approach to functional mapping of circuits in behaving zebrafish. Neuroscience 296:26–38

16. Orger MB, de Polavieja GG (2017) Zebrafish behavior: opportunities and challenges. Annu Rev Neurosci., epub ahead of print. https://doi.org/10.1146/annurev-neuro-071714-033857

17. Nusslein-Volhard C, Dahm R (eds) (2002) Zebrafish: a practical approach. Oxford University Press, Oxford

18. Westerfield M (2007) The Zebrafish Book, 5th edition: A guide for the laboratory use of zebrafish (Danio rerio). University of Oregon Press, Eugene

19. Brustein E, Saint-Amant L, Buss RR, Chong M, McDearmid JR, Drapeau P (2003) Steps during the development of the zebrafish locomotor network. J Physiol Paris 97:77–86

20. Kikuta H, Kawakami K (2009) Transient and stable transgenesis using Tol2 transposon vectors. Methods Mol Biol 546:69–84

21. Arrenberg AB, Driever W (2013) Integrating anatomy and function for zebrafish circuit analysis. Front Neural Circuits 7:74

22. Dunn TW, Mu Y, Narayan S, Randlett O, Naumann EA, Yang CT, Schier AF, Freeman J, Engert F, Ahrens MB (2016) Brain-wide mapping of neural activity controlling zebrafish exploratory locomotion. Elife 5:e12741

23. Ahrens MB, Orger MB, Robson DN, Li JM, Keller PJ (2013) Whole-brain functional imaging at cellular resolution using light-sheet microscopy. Nat Methods 10:413–420

24. Naumann EA, Fitzgerald JE, Dunn TW, Rihel J, Sompolinsky H, Engert F (2016) From whole-brain data to functional circuit models: the zebrafish optomotor response. Cell 167:947–960

25. Scalise K, Shimizu T, Hibi M, Sawtell NB (2016) Responses of cerebellar Purkinje cells during fictive optomotor behavior in larval zebrafish. J Neurophysiol 116:2067–2080

26. Berkefeld H, Fakler B, Schulte U (2010) Ca^{2+}-activated K^+ channels: from protein complexes to function. Physiol Rev 90:1437–1459

27. Bae Y-K, Kani S, Shimizu T, Tanabe K, Hojima H, Kimura Y, Higashijima S, Hibi M (2009) Anatomy of zebrafish cerebellum and screen for mutations affecting its development. Dev Biol 330:406–426

28. Hibi M, Shimizu T (2012) Development of the cerebellum and cerebellar neural circuits. Dev Neurobiol 72:282–301

29. Hashimoto M, Hibi M (2012) Development and evolution of cerebellar neural circuits. Dev Growth Differ 54:373–389

30. Hamling KR, Tobias ZJC, Weissman TA (2015) Mapping the development of cerebellar Purkinje cells in zebrafish. Dev Neurobiol 75:1174–1188

31. Hsieh J-Y, Ulrich B, Wan J, Papazian DM (2014) Rapid development of Purkinje cell excitability, functional cerebellar circuit, and afferent sensory input to cerebellum in zebrafish. Front Neural Circuits 8:147

32. Tanabe K, Kani S, Shimizu T, Bae Y-K, Abe T, Hibi M (2010) Atypical protein kinase C regulates primary dendrite specification of cerebellar Purkinje cells by localizing Golgi apparatus. J Neurosci 30:16983–16992

33. Brochu G, Maler L, Hawkes R (1990) Zebrin II: a polypeptide antigen expressed selectively by Purkinje cells reveals compartments in rat and fish cerebellum. J Comp Neurol 291:538–552

34. Hurd MW, Cahill GM (2002) Entraining signals initiate behavioral circadian rhythmicity in larval zebrafish. J Biol Rhythms 17:307–314

35. Chhetri J, Jacobson G, Gueven N (2014) Zebrafish—on the move towards ophthalmological research. Eye (Lond) 28:367–380

36. Kano M, Hashimoto K (2012) Activity-dependent maturation of climbing fiber to Purkinje cell synapses during postnatal cerebellar development. Cerebellum 11:449–450

37. Hashimoto K, Kano M (2013) Synapse elimination in the developing cerebellum. Cell Mol Life Sci 70:4667–4680

38. Westphal RE, O'Malley DM (2013) Fusion of locomotor maneuvers and improving sensory capabilities give rise to the flexible homing strikes of juvenile zebrafish. Front Neural Circuits 7:108

39. Bianco IH, Kampff AR, Engert F (2011) Prey capture behavior evoked by simple visual stimuli in larval zebrafish. Front Syst Neurosci 5:101

40. Aizenberg M, Schuman EM (2011) Cerebellar-dependent learning in larval zebrafish. J Neurosci 31:8708–8712

41. Matsui H, Namikawa K, Babaryka A, Köster RW (2014) Functional regionalization of the teleost cerebellum analyzed in vivo. Proc Natl Acad Sci U S A 111:11846–11851

42. Sengupta M, Thirumalai V (2015) AMPA receptor mediated synaptic excitation drives state-dependent bursting in Purkinje neurons of zebrafish larvae. Elife 4:e09158

43. Mundell NA, Beier KT, Pan YA, Lapan SW, Göz Aytürk D, Berezovskii VK, Wark AR, Drokhlyansky E, Bielecki J, Born RT, Schier AF, Cepko CL (2015) Vesicular stomatitis virus enables gene transfer and transsynaptic tracing in a wide range of organisms. J Comp Neurol 523:1639–1663

44. Beier KT, Mundell NA, Pan YA, Cepko CL (2016) Anterograde or retrograde transsynaptic circuit tracing in vertebrates with vesicular stomatitis virus vectors. Curr Protoc Neurosci 74:1.26–1.27

Chapter 12

Recording Extracellular Activity in the Developing Cerebellum of Behaving Rats

Greta Sokoloff and Mark S. Blumberg

Abstract

The in vivo extracellular activity of the cerebellum has been intensively investigated in adult animals to understand its roles in learning and memory and sensorimotor integration. Here we describe a method for studying extracellular activity in the cerebellum of unanesthetized, behaving infant rodents over the first 2 postnatal weeks, a time of substantial cerebellar circuit development. The study of extracellular activity during cerebellar development in behaving infants provides a unique opportunity to relate neural activity not only to cerebellar circuit development but to behavioral development as well. We propose that studying extracellular neural activity in the developing cerebellum provides a model system for examining the complex interactions between behavior and neural activity and how they contribute together to functional neural circuit development.

Key words Cerebellum, Sleep, Twitching, Infant rodent, Methods, Neural recording

1 Introduction

Beginning with the first identification of the Purkinje cell in 1837, the cerebellum has fascinated to neuroanatomists [1]. For developmental neurobiologists, the protracted postnatal development of the cerebellum in mammals, especially rodents, has spurred recent interest in a broad range of research questions addressing Purkinje cell development, synapse formation and elimination, cerebellar circuit formation, and motor learning [2–8].

Figure 1 presents representative coronal sections through the cerebellar cortex at postnatal day (P) 4, P8, and P12. It is readily apparent that there is a dramatic increase in size alone over this single week. Figure 1 also presents a subset of the cerebellar milestones that occur during the first 2 postnatal weeks in rats. At birth, Purkinje cells are immature with short dendritic processes and are not yet organized into a single Purkinje cell layer [3, 7]. Also at birth, mossy fibers form transient direct connections with Purkinje cells and innervate cells of the cerebellar nuclei (CN) [9–11]. In the first few

Roy V. Sillitoe (ed.), *Extracellular Recording Approaches*, Neuromethods, vol. 134,
https://doi.org/10.1007/978-1-4939-7549-5_12, © Springer Science+Business Media, LLC 2018

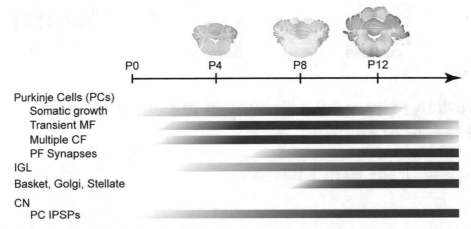

Fig. 1 A timeline illustrating some of the major developmental events in the cerebellum during the first 2 postnatal weeks. *Top*, Nissl-stained, coronal sections of the cerebellum and brainstem in P4, P8, and P12 rats. *Bottom*, timeline based on anatomical and electrophysiological studies depicting earliest observation, peak, and end (if occurring by P16) of each event. Abbreviations: *PC* Purkinje cell, *MF* mossy fiber, *CF* climbing fiber, *PF* parallel fiber, *IGL* internal granular layer, *IPSP* inhibitory postsynaptic potential

days after birth, climbing fiber innervation of Purkinje cells increases, and multiple climbing fibers innervate each Purkinje cell [8, 12–14]. Around the same time, granule cells begin to proliferate in the external granular layer (EGL) and begin their migration to the internal granular layer (IGL) [1, 2, 4].

During the second postnatal week, the Purkinje cell soma reaches its adult size, and the synapses of extraneous climbing fibers are weakened as the "winner" begins to translocate along the apical dendrite of the Purkinje cell [6, 7]. As granule cells migrate, they form connections with mossy fibers, and the transient mossy fiber connections on the Purkinje cell soma diminish as parallel fiber synapses begin to form, connecting mossy fibers to Purkinje cells via granule cells [2–4, 8, 11, 15]. Also during the second postnatal week, basket, Golgi, and stellate cells become functional [3, 8]. So, for the 2-week-old rat, the majority of the cerebellar circuit is established, although there is continued cell migration and dendritic growth and refinement through the third postnatal week and into adulthood [2–4, 7, 16].

Importantly, despite this protracted cerebellar development in rats, the Purkinje cells, neurons in the deep cerebellar nuclei (CN), as well as the primary input pathways—climbing and mossy fibers—are already functioning within the first few postnatal days [8, 10, 12, 16–19]. Specifically, climbing fibers and mossy fibers transmit sensory information from both the periphery and cerebral cortex to Purkinje cells and CN neurons by P3 [10, 17, 18]. Also, Purkinje cells at this time are able to produce inhibitory postsynaptic potentials (IPSPs) on CN neurons [10].

Molecular cues play an important role in early cerebellar development [15, 20, 21], as do activity-dependent mechanisms [22]. For example, activated climbing fibers facilitate synapse elimination and, ultimately, climbing fiber translocation at Purkinje cells [23–27]. However, it is not yet known how different sources of neural activity drive activity-dependent development of the infant cerebellum [22]. Therefore, the study of extracellular activity in the developing cerebellum is crucial for understanding the role of both spontaneous and evoked activities on cerebellar circuit formation and synapse elimination.

Due to its protracted development, the cerebellum is more susceptible than many other brain regions to perturbations that disrupt activity-dependent processes necessary for cerebellar development and connectivity with forebrain areas [28]. Importantly, the early, seminal studies of extracellular cerebellar activity were performed using anesthetized infant rats. Among its many drawbacks, anesthesia lowers the firing rates of Purkinje cells [29]. Thus, the best approach to understanding cerebellar activity and functionality in early development is to use unanesthetized subjects, thereby providing direct access to patterns of spontaneous and evoked neural activity across a period of rapid developmental change. Significantly, a better understanding of the role of neural activity on activity-dependent developmental processes may illuminate the contributions of cerebellar dysfunction to the etiology of neurodevelopmental disorders such as autism and schizophrenia [28, 30, 31].

2 Methods: Extracellular Cerebellar Neural Activity and Behavior in Infant Rodents

2.1 Head-Fix Method for Neural Recording in Behaving Infant Rodents

Using the head-fix method described below, it is possible to record from P1 to P12 in rats and from P4 to P15 in mice. Pups can be tested at younger and older ages with appropriate modifications to the apparatus and method. The reader should refer to our previously published primer which contains a comprehensive list of studies that have used this and similar methods to record extracellular activity in numerous brain regions in infant rodents [32]. Here we detail our head-fix method with special consideration for the cerebellum.

2.2 Constructing a Head-Fix

The head-fix is constructed using a metal washer, two strips of metal, hex nuts, locking washers, and machine screws. Inner and outer diameters of the washer are determined by the age (i.e., size and/or weight) of the subject and the brain area to be recorded from. For illustration purposes, Fig. 2 shows a washer size that is appropriate for week-old rats. Once all the materials are organized, the two metal strips are bent at 90° and glued (Loctite© Liquid

Super Glue, Henkel Corporation, Rocky Hill CT) or welded to opposite sides of the washer. Machine screws are inserted through holes drilled into the bent ends; the metal strips and nuts are screwed in place over locking washers. (Regular washers will work but will need to be tightened with each use.) To add stability, liquid Super Glue may be applied over the nuts and screws, but with a functioning locking washer, this should not be necessary.

Machine screws are sized to fit into the ends of ear bars that can be secured in a standard stereotaxic apparatus. Washer and metal strip thicknesses determine the rigidity of the head-fix. With older pups, a more rigid head-fix helps to minimize, or negate, movement artifact in the extracellular recording. For recordings with infant mice, the device can be fashioned from thinner metal components or even firm plastic with little loss of stability.

For extracellular recordings in the cerebellum, the head-fix is improved with modifications to the inner portion of the metal washer. Either removing a portion of the caudal part of the washer or drilling out part of the inner diameter will provide a window that is open to the occipital ridge, thereby facilitating access to the area of interest; see the inset in Fig. 2 for examples of modifications that can be made for small infant rodents (lower right) or cerebellum and hindbrain recording (upper left). Importantly, each head-fix costs less than $3.00 and is easily cleaned and repaired for reuse.

2.3 Equipment and Apparatus

Our setup for stereotaxic extracellular recording is shown in Fig. 3. A standard stereotaxic apparatus is housed within a Faraday cage atop a sturdy table. All electrical equipment within the cage are grounded. The thermal environment is controlled using a small double-walled glass chamber (constructed in a glass blowing lab) that is placed directly under the head-fixed pup. Before testing, a water circulator is turned on, and hot water is circulated through the chamber to warm the ambient environment. In the absence of having a glass blower at your institution, the Heated Hard Pad (Braintree Scientific, Braintree, MA) can be used or, alternatively, a low-noise electrical heated mat.

2.4 Surgery

On the day of testing, a pup is removed from the home cage. It is important to select a pup with a large milk band and within the appropriate weight range for its age. Using these selection criteria, you more reliably ensure overall health and adequate nutrition and hydration for the experimental session. If the duration of the experiment is expected to be long, pups can be intubated with commercial half-and-half or milk formula (3% of body weight in mL). Intubation should be done before placing the pup in the stereotaxic apparatus. We find that, without intubation, week-old rats maintain normal body temperatures and exhibit normal sleep-wake cycles following separation from the dam for up to 8 h [33].

Constructing a head-fix

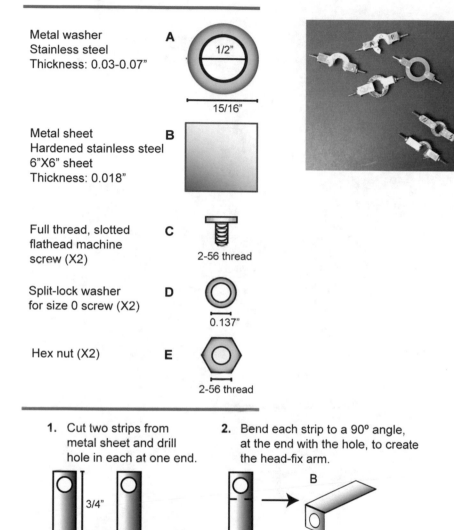

Metal washer
Stainless steel
Thickness: 0.03-0.07"

A

1/2"

15/16"

Metal sheet
Hardened stainless steel
6"X6" sheet
Thickness: 0.018"

B

Full thread, slotted
flathead machine
screw (X2)

C

2-56 thread

Split-lock washer
for size 0 screw (X2)

D

0.137"

Hex nut (X2)

E

2-56 thread

1. Cut two strips from
metal sheet and drill
hole in each at one end.

3/4"

5/16"

2. Bend each strip to a 90° angle,
at the end with the hole, to create
the head-fix arm.

B

3. Put the threaded end of the
screw through the hole in the
arm. Thread washer and nut
onto screw and tighten.

B

C

D

E

4. Using super glue, affix arms
to washer. Allow glue to cure
until set.

B   A

E D C

Fig. 2 Materials for building a head-fix and the subsequent steps for assembling it. *Inset*, examples of different sizes and modifications that can be used for the head-fix method in postnatal rodents. Specifications are standard, the head-fix washer size, screw length, and head-fix arm length will vary with the age of the pup and the testing environment. We purchase parts from McMaster-Carr (Elmhurst, IL), but any hardware supplier should stock these products

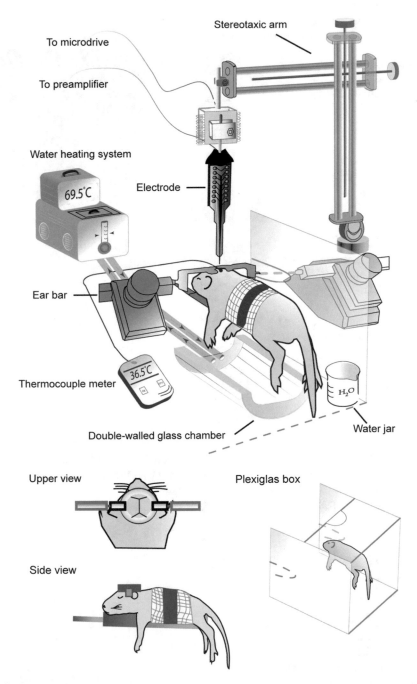

Fig. 3 An illustration of the recording setup for head-fixed infant rodents adapted from [32]. A head-fixed pup is secured to an elevated platform with tape. The head-fix attaches to ear bars that can be secured to the stereotaxic apparatus. A double-walled glass chamber is positioned under the pup so that temperature-controlled water can heat the ambient air. A Plexiglas box surrounds the pup on three sides to help maintain air temperature, and water-filled beaker inside the box helps to maintain adequate humidity. Brain temperature is monitored with a thermocouple inserted into a predrilled hole in the skull. The recording electrode is attached to a headstage and can be lowered into the cerebellum using a microdrive. Reprinted by permission of John Wiley & Sons, Inc.

Using an induction chamber filled with isoflurane (rats, 3–5%; mice, 2–3%), anesthesia is induced. Once anesthetized, the pup is placed in a prone position on a heating pad. Anesthesia is maintained with an anesthesia cone over the pup's snout. 2–3% isoflurane is typically sufficient for maintenance, but respiration should be closely monitored throughout the surgery, and pedal (foot) and pinnal (ear) reflexes should be tested regularly to insure an appropriate level of anesthesia.

To record electromyographic (EMG) activity, 27 g sterile needles are used to insert custom-made bipolar hook electrodes into the skeletal muscle (304 HPN, California Fine Wire, Grover Beach, CA). Uninsulated stainless steel wire (304, California Fine Wire) is looped transdermally on the back of the pup. EMG and ground wires are secured to the skin at insertion points with collodion. A 2″ × 2″ gauze pad is unfolded and wrapped around the pup's torso and secured with tape (e.g., 3M Micropore, St Paul, MN). The gauze is trimmed to the appropriate length to ensure that the hind limbs are well outside the wrap.

To prepare the skull for electrode implantation, the skin overlying the skull is removed using surgical scissors or a scalpel. The exposed skull is cleaned of connective tissue and dried using bleach or hydrogen peroxide. Bupivacaine is applied topically to scalp incisions to minimize pain. Loctite© Super Glue Gel Control (Henkel Corporation) is applied to the metal washer at least 15 min before it is fixed to the skull; when applied to the skull, the glue should be tacky. Before attaching the head-fix to the skull, Vetbond (3M) can be applied beneath the surface of the skin around the incision to glue it back and away from the skull, which helps ensure a secure attachment of the head-fix. Make sure to align the head-fix over the skull before positioning it so that the midline of the skull is aligned at the middle of the head-fix. Place the head-fix onto the exposed skull and apply light pressure for approximately 10 s. Ensure the head-fix is level on all planes and make any adjustments as needed. Once attached, liquid or gel Super Glue (Henkel Corporation) can be applied to the inner circumference of the head-fix to reinforce adhesion or fill in gaps between the head-fix and skull. For general analgesia, the pup is administered an NSAID (e.g., Rymadyl; 5 mg/kg s.c.). Next, the pup is temporarily wrapped in a second layer of gauze in order to restrain movement while the head-fix sets. Finally, the pup is placed in a ventilated, humidified incubator maintained at thermoneutrality (e.g., 35 °C for a P8 rat) for approximately 1 h. The entire surgical procedure takes approximately 10–15 min.

After 1 h in the incubator, the head-fix should be securely attached to the skull. At this point, for drilling holes in the skull, the pup is moved to a stereotaxic apparatus outfitted with an adaptor and anesthesia mask (Mouse and Neonatal Rat Adaptor and Gas Anesthesia Mask; Stoelting, Wood Dale, IL). The threaded end of the screws on the head-fix fits into the ear bar holders on the adaptor.

Table 1
Stereotaxic coordinates for extracellular cerebellar recordings in infant rats and mice

Age and species	AP (mm from lambda)	ML (mm)	DV (mm)	Angle (R → C)
Cerebellar cortex				
P3-P4 rat	−1.0 to −2.0	± 1.3 to 2.0	−1.2 to −3.4	8–10°
P6 rat	−1.0 to −1.5	± 1.5 to 2.0	−1.3 to −3.7	10–12°
P7-P9 rat	−1.1 to −2.4	± 1.7 to 2.0	−1.2 to −3.8	8–10°
P8-P9 mouse	−1.0 to −2.6	± 1.0 to 1.9	−1.0 to −2.3	10–14°
P11-P12 rat	−2.0 to −3.3	± 1.7 to 2.1	−1.7 to −3.1	10–12°
P14-P15 mouse	−1.1 to −2.5	± 1.1 to 1.4	−1.0 to −1.7	10–14°
Interpositus nucleus				
P7-P9 rat	−2.5 to −2.9	±2.0	−3.2 to −3.8	10–12°
P11-P13 rat	−3.7	±2.0	−3.5 to −4.5	None

AP anterior/posterior, *ML* mediolateral, *DV* dorsoventral, *R* rostral, *C* caudal, *L* lateral, *M* medial

Under light isoflurane anesthesia, stereotaxic coordinates are used to drill holes for recording electrode placement, ground and reference wires, and a thermocouple. Also at this time, cannulae can be inserted and secured to the skull for pharmacological manipulations, and/or brain lesions can be produced. Table 1 provides a list of coordinates we have used to record from cerebellar cortex and interpositus nucleus in rats and mice during the first 2 postnatal weeks.

When the pup is ready, the second layer of gauze is removed so the limbs are free to move. The pup is moved to the recording chamber and set on a narrow support platform that is attached to a metal rod that fits into the nosepiece holder of the stereotaxic apparatus. The pup is secured to the platform by wrapping a piece of tape around the pup and the platform. The nosepiece holder can be adjusted for height, and the angle can be adjusted to ensure comfortable positioning between the pup's head and body. Each ear bar is attached to the threaded end of the screws in the head-fix and then secured to the stereotaxic apparatus; take this opportunity to ensure that the skull is leveled and adjusted, if necessary. When this process is complete, the pup should appear comfortable with its legs dangling on either side of the platform; when the pup moves its limbs, you should not observe any movement of the head.

Next, a Plexiglas box is placed around the pup to help control air temperature and humidity around the pup. The Plexiglas box has three walls and fits inside the arms of the stereotaxic apparatus; the side walls have slots cutout at the height of the ear bars to allow the box to surround the pup. A separate piece of Plexiglas is used to cover the top of the box (this piece can be made to accommodate the electrode holder). The addition of a small, open jar of water placed within the box helps to humidify the air.

To measure brain temperature during acclimation and recording, we use a chromel-constantan thermocouple (Type E; #TT-E-40, Omega Engineering, Stamford, CT) and a thermocouple meter (450-AET, Omega Engineering). The thermocouple can be inserted anywhere in the brain but preferably distant from the recording location. For cerebellar recordings, we typically monitor brain temperature in the cerebral cortex. The thermocouple remains in the brain for the duration of the experiment, but the meter is turned off during data collection as it may cause electrical noise in the extracellular recording. For neurophysiological recordings, brain temperature should be maintained at 36–37 °C. Depending on the distance of the pup from the heated glass chamber, the water bath is set at 50–65 °C. It is important that the pup's limbs and tail cannot make contact with the heat source. If necessary, a heat lamp, located outside the Faraday cage, can be used to help maintain body and brain temperature; however, the lamp should never be aimed directly at the pup. It is important, especially with mice, that brain temperature not exceed 37 °C; finally, we have found that slow, steady rewarming is optimal for recovery and subsequent behavior.

2.5 Experimental Testing

Once the pup is in the recording chamber, wait at least 1 h. for recovery from anesthesia and attainment of a stable brain temperature. A well-acclimated pup should cycle regularly between sleep and wake, with sleep as the predominant behavioral state; we use the combination of EMG and behavioral observation to determine sleep and wake states. We find that EMG electrodes inserted into the nuchal muscle provide the most clear cycling between high and low muscle tones: A well-acclimated pup will exhibit clear periods of atonia (indicative of sleep) alternating with brief periods of high muscle tone (indicative of wake); chronic, low-amplitude muscle tone (as distinct from atonia) typically indicates that the pup needs more time to acclimate. Periods of active (or REM) sleep are characterized by myoclonic twitches of the limbs and tail that occur against a background of atonia. In contrast, periods of wakefulness are characterized by high-amplitude, coordinated wake movements (e.g., stepping movements, stretching) against a background of high muscle tone. A third behavioral state, quiet sleep, can be characterized as periods when the pup exhibits low muscle tone and behavioral quiescence. However, it should be noted that it can be difficult to distinguish between quiet wake and quiet sleep in infant rats younger than P11 because cortical slow waves (i.e., delta) do not emerge until then [33].

Once the pup is cycling regularly between sleep and wake, the experiment can begin. We use 16-channel silicon electrodes (NeuroNexus, Ann Arbor, MI), which give us well-resolved extracellular neural activity (site size: 177 μm^2; http://neuronexus.

com/images/Impedance.pdf). Before insertion of the electrode, we dip it in fluorescent DiI (Vybrant DiI Cell-Labeling Solution; Life Technologies, Grand Island, NY) for subsequent histological verification of electrode placement; of course, other electrode marking protocols will work. Using a pneumatic drum drive (FHC, Inc., Bowdoin, ME), the electrode is slowly lowered into the pre-drilled hole in the skull, and the reference/ground is inserted through a hole in a distant brain region. For cerebellar recordings, we typically insert the reference/ground electrode (Ag/AgCl, 0.25 mm diameter; Medwire, Mt. Vernon, NY) into the cerebral cortex. Once the desired depth is achieved and neural activity is observable, we allow the electrode to stabilize within the brain tissue for at least 10 min before starting data acquisition. During data acquisition, in addition to EMG and neurophysiological data, an observer scores behavior by pressing coded event keys.

The electrodes connect to a headstage that is connected to a battery-operated preamplifier inside the Faraday cage. All acquired signals are then sent to a data acquisition system (Tucker-Davis Technologies, Alachua, FL). For studies where time-locked behavioral or kinematic measures are required, a video camera can be used to capture synchronized video data. Hook electrodes for recording EMG connect to a bank of micro-grabbers, located inside the Faraday cage, that are connected to differential amplifier (A-M Systems, Sequim, WA) that amplifies (10,000×) and filters (300–5000 Hz bandpass) the EMG signal; a 60 Hz notch filter is also used. The neural signals are amplified (10,000×) and filtered for multiunit activity (500–5000 Hz bandpass) or, alternatively, not filtered to allow for subsequent analysis of both unit activity and LFP. Neural and EMG signals are sampled at 25 and 1 kHz, respectively.

2.6 Histology

After recording, pups are overdosed with sodium pentobarbital (1.5 mg/g i.p.) or a ketamine/xylazine cocktail (90:10; .002 mL/g i.p.) and transcardially perfused with phosphate-buffered saline followed by 4% paraformaldehyde. The recording site is visualized on 80 μm sections under a microscope with fluorescent illumination at 5–10× magnification (Leica Microsystems, Buffalo Grove, IL). Recording site locations are subsequently determined after Nissl staining using a calibrated reticle. A photomicrograph of the CN and electrode placement in the interpositus nucleus is shown in Fig. 4. For unquestionable identification of the origin of any extracellular activity in the cerebellum that is not from a Purkinje cell, juxtacellular labeling is necessary (see Chap. 1 in this volume; [34]).

2.7 Pros and Cons of Head-Fix Recording in Relation to Testing Freely Moving Pups

The head-fix method has several advantages. First, because of the small size and uncalcified skulls of rodent pups, it is difficult to secure electrodes to the skull for chronic recording in freely moving subjects. Second, recordings using the head-fix method are

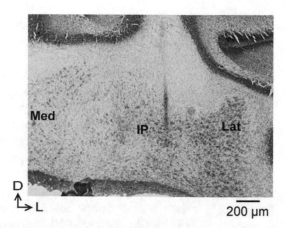

Fig. 4 Magnified Nissl-stained coronal section of the CN of a P12 rat. Red line indicates electrode location from identification of fluorescent DiI tracks in the same coronal sections. *IP* interpositus nucleus, *Lat* lateral nucleus, *Med* medial nucleus, *D* dorsal, *L* lateral

relatively free of movement artifact. Third, the head-fix method affords the experimenter the ability to adjust electrode location until the expected extracellular activity is observed and facilitates extracellular recording in combination with other manipulations, such as pharmacological inactivation [35–37]. This method is also well suited to the use of controlled peripheral stimulation in conjunction with extracellular recording (e.g., [38, 39]). Finally, for methods like two-photon imaging [40] and voltage-sensitive dye imaging [41, 42], there are no good alternatives to the head-fix method for studying brain activity in unanesthetized pups. In adult mice and rats, the head-fix method has similarly been used in conjunction with a number of different paradigms, including eyeblink conditioning ([43]; see Chap. 10 of this volume for a detailed description).

Although the head-fix method provides an environment that allows the pup to sleep comfortably, it is not ideal for pups to be separated from their mother and littermates—including the tactile, olfactory, and thermal features of the normal habitat [44]—for an extended period of time. Recently, a study examining cerebral cortical activity in head-fixed infant rats used a paradigm that included the presence of a littermate in contact with the head-fixed subject [45]. These and other modifications to the method can be developed to expand the range of questions that can be asked and resolve concerns about the artificiality of the testing procedure.

2.8 Pros and Cons of Recording Neural Activity in Unanesthetized Subjects

For those research questions in which behavior is not a priority, it may be preferable to use anesthetized subjects. However, given that anesthetics can substantially alter neural activity at the cellular and network levels, it is preferable—whenever possible—to use unanesthetized subjects. Indeed, as described below, most of the difficulties of working with behaving animals, including infants, can be avoided by considering potential pitfalls in advance.

Unlike head-fixed adult subjects, infant subjects do not require days or weeks of habituation to the restraint; on the contrary, we observe regular sleep-wake cycles within an hour or two of surgery. When head-fixed pups exhibit signs of distress, it is expressed in the form of pronounced and protracted limb movements and audible vocalizations. To resolve the problem, check to make sure that the head and neck are comfortably positioned in relation to the body and to the platform. Also, there are a number of issues to consider when testing pups after prolonged separation from the nest. Specifically, before P10, urogenital stimulation is required for micturition and defecation [46], and a pup sometimes will exhibit distress until it is voided by the experimenter (using a cotton-tipped applicator applied to the anogenital region). If a pup exhibits struggling movements or is vocalizing audibly for any extended period of time, temperature and humidity should be checked to ensure that the pup is neither too cold nor too hot and that the air is not too dry.

Mechanical vibration, including vibration resulting from body movements, can lead to artifacts in the electrophysiological recordings. EMG electrodes that are loose and dangling from the pup can result in noisy signals; securing the wires to a stable platform with tape usually resolves this problem. For a variety of reasons, including the proximity of the electrode to the neck and shoulder, cerebellar and hindbrain extracellular recordings are particularly prone to movement artifact. The attachment of the nuchal ligament and musculature to the occipital bone can be one source of movement artifact, especially in older pups. Therefore, to minimize movement artifact, an additional surgical step is to detach the nuchal ligament from its anchor point to the skull. This can be done via blunt dissection with forceps or scissor or by cauterizing at the attachment points. If movement artifact is still a problem, the occipital bone can be hardened with Vetbond.

Using head restraint with awake, behaving pups—especially at older ages—requires more diligence than with anesthetized preparations. Importantly, the second postnatal week is a time of substantial functional change in the auditory and visual systems, with implications for behavior [47–49]. For example, noise and lights that are undetectable during the first postnatal week can easily arouse a P9 mouse or a P12 rat. Lastly, the testing of tactile or proprioceptive responses becomes more difficult with age: Stimulation paradigms that are effective in young rat pups can cause profound arousal at P12.

3   Analyzing Extracellular Activity and Behavior

3.1   Recording from Purkinje Cells

As with adult rodents, infant Purkinje cell activity can be identified by the presence of complex and simple spikes. In infant rodents, however, complex spikes are unique in that they exhibit a neurophysiological signature reflective of the stage of climbing fiber synapse development [12, 17, 26] and Purkinje cell maturation [7]. Specifically, during the first 2 postnatal weeks, complex spikes present as multi-spike bursts of action potentials with inter-spike intervals of 10–50 ms depending on the age of the pup [12, 17, 50, 51]. Also, due to the multiple innervations of climbing fibers, complex spike firing rates may actually be higher in infants than in adults [26].

Because of these developmental differences in the characteristics of complex spikes, it is necessary to use an age-appropriate method for segregating complex and simple spikes. For both mouse and rat pups at P9 and younger, inter-spike interval alone can be used to identify a complex spike. Figure 5a shows representative inter-spike intervals for unit activity recorded from Purkinje cells in rats at P4, P8, and P12. Using frequency distributions of inter-spike intervals, we determined that inter-spike intervals ≤ 20 ms captured complex spike activity at P4 and inter-spike intervals ≤ 15 ms captured complex spike activity at P8 (Fig. 5b) [51]. On the basis of these inter-spike intervals and using a script implemented in Spike2 software (Cambridge Electronic Design, Cambridge, UK), complex spike bursts of two or more action potentials can be extracted from the overall unit activity and marked as single events. The remaining neural activity is designated as simple spike activity. At these ages, the firing rates for simple spikes are low in relation to adults (Table 2) [16].

By P12, complex spikes are more adult-like in form, although still variable [12, 17, 26]. Therefore, a spike sorting algorithm or template matching protocol must be used. In adults, investigators employ a number of different protocols to identify complex spikes (see [26] and [29] for slightly different examples). We sort spikes using the standard template matching algorithm in Spike2. Using a 3–5 ms window to extract each template, we can identify individual units. Once we have identified unit templates, we use the overdraw function in Spike2 to visualize the waveforms. Using a second threshold (≥ 2:1 signal-to-noise ratio), we identify subsequent depolarizations and visually verify that those depolarizations are characteristics of a complex spike (Fig. 5c). Importantly, over this 2-week period of development, the number of action potentials or components comprising complex spikes increases, while the intervals between adjacent components shorten. At the same time, simple spike firing rates increase (Fig. 5d) [12].

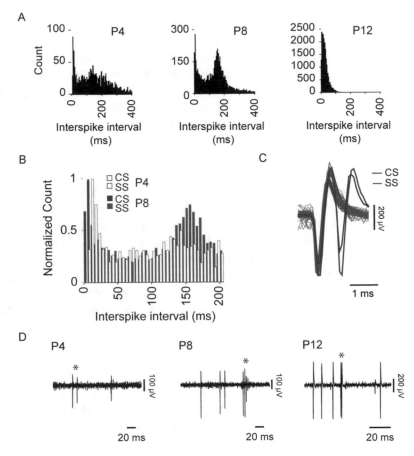

Fig. 5 Determining Purkinje cell activity in the postnatal rat. (**a**) Inter-spike interval histograms for Purkinje cells at P4, P8, and P12. Note, at P4 and P8, that the inter-spike intervals suggest two patterns of neural activity—one fast (i.e., <50 ms) and the other slower (~100–200 ms). (**b**) Determination of complex spike (**CS**) and simple spike (**SS**) activity for P4 and P8 rats. In P4 rats, CSs are defined from action potentials with inter-spike intervals ≤20 ms. At P8, inter-spike intervals are faster and CSs are defined from action potentials with inter-spike intervals ≤ 15 ms. All remaining activity is classified as SSs. (**c**) Determination of CS and SS activity in P12 rats. By P12, CS more closely resembles the adult form. Using a larger window to extract templates, CSs can be identified using the overdraw method in Spike2 to visualize templates with multiple depolarizations. (**d**) Representative multiunit activity at P4, P8, and P12 illustrating the maturation of the CS and the increase in SS activity

3.2 Recording from Other Cell Types in the Cerebellar Cortex and Cerebellar Nuclei

Purkinje cell activity during postnatal development is easily identified by high-amplitude action potentials and the presence of complex and simple spikes. The neural activity of the other cell types within the cerebellar cortex and CN, however, is harder to identify without utilizing methodologies such as juxtacellular labeling (see Chap. 1 of this volume; [34, 52]). Based on histology and patterns of extracellular activity, however, it is possible to ascertain the cell type with reasonable certainty (Fig. 6a) [34]. Furthermore, during postnatal development, certain cell types may be ruled out at certain ages (i.e., basket cells are not functional at P4) [3, 4, 8, 9].

Table 2
Mean firing rate of Purkinje cells and neurons in the interpositus nucleus during early postnatal development

	Age	Mean firing rate (Hz)
Complex spikes	P4	0.19 ± 0.0
	P8	0.64 ± 0.1
	P12	1.66 ± 0.2
Simple spikes	P4	1.87 ± 0.3
	P8	4.56 ± 0.5
	P12	11.84 ± 1.6
Interpositus nucleus	P8	3.09 ± 0.4
	P12	15.44 ± 2.6

P postnatal day. Means ± SEM

Consistent with anatomical and neurophysiological assessments of the development of the cerebellar cortex [3, 4, 8, 9], we have recorded non-Purkinje cell activity at P8 and, more commonly, at P12. Figure 6b (left) shows three channels of multiunit activity (MUA) recorded from adjacent sites within the cerebellar cortex of a P8 rat (see color coding in Fig. 6a); the corresponding inter-spike interval distributions are shown on the right. The MUA on channel 7 exhibited a typical Purkinje cell activity profile with high-amplitude action potentials and a relatively tonic firing rate (Fig. 6b, green); moreover, inter-spike intervals show the double peak typically seen at this age [16, 51]. In contrast, the MUA on channels 6 and 8 (Fig. 6b, blue and pink, respectively) only increased when the Purkinje cell was silent. These electrode sites (NeuroNexus model, A1x16-Poly2-5 mm-50s-177; site diameter = 15 μm) were situated just above and below (~43.3 μm) and lateral (± 25 μm) to the site where Purkinje cell activity was detected. This pattern of activity and inter-spike interval distributions (Fig. 6b, right) are consistent with basket or Golgi cells [34, 51].

In older infants, the IGL is large enough for the electrode to be contained entirely within it. The photomicrograph in Fig. 6c shows the IGL of cortical layers IV/V in a P14 mouse: It is apparent that all recording sites of the electrode were contained deep within the IGL. Furthermore, for one neuron, an exceptionally fast firing rate was observed. The raw MUA activity is presented below the histology image and enlarged at the right to show the bursting properties of this cell (Fig. 6c, bottom). This activity is very similar to what has been observed in extracellular recordings of granule cells in adult rabbits and mice [53].

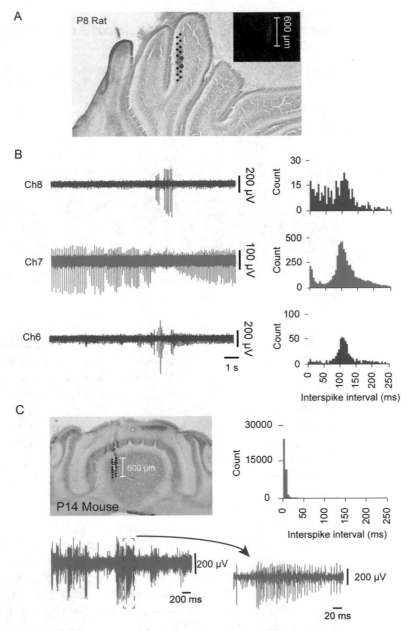

Fig. 6 Classifying cerebellar unit activity using inter-spike interval. (**a**) Nissl-stained, coronal section from a P8 rat showing electrode placement in Crus 1 of the cerebellar cortex. Filled circles illustrate electrode site locations (not drawn to scale). **Inset**, Fluorescent micrograph showing DiI fluorescence used for histological verification. (**b**) *Left*, Three channels of multiunit activity corresponding to the colored circles in (**a**). All sites were situated in closest proximity to the Purkinje cell layer. *Right*, Inter-spike interval histograms for the single unit activity from each of the MUA channels. Patterning and rate differ across the channels with channel 7 (Ch7) exhibiting the inter-spike interval pattern of a P8 Purkinje cell; the sites for channels 6 and 8 (Ch6 and Ch8) are immediately below (relative to the Purkinje cell layer) the site for Ch7 suggestive of cortical inhibitory neurons such as basket or Golgi cells. (**c**) *Upper left*, Nissl-stained, coronal section from a P14 mouse showing electrode placement in the internal granular layer of layers 4 and 5 of the cerebellar cortex. Green circle indicates approximate position of electrode site. *Upper right*, The inter-spike interval histogram on the right shows that the inter-spike interval was very brief. **Lower left,** The raw MUA record illustrates that the high firing rate occurs in bursts and is expanded, **lower right**, to show a brief period of rapid bursting in this unit, similar to what has been seen in granule cell recordings in the adult mouse and rabbit [53]

The extracellular activity of the cerebellar nuclei (CN) during motor learning in adult rodents has been well studied [54–56]. However, surprisingly little is known about CN activity across the early postnatal period [9, 10, 19]. Perhaps the best developmental research on CN activity in developing rats has focused on eyeblink conditioning during the third postnatal week [57]. In that work, extracellular activity in the interpositus nucleus exhibited developmental changes in activity that relates to the expression of conditioned responses in a way that is consistent with granule cell migration and cerebellar circuit formation.

Recently, we recorded extracellular activity in the interpositus nucleus in rats at P8 and P12. Similar to Purkinje cell activity [51], there was a substantial increase in firing rate between P8 and P12 (Table 2) [36]. Consistent with in vivo recordings in adults, extracellular activity in the developing CN exhibits heterogencity in firing rate, consistent with the heterogeneity of cell types within the nuclei [52]. We also found that inactivation of the interpositus, using the $GABA_A$ receptor agonist muscimol, reduced spontaneous activity in the red nucleus—which receives a major interpositus projection—by approximately 50% at both P8 and P12 [36]. Importantly, this finding demonstrates functional connectivity of the cerebellum with the brainstem very early in the postnatal period.

3.3 Cerebellar Activity and Behavior

In his seminal work in the 1970s on cerebellar anatomy in perinatal rats, Altman showed how the expression of skilled motor behavior occurs in lockstep with cerebellar development [1, 5, 58]. In the intervening years, much has been learned about the molecular mechanisms involved in cerebellar development [20, 21]. In addition, evidence has begun to accumulate that activity-dependent processes are also involved [21, 25]. We have sought to identify those mechanisms and how they might contribute to somatotopic organization within the cerebellar circuit [36, 50, 51].

During the first postnatal week in rats, active (or REM) sleep is the predominant behavioral state [59, 60]. The hallmark of active sleep is the muscle atonia accompanied by myoclonic twitches of skeletal muscles, resulting in hundreds of thousands of discrete jerky movements of the limbs and whiskers each day [61]. Accordingly, twitching is the predominant motor behavior of the early postnatal period. Importantly, twitch-related sensory feedback influences spinal circuit formation [62] and triggers substantial neural activity in many brain areas, including the thalamus, hippocampus, and cerebral cortex [39, 42, 63–65]. The same holds true for the cerebellum: Both Purkinje cell and interpositus activity exhibit strong sleep dependency, with more neural activity occurring during active sleep than wakefulness. Furthermore, feedback from twitches triggers increased cerebellar activity [36, 50, 51]. These observations are building a case for the importance of sleep in early development, as well as for the consideration of twitching as a self-generated form of spontaneous activity that contributes to the activity-dependent development of sensorimotor systems [36–39, 50, 51, 66].

To examine sleep-related cerebellar activity, we quantify behavioral state using EMG and behavioral scoring. First, the EMG record is rectified and smoothed (tau = 0.001 s). The mean EMG signal for periods of atonia (sleep) and high muscle tone (wake) is calculated from five representative 1 s EMG segments. The midpoint between these two values is used as a threshold for identifying sleep and wake. Once the threshold is determined, the experimental session can be coded using the EMG signal to identify periods of sleep and wake, with behavioral observation used to confirm these designations. Sleep can be further divided into periods of quiet sleep (low tone/atonia with behavioral quiescence) and active sleep (atonia with myoclonic twitches) [50, 51]. Figure 7a shows a representative Purkinje cell MUA and EMG activity from a P12 rat. The record begins with the pup in active sleep as evidenced by both nuchal and hind limb atonia. When the rat pup awakens (gray box) and moves vigorously, the neural activity decreases. Then, when the pup stops moving and becomes quiet again, the neural activity returns.

Similar to studies examining cerebellar activity during learning, perievent histograms provide a straightforward way to assess the neural response to twitches. First, we identify individual twitches as discrete EMG events with amplitudes exceeding at least three times the mean EMG level during atonia [50, 51]. With twitches converted into discrete events, it is easy to quantify the neural activity in the vicinity of a twitch; in Fig. 7b, complex spike activity increases substantially within 100 ms of twitching. The same approach can be taken for examining the effects of exafferent stimuli on cerebellar activity (Fig. 7c). In Fig. 7c, interpositus activity increases substantially after ipsilateral forelimb stimulation. Twitch-reafference and exafferent stimulation of the limbs and face can be used to investigate somatotopic organization in the developing cerebellum.

4 Conclusions: New Approaches to Recording and Analyzing Extracellular Activity

The last decade has seen a profound increase in new molecular, neurophysiological, and neuroimaging methods that are improving our ability to observe and manipulate (e.g., optogenetics) neural activity on biologically relevant timescales [43, 52]. For example, the use of transgenic lines with conditional expression of GCaMP3 has been used in infant mice with two-photon imaging to examine neural activity in the developing visual system [67]. Furthermore, mouse lines with Cre-dependent, cell-specific expression of opsins makes it possible to use optogenetic approaches at the earliest

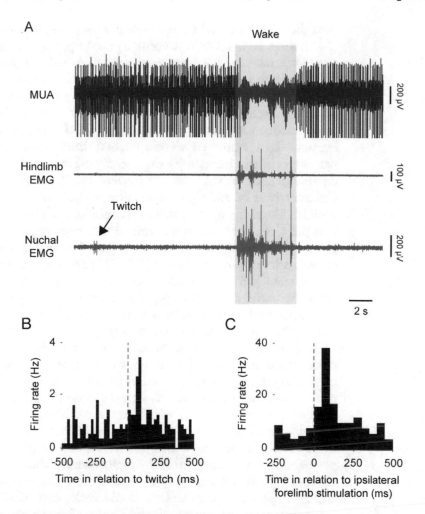

Fig. 7 Cerebellar activity and behavior in the postnatal rat. (**a**) Multiunit activity (MUA) and EMG activity from the ipsilateral hind limb and nuchal muscles for a representative P12 rat exhibiting state-dependent neural activity. The MUA exhibits a tonic, rapid rate of firing during active sleep (AS). However, when the pup wakes up and moves around the high amplitude unit, activity ceases. When the animal again becomes quiet and stops moving, the neural activity resumes. (**b**) Twitch-triggered event correlation for complex spike activity in a P8 rat. Approximately 60–80 ms following a twitch in the nuchal muscle complex spike activity increases. (**c**) Event correlation of CN activity recorded from the interpositus nucleus of a P12 rat in response to a forelimb stimulation. Over 18 stimulations, an increase in activity is observed 50–100 ms following the stimulation

postnatal ages [68]. These are just two examples of novel approaches for studying activity in the developing cerebellum using the head-fix method.

The head-fix method can also be modified for implementation in numerous testing environments. To date, no studies have used the head-fix method to look at goal-directed movements and extracellular cerebellar activity in infant rodents. In infant rats, this could provide a means to assess the effects of early motor training on cerebellar activity and vice versa. For example, an adaptation of the cylindrical treadmill from the adult mouse head-fix preparation ([43]; see Chap. 10 of this volume) could provide a measure of cerebellar responses to practiced or novel wake movements in infant rats and mice. We know from previous work that direct stimulation of the pontine nuclei in P12 rats is a potent conditioned stimulus, resulting in a conditioned eyeblink response [69]. This result clarified that the previously reported absence of motor learning is a result of developmental immaturity of sensory pathways and not immaturity of plasticity mechanisms in cerebellar circuits [5]. Given that the red nucleus receives inputs from the CN as early as P8 and that inactivation of the interpositus nucleus significantly reduces red nucleus extracellular activity, it appears that the CN exerts excitatory control over premotor areas at very young ages [36]. Perhaps the incorporation of age-appropriate sensory stimuli (i.e., vibration; [70, 71]) and/or age-appropriate motor tasks [58] would allow for the study of learning-related cerebellar activity at earlier ages than previously thought possible.

Importantly, extracellular activity in the infant rat cerebellum shows changes in patterning, timing, variety, and quantity of activity during the first 2 postnatal weeks in rats [36, 50, 51]. This corresponds to a period of rapid and dramatic development in cerebellar circuitry. Two important patterns observed during this period are the sleep- and twitch-dependent activities of the cerebellar cortex and CN, which suggest this may be a sensitive period for activity-dependent development [36, 50, 51]. Although the rodent cerebellum continues to develop beyond the first postnatal month, many critical developmental changes occur in the first 2 postnatal weeks; therefore, if these changes depend on activity-dependent processes, it would make cerebellar development susceptible to environmental factors that might affect sleep or sleep quality early in development [51]. It remains to be determined how subtle and variable deficits in cerebellar development translate into neurobehavioral disorders [28]. Therefore, a comprehensive understanding of early postnatal cerebellar activity and the susceptibility of that activity to environmental insult remains an important line of research that could benefit greatly from the head-fix method.

References

1. Addison WHF (1911) The development of the Purkinje cells and of the cortical layers in the cerebellum of the albino rat. J Comp Neurol 21:459–486

2. Altman J (1972) Postnatal development of the cerebellar cortex in the rat I. The external germinal layer and the transitional molecular layer. J Comp Neurol 145:353–398

3. Altman J (1972) Postnatal development of the cerebellar cortex in the rat II. Phases in the maturation of Purkinje cells and of the molecular layer. J Comp Neurol 145:399–464

4. Altman J (1972) Postnatal development of the cerebellar cortex in the rat III. Maturation of the components of the granular layer. J Comp Neurol 145:465–514

5. Freeman JH (2014) The ontogeny of associative cerebellar learning. In: Mauk MD (ed) International review of neurobiology: Cerebellar conditioning and learning. Elsevier, Oxford, pp 53–71

6. Hashimoto K et al (2009) Translocation of a "winner" climbing fiber to the Purkinje cell dendrite and subsequent elimination of "losers" from the soma in developing cerebellum. Neuron 63:106–118

7. McKay BE, Turner RW (2005) Physiological and morphological development of the rat. J Physiol 567:829–850

8. Shimono T, Nosaka S, Sasaki K (1976) Electrophysiological study on the postnatal development of neuronal mechanisms in the rat cerebellar cortex. Brain Res 108:279–294

9. Crepel F (1974) Excitatory and inhibitory processes acting upon cerebellar Purkinje cells during maturation in the rat; Influence of hypothyroidism. Exp Brain Res 20:403–420

10. Gardette R et al (1985) Electrophysiological studies on the postnatal development of intracerebellar nuclei neurons in rat cerebellar slices maintained in vitro. I. Postsynaptic potentials. Dev Brain Res 19:47–55

11. Kalinovsky A et al (2011) Development of axon-target specificity of ponto-cerebellar afferents. PLoS Biol. https://doi.org/10.1371/journal.pbio.1001013

12. Crepel F (1971) Maturation of climbing fiber responses in the rat. Brain Res 35:272–276

13. Crepel F, Mariani J, Delhaye-Bouchaud N (1976) Evidence for a multiple innervation of Purkinje cells by climbing fibers in the immature rat cerebellum. J Neurobiol 7:567–578

14. Hashimoto K, Kano M (2005) Postnatal development and synapse elimination of climbing fiber to Purkinje cell projection in the cerebellum. Neurosci Res 53:221–228

15. Kuwako K-I et al (2014) Cadherin-7 regulates mossy fiber connectivity in the cerebellum. Cell Rep 9:311–323

16. Woodward DJ, Hoffer BJ, Lapham LW (1969) Postnatal development of electrical and enzyme histochemical activity in Purkinje cells. Exp Neurol 23:120–139

17. Puro DG, Woodward DJ (1977) Maturation of evoked climbing fiber input to rat cerebellar Purkinje cells (I.) Exp Brain Res 28:85–100

18. Puro DG, Woodward DJ (1977) Maturation of evoked mossy fiber input to rat cerebellar Purkinje cells (II). Exp Brain Res 28:427–441

19. Gardette R et al (1985) Electrophysiological studies on the postnatal development of intracerebellar nuclei neurons in rat cerebellar slices

maintained in vitro. II. Membrane conductances. Dev Brain Res 20:97–106

20. Sillitoe RV, Joyner AL (2007) Morphology, molecular codes, and circuitry produce the three-dimensional complexity of the cerebellum. Annu Rev Cell Dev Biol 23:549–577

21. Wang VY, Zoghbi HY (2001) Genetic regulation of cerebellar development. Nat Rev Neurosci 2:484–491

22. Hashimoto K, Kano M (2013) Synapse elimination in the developing cerebellum. Cell Mol Life Sci 70:4667–4680

23. Bosman LWJ et al (2008) Homosynaptic long-term synaptic potentiation of the "winner" climbing fiber synapse in developing Purkinje cells. J Neurosci 28:798–807

24. Kakizawa S et al (2000) Critical period for activity-dependent synapse elimination in developing cerebellum. J Neurosci 20:4954–4951

25. Kano M, Hashimoto K (2012) Activity-dependent maturation of climbing fiber to Purkinje cell synapses during postnatal cerebellar development. Cerebellum 11:449–450

26. Lorenzetto E et al (2009) Genetic perturbation of postsynaptic activity regulates synapse elimination in developing cerebellum. PNAS 106:16475–16480

27. Watanabe M, Kano M (2011) Climbing fiber synapse elimination in cerebellar Purkinje cells. Eur J Neurosci 34:1697–1710

28. Wang SS-H, Kloth AD, Badura A (2014) The cerebellum, sensitive periods, and autism. Neuron 83:518–532

29. Arancillo M et al (2015) In vivo analysis of Purkinje cell firing properties during postnatal mouse development. J Neurophysiol 113:578–591

30. Ito M (2008) Control of mental activities by internal models in the cerebellum. Nat Rev Neurosci 9:304–313

31. Shevelkin AV, Ihenatu C, Pletnikov MV (2014) Pre-clinical models of neurodevelopmental disorders: Focus on the cerebellum. Rev Neurosci 25:177–197

32. Blumberg MS et al (2015) A valuable and promising method for recording brain activity in behaving newborn rodents. Dev Psychobiol 57:506–517

33. Seelke AMH, Blumberg MS (2005) The microstructure of active and quiet sleep as cortical delta activity emerges in infant rats. Sleep 31:691–699

34. Ruigrok TJH, Hensbroek RA, Simpson RI (2011) Spontaneous activity signatures of morphologically identified interneurons in the vestibulocerebellum. J Neurosci 31:712–724

35. Del Rio-Bermudez C, Sokoloff G, Blumberg MS (2015) Sensorimotor processing in the newborn rat red nucleus during active sleep. J Neurosci 35:8322–8332

36. Del Rio-Bermudez C et al (2016) Spontaneous activity and functional connectivity in the developing cerebellorubral system. J Neurophysiol 116:1316–1327

37. Tiriac A, Blumberg MS (2016) Gating of reafference in the external cuneate nucleus during self-generated movements in wake but not sleep. Elife. https://doi.org/10.7554/eLife.18749.

38. An S, Kilb W, Luhmann HJ (2014) Sensory-evoked and spontaneous gamma and spindle bursts in neonatal rat motor cortex. J Neurosci 34:10870–10883

39. Tiriac A, Rio-Bermudez CD, Blumberg MS (2014) Self-generated movements with "unexpected" sensory consequences. Curr Biol 24:2136–2141

40. Kummer M et al (2016) Column-like CA2+ clusters in the mouse neonatal neocortex reveled by three-dimensional two-photon CA2+ imagin in vivo. Neuroimage 138:64–75

41. Luhmann HJ (2016) Review of imaging network activities in developing rodent cerebral cortex in vivo. Neurophoton. https://doi.org/10.1117/1.NPh.4.3.031202

42. Tiriac A, Uitermarkt BD, Fanning AS, Sokoloff G, Blumberg MS (2012) Rapid whisker movements in sleeping newborn rats. Curr Biol 22:2075–2080

43. Heiney SA et al (2014) Cerebellar-dependent expression of motor learning during eyeblink conditioning in head-fixed mice. J Neurosci 34:14845–14853

44. Alberts JR, Cramer CP (1988) Ecology and experience: Sources of means and meaning of developmental change. In: Blass EM (ed) Developmental psychobiology and behavioral ecology. Springer, New York, pp 1–39

45. Akhmetshina D et al (2016) The nature of the sensory input to the neonatal rat barrel cortex. J Neurosci 36:9922–9932

46. Stelzner DJ (1971) The normal postnatal development of synaptic end-feet in the lumbosacral spinal cord and of responses in the hind limbs of the albino rat. Exp Neurol 31:331–357

47. Blatchley BJ, Cooper WA, Coleman JR (1987) Development of auditory brainstem response to tone pip stimuli in the rat. Dev Brain Res 32:75–84

48. Blumberg MS et al (2005) Dynamics of sleep-wake cyclicity in developing rats. PNAS 102:14860–14864

49. Routtenberg A, Strop M, Jerden J (1978) Response of the infant rat to light prior to eyelid opening: mediation by the superior colliculus. Dev Psychobiol 11:469–478

50. Sokoloff G, Uitermarkt BD, Blumberg MS (2015) REM sleep twitches rouse nascent cerebellar circuits: implications for sensorimotor development. Dev Neurobiol 75:1140–1153

51. Sokoloff G et al (2015) Twitch-related and rhythmic activation of the developing cerebellar cortex. J Neurophysiol 114:1746–1756

52. Canto CB, Witter L, De Zeeuw CI (2016) Whole-cell properties of cerebellar nuclei neurons in vivo. PLoS One. https://doi.org/10.1371/journal.pone.0165887

53. van Beugen BJ, Gao Z, Boele H-J, Hoebeek F, De Zeeuw CI (2013) High frequency burst firing of granule cells ensures transmission at the parallel to Purkinje cell synapse at the cost of temporal coding. Front Neural Circuit. https://doi.org/10.3389/fncir.2013.00095

54. Mauk MD (1997) Roles of cerebellar cortex and nuclei in motor learning: contradictions or clues? Neuron 18:343–346

55. Lee KH et al (2015) Circuit mechanisms underlying motor memory formation in the cerebellum. Neuron 86:529–540

56. Perciavalle V et al (2013) Consensus paper: current views on the role of cerebellar interpositus nucleus in movement control and emotion. Cerebellum 12:738–757

57. Freeman JH, Nicholson DA (2000) Developmental changes in eye-blink conditioning and neuronal activity in the cerebellar interpositus nucleus. J Neurosci 20:813–819

58. Altman J, Sudarshan K (1975) Postnatal development of locomotion in the laboratory rat. Anim Behav 23:896–920

59. Jouvet-Mounier D, Astic L, Lacote D (1969) Ontogenesis of the states of sleep in rat, cat, and guinea pig during the first postnatal month. Dev Psychobiol 2:216–239

60. Gramsbergen A, Schwartze P, Prechtl HFR (1970) The postnatal development of behavioral states in the rat. Dev Psychobiol 3:267–280

61. Blumberg MS, Marques HG, Iida F (2013) Twitching in sensorimotor development from sleeping rats to robots. Curr Biol 23:R532–R537

62. Petersson P et al (2003) Spontaneous muscle twitches during sleep guide spinal self-organization. Nature 424:72–75

63. Khazipov R et al (2004) Early motor activity drives spindle bursts in the developing somatosensory cortex. Nature 432:758–761

64. Mohns EJ, Blumberg MS (2008) Synchronous bursts of neuronal activity in the developing hippocampus: modulation by active sleep and association with emerging gamma and theta rhythms. J Neurosci 28:10134–10144

65. Mohns EJ, Blumberg MS (2010) Neocortical activation of the hippocampus during sleep in infant rats. J Neurosci 30:3438–3449

66. Roffwarg HP, Muzio JN, Dement WC (1966) Ontogenetic development of the human sleep-dream cycle. Science 152: 604–619

67. Ackman JB, Burbridge TJ, Crair MC (2012) Retinal waves coordinate patterned activity throughout the developing visual system. Nature 490:219–225

68. Madisen L et al (2012) A toolbox of Cre-dependent optogenetic transgenic mice for light-induced activation and silencing. Nat Neurosci 15:793–802

69. Campolattaro MM, Freeman JH (2008) Eyeblink conditioning in 12-day-old rats using pontine stimulation as the conditioned stimulus. PNAS 105:8120–8123

70. Goldsberry ME, Elkin ME, Freeman JH (2014) Sensory system development influences the ontogeny of eyeblink conditioning. Dev Psychobiol 56:1244–1251

71. Goldsberry ME, Freeman JH (2016) Sensory system development influences the ontogeny of trace eyeblink conditioning. Dev Psychobiol. https://doi.org/10.1002/dev.21468

Chapter 13

In Vivo Recordings of Network Activity Using Local Field Potentials and Single Units in Movement and Network Pathophysiology

Richard Courtemanche and Maxime Lévesque

Abstract

This chapter introduces basic and some advanced methods for recording, analyzing, and comparing local network electrophysiological activity in rodents and primates. Attention will be paid to the acquisition of network signals that consider the local field potentials (LFPs) and single-unit and multiunit activity. Analysis methods for extracting main features from LFP signals, their frequency, power, and coherence will be discussed, as well as the relation of unit activity with the LFPs. The relationship with movement and behavior will be developed, and so will the relation of these measures with network activity in specific pathologies of the local networks.

Key words Local field potentials, Unit activity, Oscillations, Synchrony

---

1 Introduction

The great quest of many neurophysiologists, to be able to record all neurons in a structure or many structures in order to understand the underlying neural code, might of course be an impossible dream. However, while current technology cannot permit to record all neurons involved in a given behavior or pathology, it is becoming increasingly more feasible to explore and address population coding by recording many single neurons, in combination with local field potentials (LFPs). Such recordings permit to address the local network processes, in combination with the modes of interaction of small populations of neurons between each other.

Local neurons are likely cooperating in small groups in order to accomplish specific functions, in addition to communicating with further distant neighbors in combining computations requiring multiple local circuits. In essence, based on the neural firing of multiple single isolated neurons, some of the local circuit properties can be assessed, complemented by LFPs, and long-range communication can be characterized by LFP–LFP relations, or by

Roy V. Sillitoe (ed.), *Extracellular Recording Approaches*, Neuromethods, vol. 134,
https://doi.org/10.1007/978-1-4939-7549-5_13, © Springer Science+Business Media, LLC 2018

spike–LFP relations [1]. These relationships provide a view of functional connectivity at the structural level, but also at the inter-structural level [2].

2 Recordings and Analysis of LFPs

2.1 Recordings

Local field potentials (LFPs) represent the integrated electrical signals recorded from a local population of neurons, and more and more is known about the underlying physiological mechanisms generating the LFP or EEG signals [3]. While in certain instances, spike activity strongly contributes to this integrated signal, the slower-frequency integrated signal mostly follows the local synaptic activity. By using low-impedance electrodes, one can record these intracellular synaptic voltage fluctuations that are integrated from a population of cells close to the electrode. The synaptic voltage changes combine together into a spatiotemporally integrated signal, that is, dynamic down to the millisecond, forming different neural generators via local circuits and long-range connections [4].

The basic type of electrode for this type of recording is a lower impedance single electrode. Many can be implanted at the same time, and the redundancy of the signal depends on the structure being recorded from and the frequency of the oscillations. With bone screws serving as ground, and a good metal conductor placed in the tissue as reference, multiple metal microelectrodes of an impedance around 1 MΩ can be placed onto a headstage [5] or in a recording chamber [6]. In certain cases, when the dipole generating the electrophysiological activity is well known, a bipolar recording arrangement is useful [7, 8]. Multiple electrode arrays can also record reliably multiple unit activity in a chronic setting, permitting the isolation of multiple spike channels at the same time and contributing to the knowledge of the circuit arrangement [9, 10]. Tetrodes, with wires brought down to a low impedance, can also be used to record stable LFPs [11] and units as well, enclosed in small microdrives [12, 13]. Tetrode recordings, by quadrangulating in a few tens of μm the various sources of voltage, along with the use of powerful software algorithms, contribute to advantageous unit detection, along with LFP recordings [14–17]. This can be done in vitro, in a rodent brain slice of approximately 450 μm [18]. Multiple single units with LFP activity can be recorded simultaneously, but pharmacological agents that will trigger neuronal network activity need to be applied. The artificial network with reduced inputs from surrounding brain regions that has been created following the slicing of the brain will not spontaneously generate single-unit and LFP activity that can be recorded.

This type of tetrode approach has been developed in macaque monkeys as well (e.g., [19]). Still, low-impedance single electrodes or wires can permit to record a strong field of activity, while keeping the capacity to resolve larger units. In terms of endpoint, the dual quality in recording is an important factor, as it permits to address the overall integration across the local network, while still keeping the capacity to isolate single neurons. Specialized apparatus also exist for recording for specific cellular configurations, for example, in the cerebellar sheet and nuclei [20–23]—here we will focus more on the general recordings of units and LFPs.

The separation between the LFP and single-unit recordings can be determined by the method itself chosen in recordings, but it can also be performed offline. Essentially, the separation is related to the filtering of the signals. Most of the power of the LFP data is situated between 0.1 and 300 Hz (up to 500 Hz in some areas), while unit activity can be perceived and recorded at higher frequencies (between 600 and 6000 Hz). Many systems offer specific recording procedures for these different types of signals, separating the signals at the source. In this case, in essence, separate amplifiers are controlled for the recording of LFPs and single units. In order to obtain good quality signals, LFP data filtered between 0.1 and 500 Hz are then amplified and can then be sampled at around 2000 Hz. This will provide the best capacity for correlating the field potential activity in the millisecond domain later on. For unit activity, separate amplifiers can extract the signal, filtering between 600 and 6000 Hz, to focus on the firing of the units, and each unit that crosses an established action potential threshold can then be detected and sampled at very high sampling rate (e.g., 32 kHz) to digitize a good spike waveform. Adjustments on spike detection can be performed throughout the recordings, and the digitized thresholded waveform can then be overlaid to verify reproducibility.

Alternatively, with the increase in the capacity to store digitized signals (i.e., size of computer hard drives and memory), the whole LFP to single-unit frequency bandwidth can be recorded in one operation. In this mode, signals are filtered 0.1 Hz–8 kHz, then differentially amplified, sampled at a high rate (e.g., 25–32 kHz), and stored to disk for offline analysis. To isolate spikes, continuous wideband extracellular recordings can be filtered offline with adapted high-pass filters (e.g., with a two-pole Butterworth 500 Hz high pass). Spikes can then be discriminated offline by thresholding the filtered trace and extracting the main parameters of their waveform. From the same stored signals, LFPs can also be extracted from the wide-band signal, with the initial signal downsampled at 2000 Hz and low-pass filtered at 500 Hz.

2.2 Data Analysis   We will not focus on the extensive methods for single-unit processing, which have been developed elsewhere and in the preceding chapters. Many different software packages now permit to perform quantitative LFP, often combined with unit signal processing [24, 25], and many laboratories have developed MATLAB (MathWorks, Natick MA) routines based on standardized functions (e.g., signal processing toolbox). Several textbooks offer a good introduction to those types of functions and algorithms [26, 27].

An essential component of analyzing LFPs concerns the analysis of time periods with oscillatory content. Network oscillations indeed represent an important dynamical organizational system that can influence communication between neural populations [28–30]. An initial process to analyze these signals can be to use a spectrogram analysis. Spectrograms can be calculated using discrete short-time Fourier transform to evaluate rhythmicity in various frequency bands. Data sampled at circa 2000 Hz can be further filtered and can be decimated if needed, and the spectrogram can be produced from separated short intervals. An example of this is given in Fig. 1. Since the time series is then split into smaller windows, a safe process is to apply a window that minimizes the potential abnormalities lying at the lateral edges, by applying a form of triangular filter, further valuing the middle of the shorter window vs. the edges [26]. For example, a Hamming window can be applied. Finally, for the iterations of windows forming the spectrogram, the windows can be overlapped by a certain fraction of this window. For each window of the spectrogram, the discrete Fourier transforms are evaluated.

From this data, one application is to use different algorithms to identify more precisely oscillatory periods. As an example, 4–12 Hz rodent cerebellar granule cell layer oscillations in vivo have been quantified, focusing on periods of high presence of oscillations and potency across the frequency band [5, 8, 31–34]. As an example, for each time window, an algorithm can identify the peak frequency and its inclusion within the searched band. The identified spindle then represents an oscillatory period that is represented in the frequency domain, with the FFT peaks staying within determined limits in frequency and in time. The parameters have to be selected through systematic data analysis and have the advantage of standardizing the detection of oscillation events over all data sets, eliminating bias. The end result is a list of "oscillation periods" throughout the recording file, and interspersed in between those, periods of weaker or no oscillation. If there is a behavior that is being concomitantly monitored, the power of oscillations as measured by the spectrogram or otherwise can be aligned with the beginning of the trials, or with a particular event [6, 35].

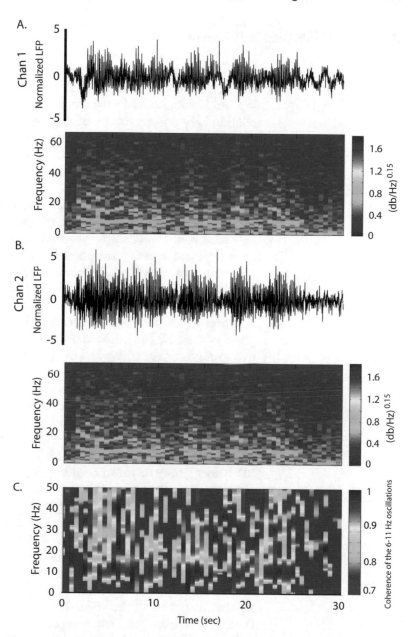

Fig. 1 Thirty seconds of two simultaneously recorded channels of rodent cerebellar granule cell layer LFPs, with oscillations around 10 Hz. (**a**) First channel LFP, along with spectrogram up to 65 Hz. (**b**) The same for second channel of LFP. (**c**) Coherence between the two channels (normalized up to 1), showing long periods of synchronous activity again around 10 Hz

This event-based approach can provide an idea of the modulation of the various rhythms across task realization, whether naturally occurring or in the context of behavioral training, or even when optogenetically stimulated [36].

3   Linking LFPs with Unit Activity

LFPs carry important information that predicts unit firing [30] and in assessing local and global network properties and influencing the likelihood of firing for groups of neurons [37, 38]. The relation between the timing of single-unit activity and LFPs can be established by cross-correlating the spike train with events representing detected LFP peaks during oscillatory periods [35] or by averaging the LFP amplitude and shape around the time of an isolated action potential.

The cross-correlation method is reminiscent of the standard methods of spike cross-correlation; the cross-correlation between the LFP peak events and the spike events can be calculated, with the LFP peak used as the reference point [6, 35, 39–41]. To establish the significance of the spike–LFP relationship, an index can be computed based on the cross-correlogram for each unit [42]. Diverse statistical approaches have been used, all related to the determination of the tightness of the temporal relationship between unit firing and the phase of the LFP [43]; essentially, the issue is to determine the relative strength of the phase-locking [5]. Here is an example to illustrate. A "chance" level of the spike–LFP relation can be computed using repeated artificial spike trains generated with randomized inter-spike delays (equal number of spikes as the original spike train, so isofrequency), to serve as a confidence interval. For each artificial spike train, an LFP peak-triggered histogram with each bin can be processed. Peaks in the mean, which are 2 SD above, or valleys 2 SD below from the mean of the spike-shuffled control histogram can then be identified in the cross-correlogram peaks. The difference in amplitude between the detected peak/valley and the next valley/peak within half a cycle can be calculated [21]. Units with a high index then show a strong spike–LFP relationship. The spike–LFP phase relationship, for each cell with a significant LFP-triggered histogram peak, can be converted to radians or degrees in proportion to the cycle duration [44]. This permits to produce a polar histogram of the spike–LFP phase relationship that can be compared across units. Two examples are given in Fig. 2. Alternatively, software packages in MATLAB also exist for evaluating the statistical significance of this phase-locking within circular variables [45].

Another method is to perform spike-triggered averages of the LFP. This method focuses on the amplitude and shape of the LFP, when it is aligned on the spike events [46]. In a manner comparable to the spike-triggered averaging of the EMG [47], this method permits to get a sense of the properties of the LFP around the time of the spike. Regularity in the shape of the pre- and post-spike LFP can provide information on the strength of the relationship, as then the peri-spike LFP can represent a signature, containing a specific frequency, or a specific combination of frequencies, in order to facilitate spiking.

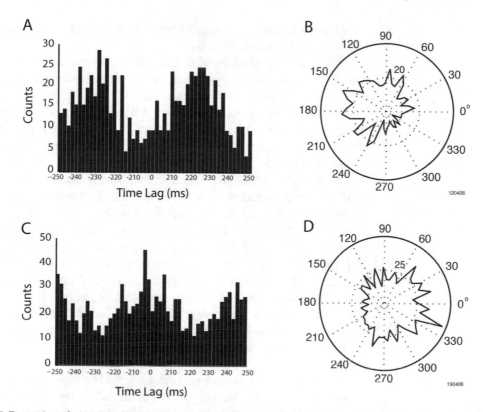

Fig. 2 Examples of phase-locking for two cerebellar cortex neurons (presumably Golgi cells). (**a**, **c**) LFP-unit cross-correlograms representing the firing rate relative to the LFP peak at zero lag. (**b**, **d**). Polar plots representing the firing of the units relative to the LFP oscillatory cycle expressed in degrees. Notice the unit in **a–b** is mostly in antiphase with the LFP (aligned on the trough), while the unit in **c–d** is mostly aligned in phase, so on the oscillatory peak

4 Task- or State-Related Network Activity via LFPs and Multiple Single Units

Behavioral state is strongly related to the organization of electrophysiological activity and in particular with the oscillatory context [48]. Recent methods permit to compute both unit activity and LFP modulation according to different behavioral events or behavioral state. As an example, in the hippocampus, population neuronal firing is coordinated relative to the theta rhythm, but also relative to local population firing at faster frequencies [49]. Multiple neural dynamics can also come into play simultaneously: an example is the capacity for theta-gamma nesting [29, 50] and its inherent capacity to influence neuronal firing [51]. Such oscillations or arrangement of oscillations, as they are implicated in certain behaviors, can guide the exploration for task-related neural coding; in fact, even if only relying on the LFP signals, direction-specific signals can be extracted and used to control a brain–machine interface [52]. In the context of task-related coding, many examples have shown a differential organization at various stages of learning of a sensorimotor task [53, 54].

A logical way of addressing the task- or state-related activity is thus to consider the firing patterns of the single units in relation to identified oscillations and between each other. The process would start by extracting from the LFP the various bands of oscillations, using filtering methods. As an example, in sensorimotor processes, obvious separations would include the delta, theta, beta, and gamma bands [55, 56]. Following, the separated filtered data could be used in order to explore the phase alignment of the units relative to the segmented rhythmic signals. The signals separated according to the frequency bands, whether coming from the "nearby" or "far away" electrodes, can be used as reference signal for the timestamps of the action potentials from various units [57–59]. In this manner, the neuron entrainment by those wave activities can be determined. Other specific neural events can also be identified using a similar process—for example, spindles in the context of state-specific anesthetized slow oscillations; spindles in the cerebellum and basal ganglia show similar 8–20 Hz nesting in slow 1 Hz oscillations during urethane anesthesia, as has also been shown in the hippocampus [33, 60].

In addition, the power vs. certain tasks is sometimes difficult to ascertain using a spectrogram. A process of assessing the relative time-dependent variations in signal content corresponding to particular oscillation in the time series has to be determined in a way that will show adapted time granularity. If the phenomenon is quickly on/quickly off and/or has an amplitude that could be submerged by neighboring phenomena, a method of analysis that aims to isolate the trace and quantify relative evolution can serve, even if in the time domain. Such a method (sometimes the "temporal spectral evolution") can use a method similar to EMG processing [47]—take the filtered wave, rectify it, and consider a smoothing procedure to capture the wave-specific variations [6]. In a specific application where the LFP signal varies quickly, for instance, oscillations between 80 and 500 Hz (called high-frequency oscillations or HFOs) are often recorded from epileptic animals and patients (see for review [61]) and last between 10 and 30 ms, which makes them difficult to visualize on a spectrogram. False positives can also be detected when using the spectrogram method alone, since HFOs can result from rapid transitions of the signal such as when sharp peaks are high-pass filtered [62, 63]. Alternative methods are then more appropriate to analyze activity in higher frequency bands. The most widely used method consists of filtering the signal between specific frequency bands (ripples: 80–200 Hz and fast ripples: 250–500 Hz), identifying oscillations with more than three consecutive cycles above a predetermined threshold (2SD), and determining whether there is an overlap between events detected in both frequency bands. For instance, genuine fast ripple oscillations would be visible only in the 250–500 Hz frequency band and not in the 80–200 Hz frequency range [62]. Such events can then be visualized using the wavelet transforms with consecutive short duration overlapping windows [64, 65].

5   Linking Distant Networks Together in Movement

A major advantage of recording LFPs is to explore their synchronization patterns in the context of brain operations. These could inform us on the spatiotemporal organization of neural areas in information processing. From the basic channel-to-channel comparison, newer methods also take into consideration the overall interactions between a number of structures.

5.1   Communication Through Coherence

Much like electroencephalography (EEG), when recording the LFP activity from multiple regions simultaneously, neurophysiologists can compare the changing patterns of activity in time, studying both the presence and characteristics of oscillations within each area and the changes in synchrony between areas. Cooperation between multiple brain areas must be optimized in order to perceive our environment, select optimal behaviors, and program our motor responses. A likely mechanism for this cooperation is "communication through coherence" (CTC) [66–68]. This mechanism permits distant brain areas to share information on the basis of coherent neuronal signals. "Riding the waves," the local networks then optimize communication by locking in neuronal activity at particular frequencies and generate effective exchange of information at the single neuronal level.

In the passage from a state of immobility to movement, multiple areas including cerebellar, striatal, and cortical oscillatory activity are strongly modulated, going from strong 7–30 Hz coherent activity to a desynchronized state [6, 35]. These brain systems are involved in the assessment of temporal and spatial cues predicting events that are likely to happen, which also triggers oscillations. Such interconnected areas collaborate in the performance of sensorimotor actions [69–72], and their functional communication can be estimated using synchronization of the field potential signals, such as in CTC. We have found previously that cerebellar and striatal oscillations in the theta range vary with time of day; here the paradigm allows to assess the impact of learning on the strength of the CTC [33].

To evaluate involvement in tasks, large-scale electrophysiological recordings of the networks can be performed, while an animal is learning and performing. Network communication can be inspected by looking at LFP activity while the learning is taking place [37]. During performance these sites can be functionally coupled to promote sensorimotor binding [73]. Global synchronization in the brain is heavily dependent on rhythmic activity [74] in areas like the cerebellum, cerebral cortex, hippocampus, and basal ganglia [30, 75–77]. Cerebellar granule cell layer (GCL) oscillations (15–20 Hz) during attentive immobility are modulated by a sensorimotor task [6]. These oscillations are maximally expressed in a state of expectancy for the animal, just prior to an imminent motor action or sensory event [6]. GCL oscillations are synchronized with primary somatosensory cortex and primary motor cortex oscillations [78], likely binding together these distant areas. The CTC within the cerebellum is dynamically modular in the sagittal plane and recruits further lateral sites during movement [79]. In the rodent, GCL oscillations at ~7 Hz, in attentive immobility, are closely related to Golgi cell firing [5]. In the basal ganglia, striatal oscillations are modulated to permit task-related firing [35]. During a T-maze task, striatal neurons become progressively phase-locked to the oscillatory activity and show a particular CTC signature, with a learning-related increase in coherence at 7 Hz vs. hippocampal oscillations [11, 80]. With repeated learning, networks across the cortex and brain refine their patterns of coherence, focusing on optimal oscillatory frequencies for communication [81, 82].

Multiple methods can be used to determine CTC signatures. From the standpoint of the analysis of the LFPs, multiple one-to-one synchronies in signal comparisons often represent the starting point. This can be done in the time domain, by doing analog cross-correlation or cross-covariance of the signals, one comparison at a time [35, 78]. In this case, the cross-correlation value can be higher in phase (around zero-lag), in antiphase (half-a-cycle apart), or relatively in out of phase (relative time lag). The time lag between areas can then help determine if there is one site that might be driving the other site. The synchrony comparison can also be done in the frequency domain, by looking at the coherence of the signals across frequency bands [4], or by looking at directed coherence, which can also establish an order of "which area triggers which" as an added interesting information [83].

5.2 Network Analysis

Another new method of determining relations is through network analysis [84]. By focusing on anatomical or physiological quantities identified in various brain areas and their interconnections, the relations between the various network elements can be identified [85]. The patterns of interactions can serve to identify local network and large network interactions [86]. The relative complexity in the anatomical relationships between different brain areas has been explored in different species [87]. Whereas anatomical relationships were mostly used originally, the study of EEG and LFP activity relationships across brain areas promises to be a determining approach in assessing functional connectivity and network organization for processing information [88]. This is a particularly interesting and promising objective, as eventually we can also relate how protein organization shapes networks along with electrical signaling and how networks are shaped by both genes and neural activity [89].

6 Applications to Network Pathophysiology

LFP signals can also be used to address when networks show abnormal or pathological neural processing. We can identify, for example, the exaggerated rhythmicity in Parkinsonian loops [77, 90–92], which is a pathological manifestation of regularly occurring oscillations in the striatum [35, 80, 93]. In this situation, a difference can be seen in the capacity of the underlying oscillatory influence to be able to drive medium spiny neurons in the striatum; in normal circuits, the oscillatory activity witnessed in the LFPs can influence the phase of the unit firing, without driving them for every cycle [35]. As such, while oscillatory activity can be seen in the LFPs, only a small portion of units shows an oscillatory discharge. However, in dopamine-deprived circuits, it appears like the oscillatory influence is stronger, capable of driving the firing of many units, seemingly bringing the basal ganglia circuits into a more global oscillatory state [90, 94, 95]. Thus, affecting a neuromodulator circuit has shown the capacity to tip the network out of balance and into a less controlled state that shows resonance throughout the nodes. Initially found in the beta range, and still of unknown origin [96, 97], a more diverse type of oscillatory profile spatially and temporally (including gamma oscillations) has appeared with and without dopamine depletion [54, 98]. Due to the network complexity, the pursuit of the network dynamics has been actively sought in network modeling [99–101]. Beta oscillations in the basal ganglia have also been used as a biomarker and target for deep brain stimulation [102].

In addition, another example could be given such as in the case of focal epileptic disorders, for which HFOs (ripples: 80–200 Hz, fast ripples: 250–500 Hz) have been observed in both animal models and patients [61]. Mesial temporal lobe epilepsy (MTLE) is the most prevalent syndrome among focal epilepsies and is characterized by seizures that recur following a latent period of up to many years after an initial brain insult such as *status epilepticus* (SE), traumatic brain injury, encephalitis, or febrile convulsions [103, 104]. These recurrent seizures originate from the hippocampus, entorhinal cortex, or amygdala [105] and are often refractory to medication [106]. The surgical removal of the seizure onset zone is often the only alternative. Epileptologists are therefore trying to find biomarkers of MTLE that will lead to a better identification of the seizure onset zone as well as to a better understanding of the disease. HFOs, and more specifically fast ripples, are recorded in the seizure onset zones of epileptic patients; they thus show a strong relationship with pathological network activity and could help guide surgical interventions [107–110]. The surgical removal of regions containing fast ripples could be associated to reduced seizure occurrence [111], but some studies have obtained similar results when regions containing ripples and fast ripples have been removed [112, 113].

In animal models, such as in the kainic acid or pilocarpine model, HFOs have also been observed in temporal lobe regions. Bragin et al. [114] have been the first to report the occurrence of HFOs in the hippocampus, entorhinal cortex, and dentate gyrus of kainic acid-treated animals. Interestingly, fast ripples in the ento- rhinal cortex could precede first seizure occurrence by 2 months and were only visible in animals that would become epileptic, indi- cating that fast ripples reflect pathological network activity that could lead to seizure generation. In the pilocarpine model, HFOs have also been observed, during the latent (before first seizure occurrence) and the chronic period (after first seizure occurrence) [115, 116]. Fast activity in the LFP may therefore be indicative of epileptogenesis or the pathological reorganization of activity that transforms a normal brain into an epileptic one. An example of detection of HFOs is given in Fig. 3.

The generation of seizures or the processes that will favor the transition from an interictal to an ictal state have also been associated to HFOs. They are interestingly associated to specific seizure onset patterns since hypersynchronous seizures, which emerge from the hippocampus and are thought to be associated to an unrestrained enhancement of excitation [117], are more likely to be associated to fast ripples [118, 119], which are also thought to depend on excitatory principal cell glutamatergic activity [61]. On the contrary, low-voltage fast-onset seizures, which are believed to depend on the activity of interneurons, are associated to ripple occurrence [118], which are generated by the activity of interneuronal networks [120, 121]. There are however some discrepancies between animal and clinical studies ([122, 123], which could be due to the fact that seizures in epileptic patients are recorded many years after the initial brain insult, whereas spontaneous seizures in animals are recorded within 2 weeks after status epilepticus. Changes such as hippo- campal sclerosis, diffuse neuronal loss, and the pathological reorganization of neuronal networks could over time affect the occurrence of HFOs recorded in the LFP.

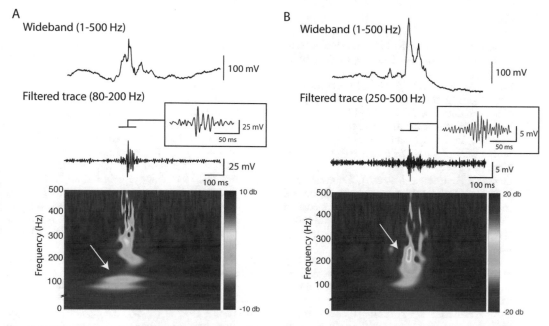

Fig. 3 High-frequency oscillations (HFOs) in the hippocampus of an epilepsy rodent model. (**a**) Interictal spike recorded from the CA3 subfield of the hippocampus of an epileptic rat that was treated with pilocarpine. Note the high power in the spectrogram between 100 and 200 Hz (*arrow*), which indicates the presence of a ripple that is visible in the filtered trace between 80 and 200 Hz. (**b**) Interictal spike recorded in the CA3 region of an epileptic rat. A fast ripple between 250 and 500 Hz is visible in the filtered trace and in the power spectrogram (*arrow*)

7   Conclusion

In this chapter, we tried to provide a sense of the various methods to gather information about neural networks of the brain using LFPs and LFPs combined with action potential coding, as well as multiple LFPs concomitantly. While EEG has been important since the 1930s [124, 125], it partly gave way to single-unit electrophysiology in the 1950s [126]; however, the integrated signal analysis of LFPs and their underlying rhythms [30] have regained a large capital of interest in trying to explain the way brain networks interact. Using recording methods that permit to record both LFPs and single units provides a powerful capacity to identify and analyze important network-level interactions in information processing, whether this is in the context of changes of state, preparation for action, or pathophysiological conditions.

Acknowledgments

Richard and Maxime wish to thank their respective former or current supervisors for their useful and prescient technical explorations. Richard thanks Maxime for being his first graduate student and for progressing so well. Special thanks are given to Roy Sillitoe for inviting us to write this chapter.

References

1. Friston KJ et al (2015) LFP and oscillations-what do they tell us? Curr Opin Neurobiol 31:1–6
2. Lowe MJ et al (2016) Modern methods for interrogating the human connectome. J Int Neuropsychol Soc 22(2):105–119
3. Buzsaki G, Anastassiou CA, Koch C (2012) The origin of extracellular fields and currents--EEG, ECoG, *LFP and spikes*. Nat Rev Neurosci 13(6):407–420
4. Frederick A, Bourget-Murray J, Courtemanche R (2013) Local field potential, synchrony of. In: Jaeger D, Jung R (eds) Encyclopedia of Computational Neuroscience: SpringerReference. Springer-Verlag, Berlin, Heidelberg; http://www.springerreference.com.
5. Dugué GP et al (2009) Electrical coupling mediates tunable low-frequency oscillations and resonance in the cerebellar Golgi cell network. Neuron 61(1):126–139
6. Courtemanche R, Pellerin JP, Lamarre Y (2002) Local field potential oscillations in primate cerebellar cortex: modulation during active and passive expectancy. J Neurophysiol 88:771–782
7. Glasgow SD, Chapman CA (2007) Local generation of theta-frequency EEG activity in the parasubiculum. J Neurophysiol 97(6):3868–3879
8. Robinson JC, Chapman CA, Courtemanche R (2017) Gap junction modulation of low-frequency oscillations in the cerebellar granule cell layer. Cerebellum 16(4):802–811
9. Ward MP et al (2009) Toward a comparison of microelectrodes for acute and chronic recordings. Brain Res 1282:183–200
10. Vetter RJ et al (2004) Chronic neural recording using silicon-substrate microelectrode arrays implanted in cerebral cortex. IEEE Trans Biomed Eng 51(6):896–904
11. DeCoteau WE et al (2007) Learning-related coordination of striatal and hippocampal theta rhythms during acquisition of a procedural maze task. Proc Natl Acad Sci U S A 104(13):5644–5649
12. Battaglia FP et al (2009) The Lantern: an ultra-light micro-drive for multi-tetrode recordings in mice and other small animals. J Neurosci Methods 178(2):291–300
13. Jeantet Y, Cho YH (2003) Design of a twin tetrode microdrive and headstage for hippocampal single unit recordings in behaving mice. J Neurosci Methods 129(2):129–134
14. Takahashi S, Anzai Y, Sakurai Y (2003) Automatic sorting for multi-neuronal activity recorded with tetrodes in the presence of overlapping spikes. J Neurophysiol 89(4):2245–2258
15. Lee CW et al (2013) The accuracy and precision of signal source localization with tetrodes. Conf Proc IEEE Eng Med Biol Soc 2013:531–534
16. Chelaru MI, Jog MS (2005) Spike source localization with tetrodes. J Neurosci Methods 142(2):305–315
17. Gao H, Solages CD, Lena C (2012) Tetrode recordings in the cerebellar cortex. J Physiol Paris 106(3–4):128–136. https://doi.org/10.1016/j.jphysparis.2011.10.005
18. Levesque M, Herrington R, Hamidi S and Avoli M (2016). Interneurons spark seizure-like activity in the entorhinal cortex. Neurobiol Dis 87:91–101. https://doi.org/10.1016/j.nbd.2015.12.011
19. Kapoor V et al (2013) Development of tube tetrodes and a multi-tetrode drive for deep structure electrophysiological recordings in the macaque brain. J Neurosci Methods 216(1):43–48
20. White JJ et al (2016) An optimized surgical approach for obtaining stable extracellular single-unit recordings from the cerebellum of head-fixed behaving mice. J Neurosci Methods 262:21–31
21. Sugihara I, Lang EJ, Llinás R (1995) Serotonin modulation of inferior olivary oscillations and synchronicity: a multiple-electrode study in the rat cerebellum. Eur J Neurosci 7:521–534
22. Welsh JP et al (1995) Dynamic organization of motor control within the olivocerebellar system. Nature 374:453–457
23. Welsh JP, Schwarz C, Nicolelis MAL (1999) Multielectrode recording from the cerebellum. In: Simon SA, Nicolelis MAL (eds) Methods for neural ensemble recordings. CRC Press, Boca Raton, FL, pp 79–100
24. Bokil H et al (2010) Chronux: a platform for analyzing neural signals. J Neurosci Methods 192(1):146–151
25. Melman T, Victor JD (2016) Robust power spectral estimation for EEG data. J Neurosci Methods 268:14–22
26. Wallisch P et al (2014) MATLAB for neuroscientists, 2nd edn. UK Academic Press, London, p 550
27. Cohen MX (2014) Analyzing neural time series data: theory and practice. MIT Press, Cambridge, MA, p 600
28. Wang XJ (2010) Neurophysiological and computational principles of cortical rhythms in cognition. Physiol Rev 90(3):1195–1268
29. Buzsaki G, Watson BO (2012) Brain rhythms and neural syntax: implications for efficient coding of cognitive content and neuropsychiatric disease. Dialogues Clin Neurosci 14(4):345–367

30. Buzsaki G (2006) Rhythms of the brain. Oxford University Press, New York, pp 1–448

31. Hartmann MJ, Bower JM (1998) Oscillatory activity in the cerebellar hemispheres of unrestrained rats. J Neurophysiol 80:1598–1604

32. O'Connor S, Berg RW, Kleinfeld D (2002) Coherent electrical activity between vibrissa sensory areas of cerebellum and neocortex is enhanced during free whisking. J Neurophysiol 87:2137–2148

33. Frederick A et al (2014) Diurnal influences on electrophysiological oscillations and coupling in the dorsal striatum and cerebellar cortex of the anesthetized rat. Front Syst Neurosci 8:145

34. Courtemanche R, Robinson JC, Aponte DI (2013) Linking oscillations in cerebellar circuits. Front Neural Circuits 7:125

35. Courtemanche R, Fujii N, Graybiel AM (2003) Synchronous, focally modulated beta-band oscillations characterize local field potential activity in the striatum of awake behaving monkeys. J Neurosci 23(37):11741–11752

36. Lu Y et al (2015) Optogenetically induced spatiotemporal gamma oscillations and neuronal spiking activity in primate motor cortex. J Neurophysiol 113(10):3574–3587

37. Buzsaki G, Draguhn A (2004) Neuronal oscillations in cortical networks. Science 304:1926–1929

38. Bullock TH (1997) Signals and signs in the nervous system: the dynamic anatomy of electrical activity is probably information-rich. Proc Natl Acad Sci U S A 94:1–6

39. Gerstein GL (1999) Correlation-based analysis methods for neural ensemble data. In: Simon SA, Nicolelis MAL (eds) Methods for neural ensemble recording. CRC Press, Boca Raton, FL, pp 157–177

40. Lamarre Y, Raynauld JP (1965) Rhythmic firing in the spontaneous activity of centrally located neurons. A method of analysis. Electroencephalogr Clin Neurophysiol 18:87–90

41. Perkel DH, Gerstein GL, Moore GP (1967) Neuronal spike trains and stochastic point processes. II. Simultaneous spike trains. Biophys J 7:419–440

42. Destexhe A, Contreras D, Steriade M (1999) Spatiotemporal analysis of local field potentials and unit discharges in cat cerebral cortex during natural wake and sleep states. J Neurosci 19(11):4595–4608

43. Hurtado JM, Rubchinsky LL, Sigvardt KA (2004) Statistical method for detection of phase-locking episodes in neural oscillations. J Neurophysiol 91(4):1883–1898

44. Perez-Orive J et al (2002) Oscillations and sparsening of odor representations in the mushroom body. Science 297:359–365

45. Berens P (2009) CircStat: a Matlab toolbox for circular statistics. J Stat Softw 31(10):1–21

46. Fries P et al (2001) Modulation of oscillatory neuronal synchronization by selective visual attention. Science 291:1560–1563

47. Loeb GE, Gans C (1986) Electromyography for experimentalists. University of Chicago Press, Chicago, p 365

48. Stamoulis C, Richardson AG (2010) Encoding of brain state changes in local field potentials modulated by motor behaviors. J Comput Neurosci 29(3):475–483

49. Harris KD et al (2003) Organization of cell assemblies in the hippocampus. Nature 424:552–556

50. DeCoteau, W.E., et al. (2004) Theta-gamma oscillations in local field potentials are prominent in the rat striatum and are coordinated with hippocampal rhythms in behaviorally selective patterns

51. Igarashi J et al (2013) A theta-gamma oscillation code for neuronal coordination during motor behavior. J Neurosci 33(47):18515–18530

52. Hwang EJ, Andersen RA (2009) Brain control of movement execution onset using local field potentials in posterior parietal cortex. J Neurosci 29(45):14363–14370

53. Howe MW et al (2011) Habit learning is associated with major shifts in frequencies of oscillatory activity and synchronized spike firing in striatum. Proc Natl Acad Sci U S A 108(40):16801–16806

54. Thorn CA et al (2010) Differential dynamics of activity changes in dorsolateral and dorsomedial striatal loops during learning. Neuron 66(5):781–795

55. MacKay WA (1997) Synchronized neuronal oscillations and their role in motor processes. Trends Cogn Sci 1(5):176–183

56. Basar E et al (2001) Gamma, alpha, delta, and theta oscillations govern cognitive processes. Int J Psychophysiol 39(2–3):241–248

57. Lubenov EV, Siapas AG (2009) Hippocampal theta oscillations are travelling waves. Nature 459(7246):534–539

58. Siapas AG, Wilson MA (1998) Coordinated interactions between hippocampal ripples and cortical spindles during slow-wave sleep. Neuron 21:1123–1128

59. Sharott A et al (2012) Relationships between the firing of identified striatal interneurons and spontaneous and driven cortical activities in vivo. J Neurosci 32(38):13221–13236

60. Clement EA et al (2008) Cyclic and sleep-like spontaneous alternations of brain state under urethane anaesthesia. PLoS One 3(4):e2004

61. Jefferys JG et al (2012) Mechanisms of physiological and epileptic HFO generation. Prog Neurobiol 98(3):250–264

62. Benar CG et al (2010) Pitfalls of high-pass filtering for detecting epileptic oscillations: a technical note on "false" ripples. Clin Neurophysiol 121(3):301–310

63. Menendez de la Prida L, Staba RJ, Dian JA. Conundrums of high-frequency oscillations (80–800 Hz) in the epileptic brain. J Clin Neurophysiol. 2015 Jun;32(3):207–19. https://doi.org/10.1097/WNP.0000000000000150.

64. Worrell GA et al (2012) Recording and analysis techniques for high-frequency oscillations. Prog Neurobiol 98(3):265–278

65. Levesque M, Behr C, Avoli M (2015) The anti-ictogenic effects of levetiracetam are mirrored by interictal spiking and high-frequency oscillation changes in a model of temporal lobe epilepsy. Seizure 25:18–25

66. Buffalo EA et al (2011) Laminar differences in gamma and alpha coherence in the ventral stream. Proc Natl Acad Sci U S A 108(27):11262–11267

67. Moser E et al (2010) Coordination in brain systems. In: von der Marlsburg C, Phillips WA, Singer W (eds) Dynamic coordination in the brain: from neurons to mind. MIT Press, Cambridge, MA, pp 193–214

68. Fries P (2005) A mechanism for cognitive dynamics: neuronal communication through neuronal coherence. Trends Cogn Sci 9(10):474–480

69. Middleton FA, Strick PL (2000) Basal ganglia and cerebellar loops: motor and cognitive circuits. Brain Res Brain Res Rev 31(2–3):236–250

70. Schmahmann JD (1997) The cerebellum and cognition—International review of neurobiology, vol 41. Academic Press, San Diego, p 665

71. Hoshi E et al (2005) The cerebellum communicates with the basal ganglia. Nat Neurosci 8(11):1491–1493

72. Bostan AC, Dum RP, Strick PL (2010) The basal ganglia communicate with the cerebellum. Proc Natl Acad Sci U S A 107(18):8452–8456

73. Roelfsema PR et al (1997) Visuomotor integration is associated with zero time-lag synchronization among cortical areas. Nature 385:157–161

74. Jacobs J, Kahana MJ (2010) Direct brain recordings fuel advances in cognitive electrophysiology. Trends Cogn Sci 14(4):162–171

75. Boraud T et al (2005) Oscillations in the basal ganglia: the good, the bad, and the unexpected. In: The Basal Ganglia VIII. Springer Science and Business Media, New York, pp 3–24

76. D'Angelo E et al (2009) Timing in the cerebellum: oscillations and resonance in the granular layer. Neuroscience 162(3):805–815

77. Gatev P, Darbin O, Wichmann T (2006) Oscillations in the basal ganglia under normal conditions and in movement disorders. Mov Disord 21(10):1566–1577

78. Courtemanche R, Lamarre Y (2005) Local field potential oscillations in primate cerebellar cortex: Synchronization with cerebral cortex during active and passive expectancy. J Neurophysiol 93(4):2039–2052

79. Courtemanche R, Chabaud P, Lamarre Y (2009) Synchronization in primate cerebellar granule cell layer local field potentials: basic anisotropy and dynamic changes during active expectancy. Front Cell Neurosci 3:6

80. DeCoteau WE et al (2007) Oscillations of local field potentials in the rat dorsal striatum during spontaneous and instructed behaviors. J Neurophysiol 97(5):3800–3805

81. Schnitzler A, Gross J (2005) Normal and pathological oscillatory communication in the brain. Nat Rev Neurosci 6(4):285–296

82. Gross J et al (2002) The neural basis of intermittent motor control in humans. Proc Natl Acad Sci U S A 99(4):2299–2302

83. Michalareas G et al (2016) Alpha-beta and gamma rhythms subserve feedback and feedforward influences among human visual cortical areas. Neuron 89(2):384–397

84. Newman MEJ (2010) Networks: an introduction. Oxford University Press, Oxford, p 772

85. Bullmore E, Sporns O (2009) Complex brain networks: graph theoretical analysis of structural and functional systems. Nat Rev Neurosci 10(3):186–198

86. Bullmore E, Sporns O (2012) The economy of brain network organization. Nat Rev Neurosci 13(5):336–349

87. van den Heuvel MP, Bullmore ET, Sporns O (2016) Comparative connectomics. Trends Cogn Sci 20(5):345–361

88. Misic B, Sporns O (2016) From regions to connections and networks: new bridges between brain and behavior. Curr Opin Neurobiol 40:1–7

89. Richiardi J et al (2015) BRAIN NETWORKS. Correlated gene expression supports synchronous activity in brain networks. Science 348(6240):1241–1244

90. Hammond C, Bergman H, Brown P (2007) Pathological synchronization in Parkinson's disease: networks, models and treatments. Trends Neurosci 30(7):357–364

91. Dostrovsky JO, Bergman H (2004) Oscillatory activity in the basal ganglia—

relationship to normal physiology and pathophysiology. Brain 127:721–722

92. Hutchison WD et al (2004) Neuronal oscillations in the basal ganglia and movement disorders: evidence from whole animal and human recordings. J Neurosci 24(42):9240–9243

93. Berke JD et al (2004) Oscillatory entrainment of striatal neurons in freely moving rats. Neuron 43(6):883–896

94. Raz A et al (1996) Neuronal synchronization of tonically active neurons in the striatum of normal and parkinsonian primates. J Neurophysiol 76:2083–2088

95. Costa RM et al (2006) Rapid alterations in corticostriatal ensemble coordination during acute dopamine-dependent motor dysfunction. Neuron 52(2):359–369

96. Beck MH et al (2016) Short- and long-term dopamine depletion causes enhanced beta oscillations in the cortico-basal ganglia loop of parkinsonian rats. Exp Neurol 286:124–136

97. Pan MK et al (2016) Neuronal firing patterns outweigh circuitry oscillations in parkinsonian motor control. J Clin Invest 126(12):4516–4526

98. Lemaire N et al (2012) Effects of dopamine depletion on LFP oscillations in striatum are task- and learning-dependent and selectively reversed by L-DOPA. Proc Natl Acad Sci U S A 109(44):18126–18131

99. Liu C et al (2017) Modeling and analysis of beta oscillations in the Basal Ganglia. IEEE Trans Neural Netw Learn Syst. https://doi.org/10.1109/TNNLS.2017.2688426

100. Belic JJ, Kumar A, Hellgren Kotaleski J (2017) Interplay between periodic stimulation and GABAergic inhibition in striatal network oscillations. PLoS One 12(4):e0175135

101. Blenkinsop A, Anderson S, Gurney K (2017) Frequency and function in the basal ganglia: the origins of beta and gamma band activity. J Physiol 595(13):4525–4548. https://doi.org/10.1113/JP273760

102. Tinkhauser G et al (2017) The modulatory effect of adaptive deep brain stimulation on beta bursts in Parkinson's disease. Brain 140(4):1053–1067

103. Cendes F et al (1993) Early childhood prolonged febrile convulsions, atrophy and sclerosis of mesial structures, and temporal lobe epilepsy: an MRI volumetric study. Neurology 43(6):1083–1087

104. French JA et al (1993) Characteristics of medial temporal lobe epilepsy: I. Results of history and physical examination. Ann Neurol 34(6):774–780

105. Spencer SS, Spencer DD (1994) Entorhinal-hippocampal interactions in medial temporal lobe epilepsy. Epilepsia 35(4):721–727

106. Engel J Jr et al (2012) Early surgical therapy for drug-resistant temporal lobe epilepsy: a randomized trial. JAMA 307(9):922–930

107. Jacobs J et al (2008) Interictal high-frequency oscillations (80–500 Hz) are an indicator of seizure onset areas independent of spikes in the human epileptic brain. Epilepsia 49(11):1893–1907

108. Staba RJ et al (2004) High-frequency oscillations recorded in human medial temporal lobe during sleep. Ann Neurol 56(1):108–115

109. Staba RJ et al (2002) Quantitative analysis of high-frequency oscillations (80–500 Hz) recorded in human epileptic hippocampus and entorhinal cortex. J Neurophysiol 88(4):1743–1752

110. Urrestarazu E et al (2007) Interictal high-frequency oscillations (100–500 Hz) in the intracerebral EEG of epileptic patients. Brain 130(Pt 9):2354–2366

111. van't Klooster MA et al (2015) High frequency oscillations in the intra-operative ECoG to guide epilepsy surgery ("The HFO Trial"): study protocol for a randomized controlled trial. Trials 16:422

112. Haegelen C et al (2013) High-frequency oscillations, extent of surgical resection, and surgical outcome in drug-resistant focal epilepsy. Epilepsia 54(5):848–857

113. Jacobs J et al (2010) High-frequency electroencephalographic oscillations correlate with outcome of epilepsy surgery. Ann Neurol 67(2):209–220

114. Bragin A et al (1999) Hippocampal and entorhinal cortex high-frequency oscillations (100–500 Hz) in human epileptic brain and in kainic acid--treated rats with chronic seizures. Epilepsia 40(2):127–137

115. Salami P et al (2014) Dynamics of interictal spikes and high-frequency oscillations during epileptogenesis in temporal lobe epilepsy. Neurobiol Dis 67:97–106

116. Levesque M et al (2011) High-frequency (80–500 Hz) oscillations and epileptogenesis in temporal lobe epilepsy. Neurobiol Dis 42(3):231–241

117. Kohling R et al (2016) Hypersynchronous ictal onset in the perirhinal cortex results from dynamic weakening in inhibition. Neurobiol Dis 87:1–10

118. Levesque M et al (2012) Two seizure-onset types reveal specific patterns of high-frequency oscillations in a model of temporal lobe epilepsy. J Neurosci 32(38):13264–13272

119. Bragin A et al (2005) Analysis of chronic seizure onsets after intrahippocampal kainic acid injection in freely moving rats. Epilepsia 46(10):1592–1598

120. Ylinen A et al (1995) Sharp wave-associated high-frequency oscillation (200 Hz) in the intact hippocampus: network and intracellular mechanisms. J Neurosci 15(1 Pt 1):30–46

121. Buzsaki G et al (1992) High-frequency network oscillation in the hippocampus. Science 256(5059):1025–1027

122. Perucca P, Dubeau F, Gotman J (2014) Intracranial electroencephalographic seizure-onset patterns: effect of underlying pathology. Brain 137(Pt 1):183–196

123. Weiss SA et al (2016) Ripples on spikes show increased phase-amplitude coupling in mesial temporal lobe epilepsy seizure-onset zones. Epilepsia 57(11):1916–1930

124. Brazier MAB (1962) The analysis of brain waves. Sci Am 207(6):1–10

125. Grey Walter W, Magoun HW (1959) Intrinsic rhythms of the brain. In: Handbook of physiology—Section 1: Neurophysiology. American Physiological Society, Washington, DC, pp 279–298

126. Hubel DH (1957) Tungsten microelectrode for recording from single units. Science 125:549–550

Chapter 14

Surgical and Electrophysiological Techniques for Single-Neuron Recordings in Human Epilepsy Patients

Juri Minxha, Adam N. Mamelak, and Ueli Rutishauser

Abstract

Extracellular recordings of single-neuron activity in awake behaving animals are one of the principal techniques used to decipher the neuronal basis of behavior. While only routinely possible in animals, rare clinical procedures make it possible to perform such recordings in awake human beings. Such human single-neuron recordings have started to reveal insights into the neural mechanisms of learning, memory, cognition, attention, and decision-making in humans. Here, we describe in detail the methods we developed to perform such recordings in patients undergoing invasive monitoring for localization of epileptic seizures. We describe three aspects: the neurosurgical procedure to implant depth electrodes with embedded microwires, electrophysiological methods to perform experiments in the clinical settings, and data processing steps to isolate single neurons. Together, this chapter provides a comprehensive overview of the methods needed to perform single-neuron recordings in humans during psychophysical tasks.

Key words Single-neuron recordings, Human intracranial, Functional neurosurgery, Stereotactic, Electrophysiology, Spike sorting

1 Introduction

Invasive intracranial EEG (iEEG) monitoring is routinely performed in patients who have localization-specific epilepsy, but for which the exact source of the seizure onset cannot be identified using noninvasive methods such as scalp EEG, MRI scans, PET and SPECT studies, or MEG. For these patients, either surface "grid" electrodes or penetrating "depth" electrodes provide a precise method to better identify seizure onset and spread patterns. "Grid" electrodes are sheets of electrodes imbedded in a thin sheet of silicone. They are typically used to identify the site and spread pattern of neocortical seizures on the brain surfaces and perform cortical functional mapping via electrical stimulation. In contrast, depth electrodes penetrate the brain surface and pass through both cortical grey and white matter. The distal end of the electrode typically rests in deep cortical or subcortical locations such as the

Roy V. Sillitoe (ed.), *Extracellular Recording Approaches*, Neuromethods, vol. 134,
https://doi.org/10.1007/978-1-4939-7549-5_14, © Springer Science+Business Media, LLC 2018

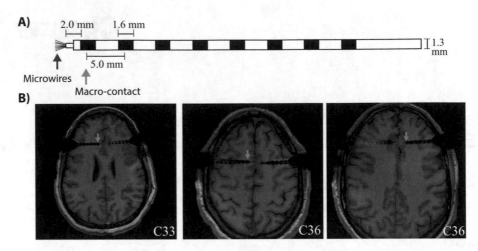

Fig. 1 Electrodes used and postoperative MRIs. (**a**) Sketch of the hybrid macro-micro depth electrode. (**b**) Example postoperative MRIs illustrating depth electrode placement

amygdala, hippocampus, primary visual cortex, or insula. Depth electrodes are typically used to identify seizure onsets in patients suspected of mesial temporal or mesial frontal seizures, although more recently the use of many small stereotactic EEG (SEEG) electrodes has been used as an alternative to grids [1].

In addition to seizure monitoring, depth electrode monitoring offers the unique opportunity to address important research questions on the function of the human nervous system [2]. For example, our current choice of electrode, called the Behnke-Fried (BF) hybrid depth electrode [3], has a standard array of 4–8 circular platinum-iridium ECoG electrodes spaced at 5 mm intervals along the electrode shaft and has a hollow core. Through the hollow core, we thread a bundle of nine microwires (40 μm diameter, platinum-iridium) (see Fig. 1a). These wires are contained in an insulated covering except on the distal end, where they come out in a "flower spray" configuration. The microwires and associated macro-electrode external sheath are FDA approved and manufactured by Ad-Tech Instrument Corp (Racine, WI). The wires extend 15 mm from the shaft. Eight of the wires are insulated, while an additional single referential ground wire is uninsulated. The assembly also has a green shrink-wrap sheath that sits over the insulation and is used during insertion to protect the wires (see below for details). We use the BF electrodes specifically to record multiunit and single-unit extracellular activity at the most medial aspect of the electrode target. At present, we have not found any other electrodes that can reliably record single-unit activity along the shaft of the main electrode, although newer technologies are being developed for this purpose. Importantly, no additional risk over standard clinical procedures is incurred by inserting microwires in addition to standard depth electrodes [4, 5].

The focus of this chapter is to provide a detailed description of the surgical methodology involved in the insertion of hybrid depth electrodes. In addition, we also briefly summarize subsequent methods to obtain reliable single-unit recording from the microwires in a clinical scenario. Our intention is that this detailed, step-by-step description will prove a useful guide to others interested in performing recordings in humans at the single-neuron level. Of note, there are several alternate techniques for inserting depth electrodes (i.e., see [6, 7]). We describe the method we have successfully employed for the last 12 years but acknowledge that other methods or modifications may be equally successful.

2   Methods

2.1   Surgical Methods

2.1.1   Target Selection

Placement of depth electrodes must always be dictated primarily by clinical concerns. Patients are undergoing depth electrode monitoring for the primary purpose of identifying a seizure focus. Because insertion of depth electrodes can carry substantial risks such as brain bleeding, stroke, infection, and even death, strict ethical standards must be maintained at all times [8]. Thus, it is unethical and unjustified to insert electrodes in nonclinically relevant areas or regions used only for research application. Failure to follow such strict ethical standards is likely to lead to potential harm to patients, which can never be justified.

In general, patients undergoing depth electrode monitoring fall into two categories. In some patients, seizures are suspected to arise from a medial temporal or limbic structure, but noninvasive monitoring and imaging tests are not sufficient to justify proceeding directly to a surgical intervention. Common examples of this include patients with suspected unilateral onset of seizures in the hippocampus or amygdala, but patients do not meet so-called skip criteria, so that depth electrode monitoring is used to confirm that all the seizures arise from one mesial temporal lobe versus having bilateral independent seizure onsets or evidence that the seizures do not arise from the mesial temporal lobe at all. Another common scenario involves patients who are believed to have localization-specific epilepsy, but noninvasive monitoring cannot reliably identify the site. In those cases depth electrode monitoring is used both to determine lateralization (i.e., what hemisphere does a seizure focus arise from) and localization (i.e., from what lobe of the brain or general region does the seizure arise from). Often in these cases patients subsequently go on to subdural grid or high density SEEG monitoring to further localize the seizures.

For most typical depth electrode cases, we rely upon orthogonal trajectories and place bilateral symmetric electrodes. The typical medial targets are amygdala, mid-body of hippocampus, medial orbitofrontal cortex (OFC), anterior cingulate cortex (ACC),

Table 1
Example electrode targets

Target	Approximate anterior–posterior location	Approximate vertical location	Other
Orbitofrontal cortex	25 mm anterior to anterior commissure		
Dorsal anterior cingulate cortex	18–30 mm anterior to anterior commissure	Mid-body of cingulate cortex	
Pre-SMA	5–8 mm anterior to anterior commissure	20–30 mm above the AC	
Amygdala	Anterior to temporal horn of ventricle	3–5 mm lateral to the uncus	Medial aspect of amygdala, basolateral nucleus
Hippocampus	10–15 mm posterior to amygdala target	Mid-body of hippocampus, slightly superior	Tip in CA1–3 rather than dentate gyrus

pre-supplementary motor area (pSMA). In addition, electrodes may often be placed in the parahippocampal gyrus, insula (frontal or temporal opercular regions), parietal cortex, or any overt structural abnormalities such as cortical dysplasias or regions of gliosis. Thus in general our patients are implanted with 5–8 electrodes in each hemisphere (see Table 1 for typical targets). Note that for all targets, the electrode tip is centered approximately 5 mm more lateral than the desired recording site, to allow room for the microwires to protrude from the end of the macro-electrode into cortical structures.

2.1.2 Stereotactic Targeting

Depth electrodes need to be inserted with a high degree of precision. Both the final target position as well as the trajectory from the surface of the brain to the target must be precisely planned to avoid injury to vascular structures such as veins and arteries on the brain surface and deep within the brain. Accidental puncture of veins and arteries is the primary cause of morbidity (brain injury) from depth electrode insertion, and these structures have substantial patient-specific variability. To accomplish this task, we rely on the use of frame-based stereotaxis. We utilize a Cosman-Roberts-Wells (CRW) stereotactic frame (Integra) and an attachment to the CRW frame that has been specifically designed for orthogonal depth electrode placements (Cosgrove Depth Electrode Insertion Kit, Ad-Tech Instrument Corp). While we are not aware of a similar insertion kit design to work with other common stereotactic frames, this design can easily be modified for those systems. Frame-based stereotaxic methods have a targeting accuracy of at least 1 mm and can be used with a variety of commercially available

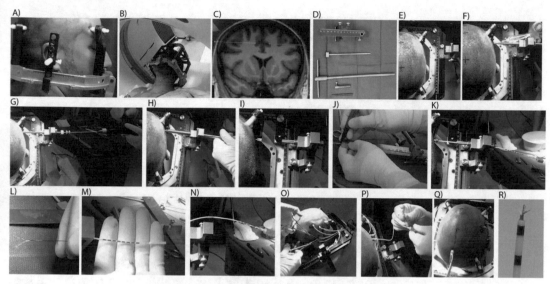

Fig. 2 Stepwise process of depth electrode insertion. (**a**) Stereotactic frame placement. (**b**) CT scan with fiducial localizer to create patient-specific coordinate system. (**c**) Screenshot from planning system (Framelink, Medtronic) used to determine stereotactic coordinates and trajectories. (**d**) Tools used for implantation (Cosgrove Depth guide, coring tool, anchor bolt wrench, reducing tubes for electrodes). (**e**) CRW base frame with Cosgrove Depth Guide mounted on patient's right side. (**f**) Coring tool inserted to skin entry point and lidocaine injection. (**g**) Handheld twist drill passed directly through coring tube to drill entry point and anchor bolt site. (**h**) Bolt insertion using hex wrench. (**i**) Anchor bolt inserted in skull, Cosgrove guide mounted with reducing tube for electrode insertion. (**j**) Marking length of electrode from end of reducing tube to target. (**k**) Inserted hybrid macro-electrode before removal of electrode stylet, inserted to length measured in (**j**). (**l**) Microwires (before cutting) together with green sheath. (**m**) Insertion of microwires through already implanted macro-electrode. (**n**) Final coupling of macro- and microelectrode. (**o**) Securing electrode by tightening anchor bolt after microwires are inserted. (**p**) Application of surgical glue to distal end to secure micro-and macro-electrode coupling. (**q**) Insertion of ground/reference strip. (**r**) Assembled hybrid macro-electrode with microwires extending (after electrode removal)

stereotactic planning software suites. These stereotactic planning systems are routinely utilized by the largest majority of neurosurgical centers around the world. With this method, a metal frame is attached to the patient's head using four disposable screws that penetrate the skin and press on but do not go into the skull (Fig. 1). We utilize ear bars inserted into the auditory canals of the patient at the time of frame placement to ensure a perfectly centered and orthogonal frame placement.

2.1.3   Surgical Protocol: Preoperative Procedures

Preoperative Area

1. Full head shave

2. Intravenous propofol sedation given by anesthesiologist

3. CRW stereotactic head frame applied using standard neurosurgical methods (Fig. 2a):

    (a) Local anesthetic injected at each pin site (Albright's solution, a pH balanced mixture of 1% lidocaine, 0.25% Marcaine, sodium bicarbonate, and epinephrine).

(b) Frame positioned with ear bars in bilateral auditory canals and held by technician. Good orthogonal position in all three planes confirmed by direct visualization by surgeon.

(c) Skull pins inserted at each of the four posts of the frame and tightened to secure the frame in a good orthogonal trajectory.

(d) Apply CT or MRI compatible stereotactic localizer.

CT or MRI Scanner

1. Patient is taken to the CT or MRI suite for scanning once sedation has worn off (typically 5–10 min). Our preference is to obtain a high-resolution 3 T MRI prior to surgery, as this is a routine part of the presurgical epilepsy work up and is therefore almost always available. On the day of the surgery, a CT scan is acquired (0° gantry angle, 1.2 mm slice thickness, axial images from the top of the frame base through the vertex, helical scan mode). Iodine contrast is also given to better visualize vascular structures. The scan is performed with the patient's head secured to a CT fixation holder to assure that the scan is orthogonal (see Fig. 2b). The CT acquisition typically takes under 1 min. In addition, a CT angiogram (CTA) sequence can be acquired to better visualize surface and deep vasculature.

2. The localizer is removed, and the patient is transferred to the operating room.

Registration of Patient-Specific Stereotactic Space[1]

1. CT images are transferred to the stereotactic planning computer.

2. The previously acquired MRI is also imported. If an MRI is not available, then the CT is not acquired, and we acquire a 1.5 T MRI with the stereotactic frame and localizer on and use these images. MRI should include at least one high-resolution multi-slice (120–180, typically 144) axial or coronal sequence and a similar high-resolution T2-weighted image for best anatomic imaging.

3. Using the planning software (Framelink, Medtronic, Inc.), the MRI and CT images are co-registered and aligned (Auto-Merge feature).

4. The fiducial rods from the localizer detected on the CT or MRI scan are then registered, generating a patient-specific stereotactic Cartesian coordinate system.

---

[1] We use the Framelink® Stereotactic Planning Software suite (Stealth Station, Medtronic) for planning at our institution. There are however many alternative solutions that are just as reliable (e.g., Brain lab, Radionics).

5. Once this step is complete, a full Cartesian coordinate system is established in which each point in space is associated with a specific Lateral (x), anterior–posterior (y), and vertical (z) coordinate that corresponds to identical x, y, and z values on the frame. A full description of stereotactic methods is beyond the scope of this chapter.

Target Calculation (See Fig. 2c)

1. The most medial point for the electrode is chose on the displayed axial, coronal, and sagittal MRI. Typical targets are listed in Table 1. The monitor cursor is placed on the target point.

2. Selecting the chosen target generates the x, y, and z coordinate values for that point on the screen. These points are recorded on an electrode targeting sheet. The "set target" button on the software is selected.

3. A point is then chosen on the lateral skin surface that is roughly parallel to the targeted A–P and vertical planes. Once selected, the x, y, and z coordinates are displayed by clicking the mouse similar to how the initial medial target was selected. The point is then adjusted with small movements of the cursor to ensure that the A–P (y) and vertical (z) values for the entry point are within 1 mm of the same values for the chosen target point. This assures that the trajectory of the electrode will be orthogonal to the insertion guide and parallel to the base frame. Once this point is established, the "set entry" button is selected, resulting in the display of the electrode trajectory from surface to medial point.

4. Using the trajectory path tool, the trajectory is then following in multiple planes to ensure that no surface or deep vessels are violated and that sulci are avoided as much as possible or feasible. If need be, the trajectory is then iteratively adjusted until the surgeon is satisfied with the trajectory.

5. This process is repeated for every electrode.

6. The insertion depth of each electrode is then recorded on the target planning sheet as 190 mm minus the absolute value of the lateral coordinate. This measurement determines the stopping point for insertion of the electrode at the lateral edge of the insertion guide (see Sect. 2.2.4 for details), as the system is designed so that the back end of the insertion guide with reducing tube in place measures exactly 190 mm from the center of the stereotactic frame.

2.1.4 Surgical Protocol: Intraoperative Procedures

Once all target coordinates and trajectories have been defined, the patient is ready to be implanted. The patient is brought into the operating. We typically perform these procedures under general anesthesia with either laryngeal mask or endotracheal tube insertion. However, the procedures can also be performed using only propofol sedation and local anesthetic if desired. In our experience, this is not necessary, slows the procedure, and increases patient discomfort. The procedural steps for insertion are as follows.

Patient Positioning and Prepping

1. Patient is placed in a semi-sitting "lounge chair" on the operating room table with the CRW frame secured in place using a standard neurosurgical head holder (Mayfield head holder with CRW adaptor plate). The head is position almost upright with the frontal eminence (top of the forehead) uppermost in the field, allowing symmetric access to both lateral sides of the head.

2. The entire head is prepped with an iodine-based antiseptic solution. Care is taken to make sure the prep extends below the zygoma (the cheek bone) on both sided and up to the frame rods.

3. The CRW base ring is attached to the head frame and secured with the locking knobs.

4. A sterile "U" drape is draped around the base of the frame but below the base ring and extended around the front of the patient with a full or three-quarter sheet placed to cover the patient body. This creates a sterile field (also see Note 1).

Equipment Required

1. Bovie electrocautery unit and cautery pencil

2. Modified electrocautery stylus (custom made – 225-cm-long insulted rod with 2 cm tip exposed at the end, inserts in coring tube).

3. Cosgrove Depth Electrode Insertion Kit (see Fig. 2d). Contains:

   (a) CRW stereotactic electrode insertion guide

   (b) Reducing tube—non-slotted

   (c) Reducing tube—slotted

   (d) Slotted electrode guide insert (inserted in slotted reducing tube)

   (e) Coring tool

4. Handheld neurosurgical twist drill

5. Disposable drill kit for depth electrode anchor bolts (Ad-Tech DDK2-2.8-30× for standard and BF depths, DDK2-2.4-30× for SEEG anchor bolts and electrodes)

6. Anchor Bolts

7. Appropriate number of BF electrodes

8. 1 × 4 or 1 × 6 contact subdural strip electrode (for ground/reference)

9. Basic surgical instrument tray with forceps, cocker clamps, etc.

10. Small gauge K-wire or Steinman pin.

Anchor Bolt Placement
The standard process of inserting BF hybrid depth electrodes using the Cosgrove Depth Insertion Kit is identical for all electrodes and insertion sites. We typically perform all right-sided insertions first, followed by all left-sided insertions.

1. The Cosgrove Depth Guide is placed on the A-P mount side of the CRW base frame and set to the A-P (y) coordinate for the electrode to be inserted (see Fig. 2e). (The CRW uses a vernier scale to allow for accuracy up to 0.1 mm.)

2. The height of the Cosgrove Guide is adjusted to match the vertical (z) coordinate of the target for the electrode to be inserted.

3. The non-slotted reducing tube in inserted into the guide until flush with it and locked in place with the tightening screw.

4. The coring tool is inserted through the reducing tube to the skin, marking the entry point. This point is injected with a 2–3 cm wheel of lidocaine 0.5% with epinephrine (see Fig. 2f).

5. The coring tool is twisted directly through the skin and muscle down to the skull. The sharp edges on the coring tool allow it to cut through the skin and muscle but will not penetrate the bone. Once making firm contact with the bone, it is locked in place with the tightening screw.

6. The modified Bovie tip is inserted directly though the hollow opening in the coring tool down to the bone and coagulation of the deep tissue carried out for several seconds. This prevents bleeding from the muscle that might occur from coring.

   (a) If no modified Bovie tip is available, the K-wire can be inserted, and the distal end of the K-wire touched to the Bovie to transmit current to the deep tissues. If this method is used care is taken to make sure the K-wire does not touch the side walls of the coring tube, resulting in an electrical short and no tissue coagulation.

7. The handheld twist drill is passed through the coring tube opening to the bone, and a twist drill hole is made in the bone. An adjustable stop on the drill is set to minimize risk of plunging into the brain. Ideally the drill penetrates the skull but stops at the dura. Standard neurosurgical techniques are applied to achieve this depth (see Fig. 2g).

8. The drill is removed, and the sharp end of the K-wire is inserted down the coring guide. This allows the surgeon to ensure that the bone has been completely breached and to palpate the dura. The dura is then punctured with the K-wire. The process can be repeated as needed to ensure complete drilling and dural opening.

9. The coring tool and reducing cannula are removed.

10. An anchor bolt is placed on the distal end of the hex wrench supplied with the Cosgrove kit (see Fig. 2h). This wrench is designed to match the hexagonal shape of the anchor bolt, with a width that is the same as the Cosgrove guide. This ensures that the insertion of the anchor bolt will be exactly in line with the drill hole and remain orthogonal at all times.

11. The hex wrench with anchor bolt is inserted through the insertion guide to the drilled hole, through the skin and hand—tightened into the calvarium. This typically requires 15–20 half turns of the wrench. Care must be taken not to turn too quickly or with too much pressure, to avoid fracturing of the anchor bolt or the underlying bone. Typically, the anchor bolt is advanced until the hexagonal aspect of the bolt is touching the skin surface. This ensures excellent purchase in the bone but not too deep a penetration to cause epidural hematoma.

12. Once in place the hex wrench is removed, and the surgeon confirms that the anchor bolt is tightly secured.

Hybrid Microwire Insertion

1. The length of the electrode that had previously been measured (step 6 in "electrode trajectory" section above) is noted. The macro-contact portion of the BF hybrid electrode is measured to this length from the distal tip of the electrode on a ruler, with the length marked using a surgical marking pen (see Fig. 2j).

2. The distal end of the microwire bundle is then cut as a single bundle using a very sharp tenotomy scissors. We typically cut the wires to be 4–5 mm long for optimal results (see Note 3 and Fig. 2l).

3. The green protective sheath is gently pulled over the microwire bundle to protect the wires during insertion (see Fig. 2l). The sheath should just cover all the wires but not be pulled up too far to avoid it coming off or bending during insertion.

4. The slotted reducing tube and slotted guide cannula are assembled so that the distal end of both pieces are flush. These are then inserted as a single assembly into the Cosgrove guide (see Fig. 2i).

5. The K-wire is again passed thought the guide cannula and the anchor bolt through the dura to ensure clear passage of the electrode.

6. The BF macro-electrode is inserted through the guide cannula and opening in the anchor bolt and passed until the marked point on the electrode just aligns with the back end of the guide cannula (see Fig. 2k). This is the target depth.

7. With the surgeon carefully holding the end of the electrode that enters the anchor bolt (so that is cannot slip), the electrode

stylet is removed. The guide cannula and slotted reducing tube are then unscrewed and gently pulled back and disassembled, leaving only the electrode exiting from the Cosgrove insertion guide. The distal electrode is then passed through the Cosgrove guide opening so that the electrode sits completely outside the Cosgrove assembly. The anchor bolt set screw is tightened one half turn.

(a) A slotted guide cannula is needed because the distal outer dimeter of the BF macro-electrode is large than the proximal end, and if it were not slotted, it could not be freed from the guide assembly.

(b) Placement of the macro-contact and holding it in place while the distal end is removed from the Cosgrove guide is the most precise portion of the procedure and the easiest place for the surgeon to accidentally move the electrode depth. Care should be taken to ensure that the surgeon has a secure grip on the electrode just as it enters the anchor bolt during this entire process to minimize risk of migration.

8. The surgeon picks up the microwire assembly and hands the distal end of the microelectrode bundle to an assistant to hold and advance, while the tip is inserted in the macro-electrode.

9. The surgeon holds the distal end of the macro-electrode, taking great care to not pull out or advance the electrode (it is not yet secured). He/she then threads the microwire assembly into the hollow opening of the macro-electrode and gently advances it until the distal connecter bushing tightly connects with the distal connector on the macro-electrode (see Fig. 2m). It is advanced until the collar on the back end of the microwire assembly aligns with and the blue line present on the macro-electrode, indicating it has been fully inserted with the distal end protruding from the distal end of the macro-electrode by the amount that was precut in step 2 (see Fig. 2n). The collar should fit snugly into the distal end of the macrowire assembly.

10. The tightening screw on the anchor bolt is then tightened, first finger tight and then further with a Kocher clamp, to lock the entire electrode assembly in place (see Fig. 2o).

11. A small drop of surgical glue (e.g., Dermabond, Indermil) is applied to the electrode coupling site at the distal end to prevent the assembly from separating (see Fig. 2p).

The process (steps 1–11) is then repeated for the next electrode, until all electrodes have been inserted.

12. Once all electrodes are inserted, a 1–2 cm incision is made in the midline scalp at the parietal vertex. A hemostat or similar

clamp is used to create a small subgaleal pocket, and the 1 × 4 subdural strip is inserted into the subgaleal space with the contacts pointing outward. This will serve as the ground and reference contacts for the recordings (see Fig. 2o). The incision is closed with a nylon suture, and the electrode tail is secured to the scalp.

13. An A-P and lateral skull X-ray are taken after all electrodes are inserted but prior to completion of the procedure. Review of the X-ray may identify electrodes that are misplaced or accidently pulled out. Any misplaced electrodes are then reinserted with final placement again confirmed by X-ray.

14. The exact number and/or color scheme for each electrode and its location are double checked with a technician, to ensure correct identification of each wire for subsequent EEG recordings.

Completion of Procedures

Once all electrodes are in place, the head must be properly dressed to prevent infection.

1. The entire head is cleaned as best as possible with several wet lap sponges, removing any dried blood and Betadine paint.

2. 1-cm-wide × 2–3-cm-long strips of Xeroform or similar bandaging material are cut and wrapped around the base of each anchor bolt.

3. The surgeon holds the head while as assistant releases the frame from the Mayfield head holder, unscrews the four skull pins, and removed the CRW frame.

   (a) Care is taken not to hit any of the protruding anchor bolts or accidentally pull on the electrodes during removal.

4. Several gauze sponges are placed on both sides of the head, and a full head dressing is applied. We use two Kerlix rolls for this purpose. The electrode tails are brought out through the top of the dressing with care taken to ensure they are not buried in it. The exiting tails are further covered with additional gauze sponges and secured with silk tape. A Spandage expandable bandage net is also applied.

5. The patient is awakened and taken to the recovery room.

Postsurgery Procedures

1. A non-contrast brain MRI is obtained within 4 hours of insertion. This confirms each electrode tip location and identifies any sites of bleeding.

   (a) A CT scan is not advisable due to metallic artifacts making any interpretation difficult.

2. A-P lateral and submental vertex plain films are taken.

3. The patient is transferred to the epilepsy monitoring unit (EMU) for recovery and monitoring.

4. Evaluation of the microwires for detecting single-unit and multiunit activity is typically first started 1–2 days after insertion, to allow the patient to recover from the procedure.

Removal of Electrodes

Once monitoring is completed, the electrodes and anchor bolts must be removed. This surgical procedure is done under propofol sedation. The basic steps are as follows:

1. In the OR patient is given propofol sedation. The head dressing is cut off, with care taken not to accidentally cut the electrode wires. The head is placed on a gel donut for support.

2. The head is not cleaned with Betadine until after the electrodes are removed.

3. Starting with the right side, the head is turned to the left. The surgeon puts on gown and gloves.

4. Using the Kocher clamp, the tightening screw on each electrode is loosened. We typically loosen all screws at one time.

5. Each electrode is then pulled out from the anchor bolt and inspected. The distal microwires should be visible (see Fig. 2r), and the entire electrode array is removed (also see Note 2).

6. Using the hex wrench unscrew each anchor bolt and remove it.

7. The entire side of the head where the bolts were removed is painted with Betadine scrub paint.

8. Each insertion site is closed with a single 2-0 or 3-0 nylon suture. The closing stich must tightly bring together the skin edges and should be inspected to ensure no egress of CSF.

9. Once one side is completed, the head is turned to the other side, and the same process (1–8) repeated for the other side.

10. Once all electrodes are removed, the head is cleaned with a moist lap sponge and water. Antibiotic ointment is then applied to each suture site. No other dressing is required.

11. The patient is awakened and returned to his/her room.

2.2 Methods for Data Acquisition and Behavioral Testing

A typical intracranial recording setup relies on three separate computers: an acquisition system, a stimulus presentation system, and an eye tracking system. Below, we briefly outline the configuration of each setup. Together, we have found this to be a very reliable setup for the use in the clinical setting.

2.2.1 Data Acquisition System

For data acquisition, we use the Atlas system from Neuralynx Inc. All signals from the microwires are preamplified on the head with small preamplifiers (head stages) that attach directly to the pigtail connector of the microelectrode. All microwire recordings are performed broadband (0.1 Hz–9 kHz band-pass filter) and are sampled at 32 kHz. In addition, this system allows the monitoring of all signals originating from macro-electrodes (depth electrodes, grids) and to pass these signals on to a clinical system running in parallel. Together, this configuration allows us to connect only a single system to the patient, which lowers noise and avoids interference (see Notes 4 and 5). While we monitor the broadband recordings throughout the experiment, all processing of the data (i.e., filtering, spike detection, spike sorting) is redone during offline analysis. We typically set the input range to ±2500 μV, resulting in <1 μV resolution. This is especially critical for spike sorting (see below), which relies on the shape of the spike waveforms themselves. Alternative products from other manufacturers (including Blackrock Microsystems Inc. and TDT Inc.) offer similar solutions to the one we described.

2.2.2 Stimulus Presentation and Eye Tracking

We implement all experimental tasks in Matlab with Psychophysics Toolbox [9, 10]. This well tested and extensively utilized toolbox has been utilized by numerous human intracranial experimenters and is well suited for this purpose. We typically show stimuli on a 19-in. screen with resolution of 1024 by 768 pixels. The screen is supported by an arm mount and also carries the camera and infrared light source for the eye tracker. We monitor monocular gaze position with a 500 Hz sampling rate with an EyeLink 1000 system (SR Research Inc.). We utilize a 9-point calibration grid to determine the eye-to-screen coordinate transformation. Throughout a typical experiment, we can monitor eye position with an accuracy of 0.42 DVA ± 0.15.

2.2.3 Response Boxes and Keyboards

To collect responses from the subjects, we primarily use the RB-740 and the RB-844 response pads (Cedrus Inc.) These response pads offer more reliable timing compared to a regular keyboard. Also, they are fully customizable, contain only a few buttons, and can be changed from experiment to experiment. As a result, we find response pads to be easier to use for patients.

2.2.4 Synchronization and Data Transfer

Since the three systems are independent, it is essential to synchronize behavioral events. We use the stimulus presentation system as the master system. Whenever a significant behavioral event occurs (stimulus onset, stimulus offset, button press), this system sends an event to both the acquisition system as well as the eye tracking system. This is achieved by utilizing the parallel port to send a signal to the transistor-to-transistor logic (TTL) input port on the data acquisition system. The same events are also sent to the eye tracking system utilizing an Ethernet IP connection and the EyeLink toolbox [11]. This way, the point of time at which each behavioral event occurred is known on all three systems, despite the underlying clocks not being synchronized. Also see [12] for further technical details on how to communicate between the three systems involved and how to utilize these connections for real-time closed-loop experiments.

2.3   Methods for Data Processing

Spike sorting is the process of extracting action potentials from the raw intracranial recording and attributing them to a particular neuron ("unit"). As is the case with all unsupervised clustering problems, one of the main challenges of spike sorting is estimating the number of neurons (i.e., clusters) that a given electrode is "listening" to. While there are many spike-sorting solutions available (see Table 2), they all execute a similar workflow: signal conditioning and filtering of the raw trace, followed by spike detection and alignment, feature extraction, and finally clustering [13, 14]. The features used for clustering are either the raw spike waveforms or derivatives thereof. To judge the quality of sorting, additional properties of the spike train associated with a given cluster have to be considered, including the distribution of inter-spike intervals, firing rates, and autocorrelations. Together, these pieces of information provide evidence for whether a given cluster can be considered representative of a single neuron or not. While the focus of this chapter is on the surgical aspects of depth electrode implantation, we briefly summarize the standard steps we utilize below. Please see [15] for further details.

2.3.1   Filtering

The first step in the processing pipeline is to remove low frequency content from the raw trace (Fig. 3a) by band-pass filtering the raw signal in the 300–3000 Hz frequency range (Fig. 3b). In order to preserve the shapes of the spike waveforms, it is important that the filtering process does not introduce phase distortions [16], which are achieved by using a zero-phase digital filter (Fig. 4). For real-time applications, such filtering is not possible because it is non-causal. In that case, the alternative is to use a linear phase FIR filter and directly account for the group delay introduced by the filter (delay $= (L-1)/2$, where L is the filter length).

2.3.2   Spike Detection and Extraction

In order to extract the action potentials from the filtered signal, we assume that the individual spikes are above the noise floor. Since the noise floor may not necessarily be stationary, we use a time-dependent threshold that is a function of the underlying noise properties of the signal. Specifically, the threshold is set to be a multiple (typically around 5) of the estimated standard deviation of the filtered trace. While this may work perfectly well for spikes that have large waveforms, it may miss some of the smaller spikes that are much closer to the noise floor. A few simple techniques can help improve the signal-to-noise ratio of spikes and therefore improve detection. One such technique is to use the energy of the signal instead of the raw trace [17]. The energy operator amplifies small differences between the spike amplitude and the noise floor, making it easier to set a threshold for spike detection. Since the processes that we are interested in (i.e., spikes) unfold over approx-

Table 2
Spike sorting algorithms commonly used for human single-neuron recordings

Name	Online	Filtering (Hz)	Automatic	Detection	Features used	Sorting algorithm	Notes
OSort [18]	Yes	300–3000	Can be but manual refinement recommended	Signal energy, computed locally with 1 ms kernel	Full, raw, decorrelated waveforms	Discriminative, with manual oversight	Does not generalize to polytrodes
Klusta [20, 39]	No	500—95% of Nyquist	Can be but manual refinement recommended	Dual threshold based on filtered trace; flood-fill algorithm used for detecting events over time and electrodes (i.e., space)	Principal components of spatiotemporal spike signature	Parametric, discriminative, maximum likelihood fitting using GMM with expectation maximization (EM)	Tailored for dense electrode arrays, activity between channels not independent
Wave_Clus [19]	No	300–3000	Yes	Amplitude threshold based on filtered trace; multiple of noise estimate	Wavelet coefficients of waveform	Nonparametric, discriminative (superparamagnetic clustering)	Can deal with polytrode data

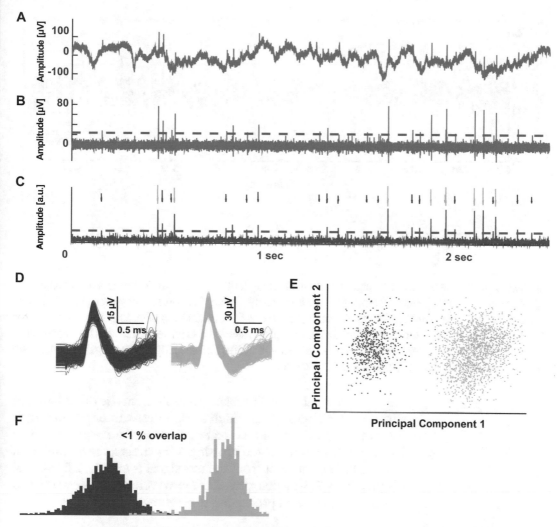

Fig. 3 Spike detection and sorting. (**a**) Example broadband recording from a microwire in the amygdala (with bipolar referencing). (**b**) The same trace as in (**a**) but band-pass filtered in the 300–3000 Hz band. (**c**) Signal used for spike detection (*green line* is the threshold). Signal shown is the local energy (with a 1 ms kernel), which improves the SNR of spikes with respect to the baseline. (**d**) After spike detection and alignment, two prominent waveform shapes, each one belonging to a different underlying unit (*green* and *red*), were identified on this channel. (**e**) The individual waveforms (256 samples per electrode) from the two clusters projected into principal component space. The clusters are well separated. (**f**) Projection test to validate the separation between the two putative single units (clusters). Shown are two overlapping histograms, each corresponding to one cluster. There was less than 1% overlap

imately 1 ms, we compute the local energy of the signal at that time scale by convolving it with a rectangular kernel of 1 ms width. This results in an "energy signal" (Fig. 3c), which is then thresholded. The threshold is set to a multiple of the standard deviation of the energy signal (here, 5× s.d.). The threshold parameter (i.e., the multiple of the standard deviation) can vary case by case. For a channel that has very few spikes, we may need to set this parameter

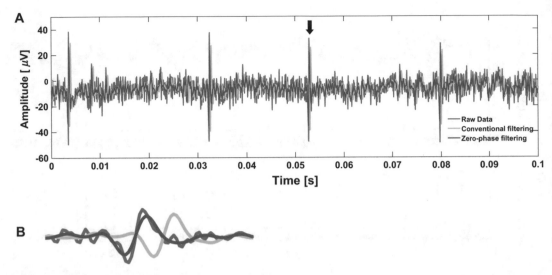

Fig. 4 Zero-phase filtering. (**a**) Band-pass filtering (here, 300–3000 Hz) is a common first step in improving the signal-to-noise ratio of spike waveforms. The way in which the filtering is performed however, greatly changes results. Shown is the result of filtering raw data (*blue trace*, Fs = 32,000 Hz) with no phase distortion (*red trace*) and with a conventional filter (*yellow trace*). Zero-phase filtering was implemented with the Matlab function *filtfilt*. (**b**) While both methods largely preserve waveform shape (but see [16]), conventional filtering delays the spike waveform, while zero-phase filtering does not

higher than usual to avoid picking up noise. On the other hand, in situations with cells of very high firing rates with large amplitude waveforms, the threshold has to be set lower than usual (i.e., 3 or 4× s.d.). For each threshold crossing, we extract a fixed number of samples before and after from the raw signal (typically, 2.5 ms total length). This 2.5-ms-long trace is the waveform of the spike and forms the basis for all processing that follows.

2.3.3 Spike Sorting (Clustering)

Spike sorting involves two steps: identification of features from each waveform followed by unsupervised clustering of these features. The most commonly used feature for clustering is the spike waveform. The goal is to identify features of the waveform that maximally separate different cells. On one extreme, we could use a single scalar, such as the peak-to-trough amplitude or the spike width. On the other extreme, we could use the entire waveform and clustered in this N-dimensional space (where N is the number of samples that make up the waveform). An alternative approach is to capture as much of the variance in the waveforms as possible using only a few dimensions. This can be achieved by utilizing a dimensionality reduction technique such as principal component analysis (PCA).

Once spike waveform features have been identified, an unsupervised clustering algorithm is used to partition the space. In practice, a variety of algorithms have been used for human single-neuron recordings, including OSort, Wave_clus, and variants of Klustakwik/Klusta (see Table 2). Some are parametric and make assumptions about the underlying distribution of the data (such as Gaussian mixture models). Others are nonparametric and rely on heuristics computed directly from the data. We utilize the OSort [18] algorithm for spike detection and clustering. OSort uses a distance metric between the raw waveforms for clustering, runs spike-by-spike (online), and determines the number of clusters automatically. Other spike sorting approaches used for human single-neuron recordings are example pipelines include Wave_clus [19] and Klusta [20].

2.3.4   *Quality Metrics*

We rely a list of quantitative metrics to assess how likely a given cluster represents a single neuron and to assess whether a given cluster is over- or under-merged.

Stable waveform: A key metric is the peak-to-trough amplitude of the spikes associated with a cluster as a function of time. Ideally, the amplitude of the spike should remain constant throughout the experiment (see Fig. 5b). Large deviations in the shape of the waveform are usually indicative of electrode movements (the electrode changes position with respect to the cell) or an artifact of the spike-sorting algorithm (e.g., two clusters were merged when they should not have been, or a single cluster was split into two when it should not have been). This can often be corrected by manually merging clusters.

Stable firing rate (on long enough time scales): A second metric to tracking the stability of a cell is firing rate as a function of time. While there may be task-dependent modulation at finer time scales, on long enough time scales the average firing rate should be relatively stable. As in the case of the peak-to-trough amplitude, a large deviation in the average firing rate of a cell is usually indicative of electrode movement or over-splitting in spike sorting (see Fig. 5b for an example). This can often be corrected by manually merging clusters.

Inter-spike interval histogram: One of the unmistakable features of a cell is the distribution of its inter-spike intervals (ISIs). The ISI histogram can be used in two ways: (1) tell the difference between cells that have similar waveform shapes but might be functionally different and (2) verify that there are no violations of the refractory period (i.e., there are few ISIs < 3 ms). Violations of the refractory period are indicative that a cluster represents multiunit activity.

Alignment check: An important check of the quality of a cluster is the distribution of the peak amplitude across all the waveforms. This distribution should be unimodal and tightly clustered

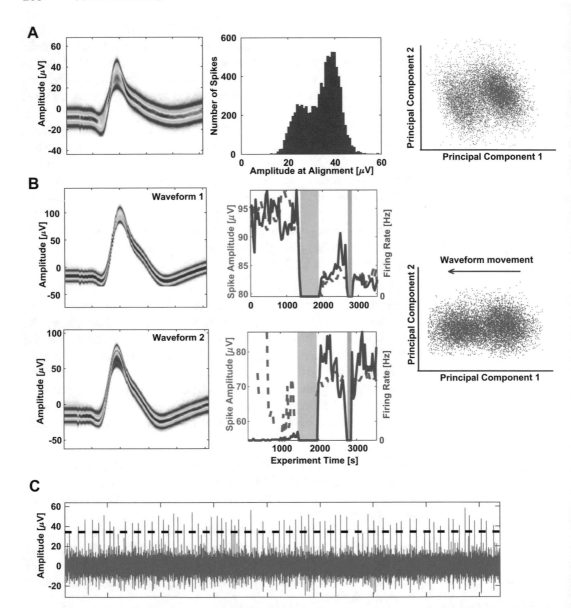

Fig. 5 Typical problems in spike sorting. (**a**) One of the most common problems in spike sorting is over-merging of two clusters. This can happen if the within-group differences between spike waveforms are larger than that across groups. In this case, the algorithm (in this case OSort) will merge the two clusters into a single cluster. There are a few simply ways to detect this kind of phenomenon. Illustrated here is the distribution of amplitudes at the alignment point (*middle panel*). If bimodal distribution, this is an indication that this cluster is a composite of two (or more) other clusters. A projection of the individual spikes into PCA space confirms this (notice the two clusters in the third panel). (**b**) An example of under-merging of two clusters due to non-stationarities through the experiment (usually due to electrode movement). The firing rate and amplitude (second column) of the waveforms as a function of experiment time is a useful tool in diagnosing such problems. Note that during grey periods, recording was off. (**c**) High firing rates can bias the automatic threshold selection (*dashed line*), leading to missing spikes with lower amplitudes. For channels with high firing rates of high-amplitude spikes, the threshold has to be lowered manually

around the true mean. A multimodal distribution is indicative of over-clustering (two or more clusters have been mistakenly merged into one) or misalignment of the waveforms (see Fig. 5a).

Cluster quality checks: We typically provide, for each cluster included in a paper, histograms of a number of spike isolation quality metrics to allow an assessment of how well separated the cells included in a study were (e.g., see supplementary figures in [21, 22]). These include projection tests between all possible pairs of clusters on the same wire [18, 23, 24], isolation distance (for each cluster vs. all other detected spikes on a wire), L-ratio [25, 26], %ISI violations <3 ms, and signal-to-noise ratio of the mean waveform of a cluster.

3 Single-Neuron Analysis, a Case Study

In this section, we demonstrate examples of results that can be obtained using the methods described in this chapter. The example experiment shown involves all components described, including the implantation surgery, eye tracking, and single-neuron recordings. For simplicity, we focus this case study on face-selective neurons recorded in the amygdala [27], a structure important in processing stimuli with ecological significance, in particular faces but also rewards and punishments [28, 29]. Face-selective responses have been identified in many brain areas [30], including the human amygdala [31, 32]. However, little so far is known about what specifically the functional role is of face responses in the amygdala [33]. The study presented here is an effort to understand the nature of these face responses in the amygdala during natural vision and their potential modulation by attention. In the context of natural vision, the amygdala is of particular interest because amygdala lesions are known to interfere with the efficient visual exploration of faces [34, 35]. Moreover, lesions of the amygdala produce a complex constellation of impairments in social behavior [36]. The amygdala is thus a prime candidate for mediating between the perceptual representations of faces in the cortical face patch system [30] and the mediation of social behaviors based on such perception. Elucidating this role, however, requires both a more naturalistic presentation of stimuli and a better quantification of how they are attended. In this study, we achieved both of these imperatives by allowing subjects to freely view a complex array of images that competed for attention, while we monitored eye movements.

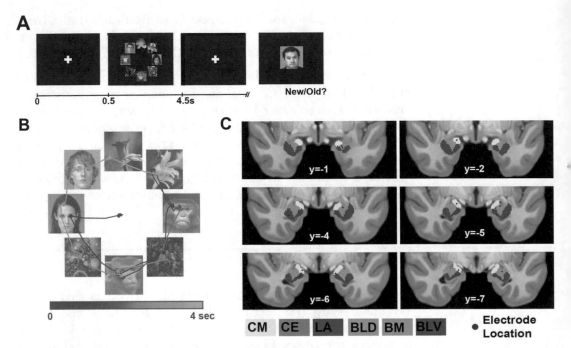

Fig. 6 Task setup and location of amygdala electrodes for case study. (**a**) All subjects (*n* = 12) performed a free-viewing task, embedding in a memory task. The subjects were instructed to scan the array of images for a later memory test. The array contained two monkey and two human faces as well as four non-face distractors. (**b**) Example scan paths from a single trial. Dots are fixations, and lines are saccades. Note that the first saccade (starting at the central fixation point) targets a human face. Trial time is encoded by color as indicated. (**c**) Recording sites in all human subjects in MNI152 space, illustrated using the Atlas provided by [37]. Amygdala nuclei are indicated in color. Each electrode location is indicated (*red dot*) at its appropriate location on the *y*-axis. Abbreviations for the anatomical areas are *LA* lateral nucleus, *BLD* dorsal basolateral, *BM* basomedial, *CE* central, *CM* cortical and medial nuclei; *BLV* ventral basolateral

Task: Each trial (Fig. 6a) consisted of a circular array of eight images chosen at random from two face categories (human and monkey faces) and two non-face categories (flowers and fractals). Each image array was displayed for 3–4 s, and subjects were free to view any location. Figure 6b shows an example scan path, illustrating preference for looking at faces.

Fixation-target sensitive neuronal responses. We isolated in total 422 putative single neurons from the human amygdala (on average, 19 per session, 1.2 cells/electrode; see Fig. 6c for recording locations illustrated using the Amygdala Atlas of Tyszka et al. [37]). We first determined whether the responses of amygdala neurons were modulated by the identity of the fixated stimuli. For each neuron, we tested whether the firing rate following fixation onset co-varied with the identity of the fixated images. When different locations within an image were successively fixated, time of fixation onset was determined by the first fixation that fell within that image's region. We found that 20% ($n = 85/422$) of human neurons significantly modulated their firing rate after fixation onset. These "fixation-target-sensitive" responses appeared transiently shortly after fixation onset (Fig. 7a, b shows an example neuron).

Determining response latency: Trial-onset latencies of visual selectivity in the amygdala have been well characterized [38], but no such numbers existed for fixation-aligned responses. In order to determine the post-fixation window of analysis, we computed the mutual information between the spike counts (S) and the image category (C) using:

$$I(S,C) = \sum_{S,C} P(S,C) \cdot \log_2 \frac{P(S|C)}{P(S)}$$

where C is a discrete variable that can take 1 of 4 possible values, and S is also a discrete variable that can take 1 of N possible values, depending on the maximum firing rate of the cell. The mutual information was computed for each cell and at each point along the PSTH (from -0.5 to 1 s around the fixation). The mutual information for each cell was then averaged to produce the mean trace. The location of the center of the fixed window for all follow-up analysis was set to the point of time at which MI was maximal. From this analysis, we determined that the optimal analysis window should be centered at 325 ms after the onset of the fixation (see [27] for details).

Selection of units and population averages: We determined whether a cell's response is sensitive to the identity of fixated stimuli using a 1×4 ANOVA of the spike counts during a 250-ms-long time window centered on the point of time at which MI was maximal. In order to avoid leakage from one category response to another, we excluded successive fixations that fall on the same category. If the ANOVA was significant ($p < 0.05$), we determined the category with the largest mean response in the same time window. This category was used as the preferred category of the cell. The majority ($n = 27/85$ tuned cells) of fixation-target-sensitive neurons preferred faces of conspecifics (i.e., human faces), and 71% of all fixation-sensitive neurons preferred faces to non-faces. Figure 7c, d, shows the proportion of cells that was tuned for each category as well as the average PSTHs across all cells that responded selectively to human faces (yellow) and monkey faces (purple).

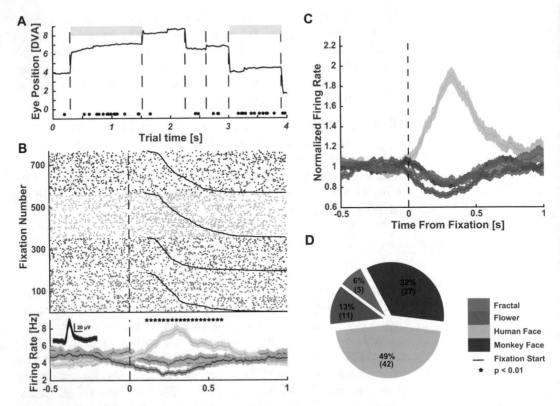

Fig. 7 Single neuron fixation-related activity in the amygdala. (**a**) Example trial from the face-selective neuron shown in (**b**). Spikes are indicated by black dots. Note that whenever gaze falls onto a face of a conspecific (*colored patch*), the neuron increased its firing rate. (**b**) Raster (*top*) and mean firing rate (PSTH, *bottom*) for an example neuron recorded in the amygdala. *Dotted vertical line* ($t = 0$) marks fixation onset. Trials were sorted by category of the fixated image (*color code*) and fixation duration (*black line*). Stars above the PSTHs indicate bins of neural activity (of 250 ms duration) with a significant (1×4 ANOVA, $p < 0.01$) difference in firing rate. Scale bar for waveforms is 0.2 ms. (**c**) Average normalized PSTH for all cells ($n = 42$) that were selective for human faces. (**d**) Preferred stimulus of all recorded fixation sensitive visually selective cells. Each selective cell (assessed with a 1×4 ANOVA) was assigned to one of the four visual categories based on the category to which it had the highest peak firing rate. The largest proportion of neurons responded maximally to faces of conspecifics ($n = 42$ cells, 49%)

4   Notes

1. The CRW frame rods remain exposed during the procedure and are not sterile. Thus, the surgeon must take care not to accidentally touch them during the procedure. If touched, or even if there is a concern of possible touching, the surgeon gloves are changes. We typically change gloves at least 3–4 times during a procedure.

2. During removal, if an electrode does not easily slide out from the anchor bolt and "hangs up," the macro-contact may be caught on the dura. Do not pull hard on the electrode, as this

may shear the electrode and leave a residual in the brain. Rather, cut the electrode as close to the anchor bolt as possible. Then when the anchor bolt itself is removed, the electrode will come out with it in the largest majority of cases.

3. Although we have experimented with various lengths ranging from 3 to 8 mm. 4–5 mm seems to give optimal recordings. Care is taken to cut quickly and to not crush the distal ends of the microwires. A sharp scissor helps with this. The microwire bundle is examined and can be slightly fashioned with the scissor blades to make sure the wires protrude in a "flower spray" configuration.

4. The choice of grounds and reference electrodes is critical for stable single-neuron recordings with high signal-to-noise ratios and an absence of movement artifacts. Strips implanted below the scalp (see surgical methods), with the exposed contacts pointing away from the brain, provide the best ground and reference contacts. For single-neuron recordings alone, the best reference is local, i.e., one of the eight microwires serves as a reference for the other seven microwires. Such bipolar recordings have the highest signal-to-noise ratio because that way, all wires have approximately the same impedance, their tips are located within a few mm of each other and common low-frequency activity cancels out. However, this configuration cancels out much of the local field potential (LFP). Thus, if LFP is important, it is advisable to use either the local reference wire or a remote reference.

5. Common sources of recording noise, including line noise, are devices connected to the patient. Before starting a recording, plug out all devices that are directly connected to the patient or which are being touched by the patient. All such devices should run on battery and be disconnected from the wall. This, in particular, includes IV pumps, leg warmers, remote controllers, cell phones, computers, etc.

References

1. Mullin JP, Shriver M, Alomar S, Najm I, Bulacio J, Chauvel P, Gonzalez-Martinez J (2016) Is SEEG safe? A systematic review and meta-analysis of stereo-electroencephalography-related complications. Epilepsia 57(3):386–401

2. Fried I, Rutishauser U, Cerf M, Kreiman G (2014) Single neuron studies of the human brain: probing cognition. MIT Press, Boston

3. Fried I, Wilson CL, Maidment NT, Engel J, Behnke E, Fields TA, MacDonald KA, Morrow JW, Ackerson L (1999) Cerebral microdialysis combined with single-neuron and electroencephalographic recording in neurosurgical patients—technical note. J Neurosurg 91(4):697–705

4. Schmidt RF, Wu C, Lang MJ, Soni P, Williams KA Jr, Boorman DW, Evans JJ, Sperling MR, Sharan AD (2016) Complications of subdural and depth electrodes in 269 patients undergoing 317 procedures for invasive monitoring in epilepsy. Epilepsia 57(10):1697–1708

5. Hefft S, Brandt A, Zwick S, von Elverfeldt D, Mader I, Cordeiro J, Trippel M, Blumberg J, Schulze-Bonhage A (2013) Safety of hybrid electrodes for single-neuron recordings in humans. Neurosurgery 73(1):78–85; discussion 85

6. Mehta AD, Labar D, Dean A, Harden C, Hosain S, Pak J, Marks D, Schwartz TH (2005) Frameless stereotactic placement of depth electrodes in epilepsy surgery. J Neurosurg 102(6):1040–1045

7. Misra A, Burke JF, Ramayya AG, Jacobs J, Sperling MR, Moxon KA, Kahana MJ, Evans JJ, Sharan AD (2014) Methods for implantation of micro-wire bundles and optimization of single/multi-unit recordings from human mesial temporal lobe. J Neural Eng 11(2):026013

8. Mamelak A (2014) Ethical and practical considerations for human microelectrode recording studies. In: Fried I et al (eds) Single neuron studies of the human brain. MIT Press, Boston

9. Pelli DG (1997) The VideoToolbox software for visual psychophysics: transforming numbers into movies. Spat Vis 10:437–442

10. Brainard DH (1997) The psychophysics toolbox. Spat Vis 10:433–436

11. Cornelissen FW, Peters EM, Palmer J (2002) The Eyelink Toolbox: eye tracking with MATLAB and the Psychophysics Toolbox. Behav Res Methods Instrum Comput 34(4):613–617

12. Rutishauser U, Kotowicz A, Laurent G (2013) A method for closed-loop presentation of sensory stimuli conditional on the internal brain-state of awake animals. J Neurosci Methods 215(1):139–155

13. Gibson S, Judy JW, Markovic D (2012) Spike sorting the first step in decoding the brain. IEEE Signal Process Mag 29(1):124–143

14. Lewicki MS (1998) A review of methods for spike sorting: the detection and classification of neural action potentials. Network 9(4):R53–R78

15. Rutishauser U, Cerf M, Kreiman G (2014) Data analysis techniques for human microwire recordings: spike detection and sorting, decoding, relation between neurons and local field potential. In: Fried I et al (eds) Single neuron studies of the human brain. MIT Press, Boston, pp 59–98

16. Quiroga RQ (2009) What is the real shape of extracellular spikes? J Neurosci Methods 177(1):194–198

17. Bankman IN, Johnson KO, Schneider W (1993) Optimal detection, classification, and superposition resolution in neural waveform recordings. IEEE Trans Biomed Eng 40(8):836–841

18. Rutishauser U, Schuman EM, Mamelak AN (2006) Online detection and sorting of extracellularly recorded action potentials in human medial temporal lobe recordings, in vivo. J Neurosci Methods 154(1):204–224

19. Quiroga RQ, Nadasdy Z, Ben-Shaul Y (2004) Unsupervised spike detection and sorting with wavelets and superparamagnetic clustering. Neural Comput 16(8):1661–1687

20. Rossant C, Kadir SN, Goodman DF, Schulman J, Hunter ML, Saleem AB, Grosmark A, Belluscio M, Denfield GH, Ecker AS, Tolias AS, Solomon S, Buzsaki G, Carandini M, Harris KD (2016) Spike sorting for large, dense electrode arrays. Nat Neurosci 19(4):634–641

21. Rutishauser U, Ross IB, Mamelak AN, Schuman EM (2010) Human memory strength is predicted by theta-frequency phase-locking of single neurons. Nature 464(7290):903–907

22. Kaminski J, Sullivan S, Chung JM, Ross IB, Mamelak AN, Rutishauser U (2017) Persistently active neurons in human medial frontal and medial temporal lobe support working memory. Nat Neurosci 20(4):590–601

23. Pouzat C, Mazor O, Laurent G (2002) Using noise signature to optimize spike-sorting and to assess neuronal classification quality. J Neurosci Methods 122(1):43–57

24. Rutishauser U, Ye S, Koroma M, Tudusciuc O, Ross IB, Chung JM, Mamelak AN (2015) Representation of retrieval confidence by single neurons in the human medial temporal lobe. Nat Neurosci 18(7):1041–1050

25. Hill DN, Mehta SB, Kleinfeld D (2011) Quality metrics to accompany spike sorting of extracellular signals. J Neurosci 31(24):8699–8705

26. Schmitzer-Torbert N, Jackson J, Henze D, Harris K, Redish A (2005) Quantitative measures of cluster quality for use in extracellular recordings. Neuroscience 131(1):1–11

27. Minxha J, Mosher C, Morrow JK, Mamelak AN, Adolphs R, Gothard KM, Rutishauser U (2017) Fixations gate species-specific responses to free viewing of faces in the human and macaque amygdala. Cell Rep 18(4):878–891

28. Adolphs R (2010) What does the amygdala contribute to social cognition? Ann N Y Acad Sci 1191(1):42–61

29. Paton JJ, Belova MA, Morrison SE, Salzman CD (2006) The primate amygdala represents the positive and negative value of visual stimuli during learning. Nature 439(7078):865–870

30. Tsao DY, Freiwald WA, Tootell RB, Livingstone MS (2006) A cortical region consisting entirely of face-selective cells. Science 311(5761):670–674

31. Rutishauser U, Tudusciuc O, Wang S, Mamelak AN, Ross IB, Adolphs R (2013) Single-neuron correlates of atypical face processing in autism. Neuron 80(4):887–899

32. Rutishauser U, Tudusciuc O, Neumann D, Mamelak AN, Heller AC, Ross IB, Philpott L, Sutherling WW, Adolphs R (2011) Single-unit responses selective for whole faces in the human amygdala. Curr Biol 21(19):1654–1660

33. Rutishauser U, Mamelak AN, Adolphs R (2015) The primate amygdala in social perception—insights from electrophysiological recordings and stimulation. Trends Neurosci 38(5):295–306

34. Adolphs R, Gosselin F, Buchanan TW, Tranel D, Schyns P, Damasio AR (2005) A mechanism for impaired fear recognition after amygdala damage. Nature 433(7021):68–72

35. Dal Monte O, Costa VD, Noble PL, Murray EA, Averbeck BB (2014) Amygdala lesions in rhesus macaques decrease attention to threat. Nat Commun 6:10161–10161

36. Adolphs R, Tranel D, Damasio AR (1998) The human amygdala in social judgment. Nature 393(6684):470–474

37. Tyszka, MJ, Pauli, WM (2016) A high resolution in vivo MRI atlas of the adult human amygdaloid complex. Submitted for publication

38. Mormann F, Kornblith S, Quiroga RQ, Kraskov A, Cerf M, Fried I, Koch C (2008) Latency and selectivity of single neurons indicate hierarchical processing in the human medial temporal lobe. J Neurosci 28(36): 8865–8872

39. Harris KD, Henze DA, Csicsvari J, Hirase H, Buzsaki G (2000) Accuracy of tetrode spike separation as determined by simultaneous intracellular and extracellular measurements. J Neurophysiol 84(1):401–414

INDEX

Roy V. Sillitoe (ed.), *Extracellular Recording Approaches*, Neuromethods, vol. 134,
https://doi.org/10.1007/978-1-4939-7549-5, © Springer Science+Business Media, LLC 2018

Printed in the United States
By Bookmasters